T0135500

Technische Universität München

Zentrum Mathematik

Time-Frequency and Wavelet Analysis of Functions with Symmetry Properties

Holger Rauhut

Vollständiger Abdruck der von der Fakultät für Mathematik der Technischen Universität München zur Erlangung des akademischen Grades eines

Doktors der Naturwissenschaften

genehmigten Dissertation.

Vorsitzender: Univ.-Prof. Dr. P. Gritzmann
Prüfer der Dissertation:

       1. Univ.-Prof. Dr. R. Lasser
       2. Prof. Dr. H.G. Feichtinger,
          Universität Wien / Österreich
       3. Univ.-Prof. Dr. St. Dahlke,
          Philipps-Universität Marburg

Die Dissertation wurde am 16. Juni 2004 bei der Technischen Universität München eingereicht und durch die Fakultät für Mathematik am 4. Dezember 2004 angenommen.

Bibliografische Information Der Deutschen Bibliothek

Die Deutsche Bibliothek verzeichnet diese Publikation in der Deutschen
Nationalbibliografie; detaillierte bibliografische Daten sind im Internet über
http://dnb.ddb.de abrufbar.

ISBN 3-8325-0778-7

Logos Verlag Berlin
Comeniushof, Gubener Str. 47,
10243 Berlin
Tel.: +49 030 42 85 10 90
Fax: +49 030 42 85 10 92
INTERNET: http://www.logos-verlag.de

Contents

Introduction

Since the fundamental work of Meyer, Daubechies, Mallat, Dahmen and others in the mid 1980's wavelet analysis and time-frequency analysis have become rapidly growing research fields. Both theories have applications in various areas of pure and applied mathematics, as well as physics and electrical engineering. Wavelet analysis has proven to be very useful for the numerical solution of partial differential equations [LMR98], for the study of function spaces and related operators [FJ90, FJW91], and for signal and image analysis [Mal88]. In particular, wavelets have found their way into the new JPEG 2000 standard for image compression. Time-frequency analysis is widely used in audio signal processing. Moreover, it has applications in wireless communication [Böl03], and also turned out to be an appropriate tool for the study of pseudodifferential operators [Grö01].

The basic object in wavelet analysis is the continuous wavelet transform (CWT) on $\mathbb{R}^d$ defined by

$$\text{CWT}_\psi f(x, a, R) = a^{-d/2} \int_{\mathbb{R}^d} f(t) \overline{\psi(a^{-1}R^{-1}(t - x))} dt, \ x \in \mathbb{R}^d, a \in (0, \infty), R \in SO(d).$$

Depending on the wavelet ψ it analyzes a given function f at various scales a, space locations x and orientations R. Whenever $\int_{\mathbb{R}^d} |\hat{\psi}(\xi)|^2 |\xi|^{-d} d\xi < \infty$ holds true, with $\hat{\psi}$ denoting the Fourier transform of ψ, then ψ is called an admissible wavelet. In this case, a stable inversion formula holds for the wavelet transform CWT_ψ on $L^2(\mathbb{R}^d)$. The counterpart in time-frequency analysis is represented by the short time Fourier transform (STFT) defined as

$$\text{STFT}_g f(x, \omega) = \int_{\mathbb{R}^d} f(t) \overline{g(t - x)} e^{-2\pi i \omega \cdot t} dt, \quad (x, \omega) \in \mathbb{R}^d \times \mathbb{R}^d.$$

Considering the function g as a window function located near the origin $t = 0$, the STFT at (x, ω) cuts out the part of f near x and then takes the Fourier transform at ω. Thus, the value $\text{STFT}_g f(x, \omega)$ can be interpreted as the content of f at the position (time) x and the frequency ω. The short time Fourier transform STFT_g is a bounded operator from $L^2(\mathbb{R})$ into $L^2(\mathbb{R}^d \times \mathbb{R}^d)$ and has a stable inversion formula for all $g \in L^2(\mathbb{R}^d)$.

A key result of time-frequency and wavelet analysis states that both the CWT and the STFT allow discretizations. Actually, it is this fact which makes wavelet and time-frequency analysis so effective in various applications. By a discretization of the CWT one means the construction of an orthonormal basis for $L^2(\mathbb{R}^d)$, or more generally a Riesz basis or a frame, which consists of dilations, rotations and translates of one single function ψ. This function ψ is usually called the mother wavelet. In particular, any L^2-function f can be expanded into a series of the form

$$f(t) = \sum_{i \in I} c_i(f)\, a_i^{-d/2} \psi(a_i^{-1} R_i^{-1}(t - x_i)), \qquad t \in \mathbb{R}^d, \qquad (0.1)$$

with $x_i \in \mathbb{R}^d, a_i \in (0, \infty), R_i \in SO(d)$ and some discrete index set I. (Hereby, convergence is in the L^2-sense.) So instead of using the whole parameter space $\mathbb{R}^d \times (0, \infty) \times SO(d)$ it suffices to work with a discrete subset. The concept of multiresolution analysis (MRA) invented by Mallat [Mal89] provides a fundamental tool for the explicit construction of orthonormal bases for $L^2(\mathbb{R})$ of the form

$$\{ \psi_{j,k}(t) := 2^{j/2} \psi(2^j t - k),\ j, k \in \mathbb{Z} \}. \qquad (0.2)$$

It moreover allows to formulate a fast algorithm, the so-called discrete wavelet transform. Daubechies' construction of smooth compactly supported wavelets ψ generating an orthonormal basis of the form (0.2) is one of the most celebrated results in wavelet theory [Dau92].

A Gabor system – the counterpart in time-frequency analysis – involves translates and modulations of one single function. A common choice are regular Gabor systems

$$\{ g_{j,k}(t) := e^{-2\pi i bk \cdot t} g(t - aj),\ j, k \in \mathbb{Z}^d \}, \quad a, b > 0.$$

Provided the parameters a, b and the function g satisfy certain properties, such a system forms a frame for $L^2(\mathbb{R}^d)$. Essentially, a frame is a complete redundant system allowing stable reconstruction, and it can therefore be considered as a generalized orthonormal basis. Thus, any $f \in L^2(\mathbb{R}^d)$ has a series expansion

$$f(t) = \sum_{j,k \in \mathbb{Z}^d} c_{j,k}(f) e^{-2\pi i bk \cdot t} g(t - aj), \qquad t \in \mathbb{R}^d, \qquad (0.3)$$

with certain coefficients $c_{j,k}(f) \in \mathbb{C}$.

Grossmann, Morlet and Paul observed in [GMP85, GMP86] that time-frequency analysis and wavelet analysis are closely connected to the theory of square-integrable group representations. Let π be such a representation of a locally compact group $\mathcal{G}$ on some Hilbert space $\mathcal{H}$. The (abstract) wavelet transform – also called voice transform – is then defined by $V_g f(x) = \langle f, \pi(x)g \rangle_{\mathcal{H}}$ for $x \in \mathcal{G}$, $f, g \in \mathcal{H}$. Taking a certain representation of the similitude group $\mathbb{R}^d \rtimes (SO(d) \times (0, \infty))$ on $\mathcal{H} = L^2(\mathbb{R}^d)$, the CWT can be obtained in this way. On the other hand, using the Schrödinger representation of the Heisenberg group we essentially get the STFT. This observation allows to treat some aspects of both time-frequency and wavelet analysis simultaneously under the common framework of representation theory.

Indeed, Feichtinger and Gröchenig developed a discretization method which works in this abstract setting [FG88, FG89a, FG89b, FG92c, Grö91]. Based on a careful analysis of convolution operators on the group $\mathcal{G}$ they constructed frames for $\mathcal{H}$ of the form $\{\pi(x_i)g\}_{i\in I}$. Their approach does not only work in the Hilbert space $\mathcal{H}$ but also in certain Banach spaces associated to π, called coorbit spaces. In the special case of wavelet analysis these spaces coincide with the Besov and Triebel-Lizorkin spaces. Their counterpart in time-frequency analysis are the modulation spaces. Based on the discretizations it is possible to study certain properties of the coorbit spaces and, in particular, embedding theorems. This theory is nowadays known as coorbit space theory or Feichtinger-Gröchenig theory.

This thesis is concerned with an aspect of time-frequency and wavelet analysis that has not yet been treated thoroughly in the literature. We consider functions on $\mathbb{R}^d$ that are invariant under some symmetry group. Here, a symmetry group is understood to be a closed subgroup $\mathcal{A}$ of the orthogonal group $O(d)$. Invariance of a function f under $\mathcal{A}$ means that $f(A^{-1}x) = f(x)$ for all $A \in \mathcal{A}$ and $x \in \mathbb{R}^d$. We treat the question whether it is possible to exploit the symmetry in some way when doing time-frequency or wavelet analysis with invariant functions. Indeed, it sounds reasonable that one can take advantage of the additional information that the function under consideration possesses symmetries. For instance, one should be able to reduce complexity or improve the approximation quality. This is in spirit of classical mathematical physics, where symmetries are introduced in order to reduce multidimensional partial differential equations to lower-dimensional ones, or even to ordinary differential equations.

The functions that are invariant under $\mathcal{A} = O(d)$ are called radial functions. For instance, a purely radial setting appears naturally when separating variables in polar coordinates and treating the radial and the spherical part separately. There have already been a few attempts to treat wavelet analysis of radial functions. Trimèche [Tri95, Tri97] and Rösler [Rös97] independently introduced a continuous radial wavelet transform. A discrete radial wavelet transform was developed by Epperson and Frazier in [EF95]. Recently, two different approaches for the short time Fourier transform of radial functions were proposed [CG03b, Dac01, Tri97]. However, they both lack important properties.

We take a systematic approach to time-frequency and wavelet analysis of functions which are invariant under general symmetry groups. We work in the abstract setting of square-integrable group representations. First, we develop a satisfying theory for the (abstract) wavelet transform of invariant functions. Based on this we work out a generalized coorbit space theory. This allows us to formulate discretization theorems which are adapted to invariant functions. As in classical coorbit space theory, these theorems are not only valid for the corresponding Hilbert space, but also for certain Banach spaces, like Besov and Triebel-Lizorkin or modulation spaces of invariant functions. Furthermore, we adapt the concept of multiresolution analysis to the special case of radial functions on $\mathbb{R}^3$. This enables us to construct an orthonormal radial wavelet basis for $L^2_{rad}(\mathbb{R}^3)$, the space of radial square-integrable functions.

Our theory leads to a natural approach to time-frequency analysis of radial functions. It avoids the drawbacks of the previous attempts [CG03b, Dac01, Tri97]. In particular, we provide an explicit construction method for radial Gabor frames. Results of this kind have not yet appeared. In the special case of wavelet analysis of radial functions we recover the radial CWT of Rösler and Trimèche. Our discretization theorems give results similar to those derived by Epperson and Frazier [EF95]. However, we can work under less restrictive assumptions.

When trying to exploit symmetry properties in time-frequency and wavelet analysis, one immediately makes the following easy but crucial observation. It goes without saying that the wavelet expansion (0.1) and the Gabor expansion (0.3) also apply to invariant functions f. But even if a function g is invariant under some subgroup $\mathcal{A}$ of $O(d)$, none of its translates is invariant again. Therefore both expansions (0.1) and (0.3) represent a series approximation of f by non-invariant functions. In particular, an approximation of f by a partial sum of the form (0.1) or (0.3), respectively, is in general not invariant under $\mathcal{A}$.

In light of this observation our problem can be rephrased as follows. Is it possible to construct Gabor-like and wavelet-like expansions of the form

$$f = \sum_{i \in I} c_i \widetilde{\pi}(x_i) g, \qquad (0.4)$$

where all elements $\widetilde{\pi}(x_i)g$ are invariant under $\mathcal{A}$? So our first task is to find an operator $\widetilde{\pi}(x)$ depending on some parameter x which maps invariant functions onto invariant ones. This operator should be designed such that the expansion (0.4) can still be interpreted as some kind of wavelet or Gabor expansion.

We show that a natural definition of such an operator $\widetilde{\pi}(x)$ can be obtained in the abstract setting of representation theory. Once again we assume π to be a square-integrable representation of some locally compact group $\mathcal{G}$ on some Hilbert space $\mathcal{H}$. The symmetry group is realized as a compact automorphism group $\mathcal{A}$ of $\mathcal{G}$. It acts on $\mathcal{H}$ by means of another representation σ. In addition, we have to require that the representations π and σ satisfy the relation

$$\pi(A(x))\sigma(A) = \sigma(A)\pi(x) \qquad \text{for all } A \in \mathcal{A}, x \in \mathcal{G}. \qquad (0.5)$$

The invariant elements of $\mathcal{H}$ are those that satisfy $\sigma(A)f = f$ for all $A \in \mathcal{A}$. We show that the desired operator $\widetilde{\pi}$ is given by

$$\widetilde{\pi}(x) = \int_{\mathcal{A}} \pi(A(x))dA, \quad x \in \mathcal{G},$$

where dA denotes the Haar measure of $\mathcal{A}$. Indeed, if g is invariant under $\mathcal{A}$ then also $\widetilde{\pi}(x)g$ is invariant under $\mathcal{A}$. It further turns out that

$$V_g f(x) = \langle f, \pi(x)f \rangle = \langle f, \widetilde{\pi}(x)g \rangle \tag{0.6}$$

for invariant elements $f, g \in \mathcal{H}$. We derive a new inversion formula for the restriction $\widetilde{V}_g$ of the wavelet transform V_g to the subspace $\mathcal{H}_{\mathcal{A}}$ of invariant elements. Furthermore, we prove a covariance principle for $\widetilde{V}_g$.

With formula (0.6) we already achieve a reduction of complexity. For instance, let $\mathcal{H} = L^2(\mathbb{R}^d)$ and assume that we have radial symmetry. Then the inner product in $L^2(\mathbb{R}^d)$ can be computed by an integral over the positive half-line $\mathbb{R}_+$, which means that the dimension is reduced from d to 1.

Applied to wavelet analysis of radial functions the operator $\widetilde{\pi}$ coincides with the one that was used in [Tri97, Rös97, EF95]. Here, the usual translation on $\mathbb{R}^d$ is replaced by a generalized translation τ. This operator τ performs a usual translation followed by a projection onto the radial functions by averaging over spheres.

In the special case of time-frequency analysis of radial functions we essentially obtain the operator

$$\widetilde{\pi}(x, \omega)f(t) = \int_{SO(d)} e^{-2\pi i A\omega \cdot t} f(t - Ax)dA. \tag{0.7}$$

The corresponding "radial STFT" has not yet been considered in the literature. It can be written as an integral over $\mathbb{R}_+ := [0, \infty)$.

There is a remarkable connection to hypergroup theory. Hypergroups are certain generalizations of locally compact groups. Radial functions are closely connected to the so-called Bessel-Kingman hypergroups. Moreover, for invariant $f, g \in \mathcal{H}_{\mathcal{A}}$ the transform $\widetilde{V}_g f$ is a function on the group $\mathcal{G}$ that is invariant under $\mathcal{A}$. Therefore, it can be considered as a function on the space $\mathcal{G}^{\mathcal{A}}$ of all orbits $\mathcal{A}x$ of $\mathcal{G}$ under $\mathcal{A}$. Again, $\mathcal{G}^{\mathcal{A}}$ possesses the structure of a hypergroup. We prove that the operators $\widetilde{\pi}(x)$ form an irreducible representation of the hypergroup $\mathcal{G}^{\mathcal{A}}$ on $\mathcal{H}_{\mathcal{A}}$. Moreover, we discover a new non-commutative hypergroup. It appears as the resulting orbit space when we choose $\mathcal{G}$ as the Heisenberg group and $\mathcal{A}$ as the special orthogonal group $SO(d)$.

The Fourier transform of a radial function can be computed by the Hankel transform. Its kernel – called the spherical Bessel function – is closely related to the generalized translation τ by means of a product formula. In the special case of $\mathbb{R}^3$ the spherical

Bessel function coincides with the sinc-function $t \mapsto \frac{\sin(t)}{t}$. Its explicit form allows us to develop a multiresolution analysis for radial functions on $\mathbb{R}^3$. Thus, we arrive at the construction of an orthonormal radial wavelet basis of $L^2_{rad}(\mathbb{R}^3)$. We also formulate fast algorithms. Our construction heavily depends on the fact that the zeros of the sinc-function are equidistant. Consequently, there seems to be no hope to generalize this construction to arbitrary dimensions. This part of the thesis was developed in cooperation with M. Rösler and will also appear in a joint paper [RR03].

In order to treat the discretization problem in the general abstract setting, we develop a generalized coorbit space theory. This enables us to construct frames for $\mathcal{H}_{\mathcal{A}}$ of the form $\{\widetilde{\pi}(x_i)g\}_{i\in I}$. Moreover, our results do not only hold for $\mathcal{H}_{\mathcal{A}}$ but also for more general coorbit spaces. Classical coorbit space theory requires quite a number of tools. Each of them has to be adapted to our situation. For instance, we prove that there exist certain coverings of the group $\mathcal{G}$ involving sets which are invariant under $\mathcal{A}$. Also, we investigate some properties of Wiener amalgam spaces on $\mathcal{G}$ consisting of invariant functions.

With our abstract discretization theorems we give an answer to the initial question. An invariant function f can be developed into a series expansion of the form (0.4). Hereby, each building block $\widetilde{\pi}(x_i)g$ is itself invariant and, therefore, the same is true for any approximation of f by a partial sum. Applying the abstract theorems to time-frequency analysis of radial functions we are able to construct explicitly a "radial Gabor frame". We emphasize again that any element of this frame is a radial function and is constructed from a single function g by applying a generalized time-frequency shift of the form (0.7). Our result holds not only for $L^2_{rad}(\mathbb{R}^d)$ but also for general radial modulation spaces. A construction of this kind has not yet been presented.

Our abstract discretization theorems also allow us to construct radial wavelet frames explicitly. Once again, each element of the frame is a radial function. Furthermore, all elements are obtained from a single mother wavelet ψ by applying a dilation and a generalized translation τ. Our result extends from $L^2_{rad}(\mathbb{R}^d)$ to general Besov and Triebel-Lizorkin spaces of radial functions. Our wavelet frames are of similar type as in [EF95]. However, we allow more freedom in the choice of the mother wavelet and of the discretization lattice.

We expect that our results can be used to study properties of function spaces of invariant functions. Sickel and Skrzypczak observed in [SS00] that certain non-compact embeddings of Besov and Triebel-Lizorkin spaces become compact when restricting the latter to radial functions. It seems that our results can be used to give an abstract explanation for this phenomenon. Furthermore, we conjecture that our radial Gabor frames yield an appropriate tool to prove similar results for the modulation spaces of radial functions. Results of this kind would be new to the literature.

This thesis is organized as follows. In Chapter 1 we collect some preliminaries and introduce notation. We treat the harmonic analysis of functions on locally compact groups that are invariant under some symmetry group. In particular, we give a detailed account of Fourier analysis of radial functions. Furthermore, basics on hypergroup theory are introduced. We discuss the short time Fourier transform and the continuous wavelet transform. We also give some information on the theory of square-integrable group representations and explain how they are related to time-frequency and wavelet analysis.

In Chapter 2 we treat the (abstract) wavelet transform of invariant functions. Among other properties we prove a new inversion formula. The special cases of time-frequency analysis and wavelet analysis of radial functions are discussed in detail. Parts of this chapter will appear in [Rau03b].

Chapter 3 is devoted to the radial multiresolution analysis in $\mathbb{R}^3$. It is part of a joint paper with Rösler [RR03]. We carry out our construction similarly to the classical case and show how to obtain a radial orthonormal wavelet basis for $L^2_{rad}(\mathbb{R}^d)$. We point out a relation of radial wavelets to classical wavelets on $\mathbb{R}$ and present fast wavelet algorithms.

In Chapter 4 we develop our generalized coorbit space theory. First we study properties of certain translation invariant function spaces. We show the existence of a particular covering of a general locally compact group that involves invariant sets and construct an associated partition of unity. Then we study certain sequence spaces related to this covering and to a function space on the group. Moreover, we introduce Wiener amalgam spaces and investigate properties of their subspaces of invariant elements. After this preparation we introduce the coorbit spaces and their subspaces of invariant elements. We show basic properties of these spaces. Next we state the abstract discretization theorems and provide their proofs. Finally, we carry out in detail how to apply coorbit space theory to time-frequency analysis and wavelet analysis of radial functions. In particular, we construct radial Gabor frames and radial wavelet frames explicitly. Parts of this chapter were published as a preprint [Rau03a].

Appendix A collects some information about integration on spheres and on the special orthogonal group $SO(d)$. Finally, Appendix B discusses previous approaches to the short time Fourier transform for radial functions, or for even more general hypergroups [CG03b, Dac01, Dac03, Tri97].

I would like to thank my supervisor Prof. Rupert Lasser for his support in carrying out my research. I am also very grateful to Prof. Hans G. Feichtinger for valuable comments and interesting discussions. Parts of this thesis were developed during a stay with his research group NuHAG at the University of Vienna in spring 2003. I would like to express my gratitude to Frank Filbir for his useful hints and proof-reading of parts of this thesis. He always supported me and in particular initiated my stay in Vienna. I would also like to thank Margit Rösler who provided good ideas and was my coauthor of the joint paper [RR03]. Special thanks go to Massimo Fornasier for lots of exciting discussions. It was also nice to have valuable comments from Hartmut Führ, Prof. Karlheinz Gröchenig, Prof. Stephan Dahlke and Wolfgang zu Castell-Rüdenhausen. I particularly enjoyed a stay in Marburg with the group of Prof. Stephan Dahlke. Furthermore, I would like to thank the graduate program "Applied Algorithmic Mathematics" funded by the Deutsche Forschungsgemeinschaft for financial support. My stay in Vienna was partially supported by the European Union's Human Potential Programme under contract HPRN-CT-2002-00285 (HASSIP). I am grateful to all members of the graduate program and of NuHAG for providing a nice environment for my research. My very special thanks go to my dear Daniela Hobst.

Munich, June 14, 2004   Holger Rauhut

Chapter 1

Harmonic Analysis in Various Settings

In this chapter we discuss the preliminaries that are necessary for our investigations. In particular, various aspects of harmonic analysis will be important for us.

In Section 1.1 we fix some notation and introduce important function spaces, like L^p-spaces and spaces of continuous functions. Since our main interest lies in functions with symmetry properties, we treat a very general situation in Section 1.2. Namely, we consider functions on a locally compact group $\mathcal{G}$, which are invariant under some compact automorphism group of $\mathcal{G}$. In particular, we study the convolution of invariant functions. Our main example consists of radial functions on $\mathbb{R}^d$. The harmonic analysis of such functions is treated in detail in Section 1.3. Motivated by the observation that invariant functions on groups, and in particular radial functions, can be interpreted as functions on certain hypergroups, we discuss some basics of hypergroup theory in Section 1.4. As our main interest lies in time-frequency analysis and wavelet analysis we discuss the short time Fourier transform and the continuous wavelet transform in Section 1.5. Moreover, we explain their relationship to square-integrable group representations.

1.1 Notation

Assume that X is a locally compact Hausdorff space. Then $C(X)$ denotes the space of complex-valued continuous functions on X, and $C^b(X), C_0(X), C_c(X)$ its subspace of bounded functions, functions vanishing at infinity and compactly supported functions, respectively. For some compact set $K \subset X$ we let $C(X, K) := \{f \in C_c(X), \operatorname{supp} f \subset K\}$, where $\operatorname{supp} f$ denotes the support of f. Together with the norm $\|f\|_\infty := \sup_{x \in X} |f(x)|$ the spaces $C^b(X), C_0(X), C(X, K)$ are Banach spaces. Let $(K_i)_{i \in I}$ denote a family of compact sets of X such that any arbitrary compact set K is contained in at least one of the K_i. The space $C_c(X)$ is endowed with the inductive limit topology of the family of spaces $C(X, K_i)$, i.e., the strongest locally convex topology that makes all embedding mappings $C(X, K_i) \to C_c(X)$, $i \in I$, continuous, see also [Edw65, Chapter 6.3]. If their is no danger of confusion we sometimes write C, C^b, C_0 or C_c for short. By the Riesz representation theorem [Rud87, Theorem 6.19] the dual space of $C_0(X)$ can be identified with the space $M(X)$ of all regular complex Borel measures on X. Together with the total variation norm $M(X)$ is a Banach space. Moreover, the topological dual of $C_c(X)$ can be identified with the space of all Radon measures on X, see [Edw65, Chapter 4.3]. The product measure of two Radon measures ρ, μ is denoted by $\rho \otimes \mu$.

Assume further that we have given a positive Radon measure μ on X. Then for $1 \leq p < \infty$ the Lebesgue space $L^p(X, \mu)$ is defined as the collection of all μ-measurable functions whose norm

$$\|f|L^p(X,\mu)\| := \|f\|_p := \left(\int_X |f(x)|^p d\mu(x) \right)^{1/p}$$

is finite. Furthermore, $L^\infty(X, \mu)$ consists of all μ-measurable functions f, for which

$$\|f|L^\infty(X,\mu)\| := \|f\|_\infty := \operatorname{ess\,sup}_{x \in X} |f(x)| < \infty,$$

where the essential supremum is taken with respect to μ. As usual one identifies functions in $L^p(X, \mu)$, which differ only on a set of measure zero. If $p = 2$ then $L^2(X, \mu)$ is a Hilbert space with inner product

$$\langle f, g \rangle_{L^2(X,\mu)} = \int_X f(x)\overline{g(x)} d\mu(x).$$

We sometimes omit the subscript at the inner product if no confusion can arise. Also, we will write $L^p(X)$ or simply L^p if the measure μ and the space X are clear from the context. In particular, if $X = \mathbb{R}^d$, $d \in \mathbb{N}$, then the measure μ will always be the Lebesgue measure and we write $L^p(\mathbb{R}^d)$ for the corresponding spaces. More generally, if $\mathcal{G}$ is a locally compact group then μ denotes its Haar measure and $L^p(\mathcal{G}) = L^p(\mathcal{G}, \mu)$ the corresponding Lebesgue space.

For some strictly positive measurable weight function w on X we define the corresponding weighted Lebesgue spaces by

$$L_w^p(X, \mu) := \{f \text{ measurable}, fw \in L^p(X, \mu)\}, \quad 1 \leq p \leq \infty.$$

Its norm is given by $\|f|L_w^p(X, \mu)\| := \|fw|L^p(X, \mu)\|$. Again, we use the abbreviations $L_w^p(X)$ or L_w^p. In case that $X = I$ is a discrete space endowed with the counting measure then the corresponding Lebesgue space is denoted by $l_w^p(I)$, or simply $l^p(I)$ if the weight is trivial.

The space $L_{loc}^1(X, \mu)$ of locally integrable functions, is defined to be the collection of μ-measurable functions f, for which

$$p_K(f) := \int_K |f(x)| d\mu(x) < \infty \qquad (1.1)$$

for all compact subsets $K \subset X$. We endow $L_{loc}^1(X, \mu)$ with the locally convex topology generated by the collection of semi-norms $\{p_K, K \subset X, K \text{ compact}\}$. The spaces $C^b(\mathcal{G}), C_0(\mathcal{G})$ and $L^p(X, \mu)$, $1 \leq p \leq \infty$, are continuously embedded into $L_{loc}^1(X, \mu)$. Moreover, a function $f \in L_{loc}^1(X, \mu)$ can be identified with a Radon measure by $\mu_f(h) = \int_X h(x) f(x) d\mu(x)$, $h \in C_c(X)$. In particular, $L_{loc}^1(X, \mu)$ is a subspace of the dual space of $C_c(X)$ and, hence, any $f \in L_{loc}^1(X, \mu)$ is completely described by its action on $C_c(X)$. Moreover, $L^1(X, \mu)$ can be identified with a closed subspace of $M(X)$ in this way.

More generally, if B is some Banach space of functions (measures) on X, then B_{loc} is defined to be the collection of elements f, for which

$$p_{K,B}(f) := \inf\{\|g|B\|, g \in B, f|_K = g|_K\} < \infty$$

for all compact $K \subset X$. Again, B_{loc} carries the topology generated by the semi-norm $p_{K,B}$.

Suppose further that $\mathbb{S}(X)$ is some locally convex vector space of (test) functions on X. (Hereby, we assume any vector space topology to be Hausdorff throughout this thesis.) We agree to call the elements of its topological dual $\mathbb{S}'(X)$ distributions on X. We endow the space $\mathbb{S}'(X)$ with the weak-$*$ topology, i.e., the $\sigma(\mathbb{S}'(X), \mathbb{S}(X))$-topology. Whenever $\mathbb{S}(X) \subset L^2(X, \mu)$ with dense and continuous embedding then elements of $\mathbb{S}(X)$ may be identified with elements of its dual by $f(g) = \int_X f(x)g(x) d\mu(x)$, $f, g \in \mathbb{S}(X)$, and, hence, $\mathbb{S}(X) \subset \mathbb{S}'(X)$.

In particular, let $X = \mathbb{R}^d$. We denote by $\mathcal{S}(\mathbb{R}^d)$ the **Schwartz space** of rapidly decreasing infinitely differentiable functions, which becomes a Fréchet space with the topology generated by the semi-norms

$$p_{\alpha, \beta}(f) := \sup_{x \in \mathbb{R}^d} |x^\alpha D^\beta f(x)|, \quad \alpha, \beta \in \mathbb{N}_0^d,$$

where $x^\alpha = x_1^{\alpha_1} \cdots x_d^{\alpha_d}$ and $D^\beta f = \frac{\partial^{|\beta|}}{\partial^{\beta_1} x_1 \cdots \partial^{\beta_d} x_d} f$. Elements of its topological dual $\mathcal{S}'(\mathbb{R}^d)$ are called tempered distributions. We extend the inner product of $L^2(\mathbb{R}^d)$ to a sesquilinear mapping $\mathcal{S}'(\mathbb{R}^d) \times \mathcal{S}(\mathbb{R}^d) \to \mathbb{C}$ by

$$\langle \phi, f \rangle := \phi(\overline{f}), \quad \phi \in \mathcal{S}'(\mathbb{R}^d), f \in \mathcal{S}(R^d),$$

where, as usual, $\overline{f}$ denotes the complex conjugate of the function f.

If T is a bounded operator from some Banach space B_1 into another Banach space B_2 we denote its operator norm by

$$\|T|B_1 \to B_2\| = \sup_{f \in B_1 \backslash \{0\}} \frac{\|Tf|B_2\|}{\|f|B_1\|}.$$

If $B = B_1 = B_2$ then we will use the abbreviation $\|T|B\|$.

The scalar product of $x, y \in \mathbb{R}^d$ is denoted by $x \cdot y$ and the Euclidean norm by $|x| = \sqrt{x \cdot x}$. Further, we agree to write $A \asymp B$ if there exist constants $C_1, C_2 > 0$ (independent of some parameters on which A and B might depend) such that $C_1 A \leq B \leq C_2 A$. Finally, we denote $F(x) \sim G(x)$ if $\lim_{x \to \infty} \frac{F(x)}{G(x)} = 1$ (provided $x \to \infty$ has some reasonable meaning).

1.2 Invariant Functions on Groups

In this thesis we are mainly interested in the analysis of functions with symmetry properties. A very general setup for such investigations are functions on a locally compact group $\mathcal{G}$, which are invariant under some compact automorphism group of $\mathcal{G}$. So we assume $\mathcal{G}$ to a locally compact group with unit element e, left invariant Haar measure μ and modular function Δ. In order to avoid technicalities at some places we additionally suppose $\mathcal{G}$ to be σ-compact. Further, let $\mathcal{A}$ be a compact automorphism group of $\mathcal{G}$ (a symmetry group) such that $\mathcal{A}$ acts continuously on $\mathcal{G}$, i.e., the mapping $\mathcal{G} \times \mathcal{A} \to \mathcal{G}, (x, A) \mapsto A(x)$ is continuous. Hereby, the product topology is taken on $\mathcal{G} \times \mathcal{A}$. The Haar measure of $\mathcal{A}$ is denoted by ν and we assume ν to be normalized, i.e., $\nu(\mathcal{A}) = 1$. Clearly, ν is also right invariant by compactness of $\mathcal{A}$. For simplicity, we will usually use the short forms $\int_{\mathcal{G}} \cdots dx$ and $\int_{\mathcal{A}} \cdots dA$ for integrals with respect to μ and ν, respectively, if this causes no confusion.

Given a function f on $\mathcal{G}$ and $A \in \mathcal{A}$ we let

$$U_A(f)(x) := f_A(x) := f(A^{-1}x).$$

More generally, suppose $\mathbb{S}(\mathcal{G})$ is some locally convex vector space of test functions on $\mathcal{G}$ satisfying $U_A(\mathbb{S}(\mathcal{G})) \subset \mathbb{S}(\mathcal{G})$. Then for a distribution $\phi \in \mathbb{S}'(\mathcal{G})$ we define the action of $A \in \mathcal{A}$ on $\mathbb{S}'(\mathcal{G})$ by

$$(U_A\phi)(f) := \phi_A(f) := \phi(f_{A^{-1}}) \qquad \text{for all } f \in \mathbb{S}(\mathcal{G}), \phi \in \mathbb{S}'(\mathcal{G}).$$

Taking in particular $\mathbb{S}(\mathcal{G}) = C_c(\mathcal{G})$ this definition applies to Radon measures on $\mathcal{G}$.

Definition 1.2.1. A function (measure, distribution) f is called invariant under $\mathcal{A}$, if $f_A = f$ for all $A \in \mathcal{A}$.

The following lemma will be useful in our context.

Lemma 1.2.1. *The Haar measure μ and the modular function Δ of $\mathcal{G}$ are invariant under the action of $\mathcal{A}$, i.e.,*

$$\int_{\mathcal{G}} f_A(x)dx = \int_{\mathcal{G}} f(x)d\mu(x) \qquad and \qquad \Delta(A^{-1}x) = \Delta(x)$$

for all $A \in \mathcal{A}, x \in \mathcal{G}$ and $f \in L^1(\mathcal{G})$.

Proof: Since

$$\int_{\mathcal{G}} f(A(yx))d\mu(x) = \int_{\mathcal{G}} f(A(y)A(x))d\mu(x) = \int_{\mathcal{G}} f(A(x))d\mu(x)$$

for all $f \in C_c(\mathcal{G})$ the measure μ_A is left invariant. By uniqueness of the Haar measure we conclude $\mu_A = \lambda(A)\mu$ with some number $\lambda(A) > 0$. We choose a non-zero positive function $f \in C_c(\mathcal{G})$ which is invariant under $\mathcal{A}$. Such a function exists: Choose any function $g \geq 0, g \neq 0$ on $\mathcal{G}$ with compact support and put $f(x) = \int_A g(A^{-1}x)dA$. The integral is finite for each x, $f \geq 0$, $f \neq 0$ and f has compact support by compactness of $\mathcal{A}$. Furthermore, $f(B^{-1}x) = \int_A g((BA)^{-1}(x))dA = \int_A g(A^{-1}x)dA = f(x)$ by invariance of the Haar measure of $\mathcal{A}$. With this function f it holds $0 \neq \mu_A(f) = \mu(f_{A^{-1}}) = \mu(f)$ and, hence, $\lambda(A) = 1$ for all $A \in \mathcal{A}$.

The invariance of the modular function is now an easy consequence. For any function $f \in C_c(\mathcal{G})$ which is invariant under $\mathcal{A}$ it holds

$$\Delta(y) \int_{\mathcal{G}} f(x)dx = \int_{\mathcal{G}} f(xy^{-1})dy = \int_{\mathcal{G}} f(A^{-1}(x)y^{-1})dx$$
$$= \int_{\mathcal{G}} f(A^{-1}(xA(y^{-1})))dx = \int_{\mathcal{G}} f(xA(y^{-1}))dx = \Delta(Ay) \int_{\mathcal{G}} f(x)dx.$$

Hence, $\Delta(Ay) = \Delta(y)$ for all $A \in \mathcal{A}$. $\qquad\square$

We remark that the proof relied on the fact that $\mathcal{A}$ is compact. It follows also that the mapping $f \mapsto f_A$ is an isometry of $L^p(\mathcal{G})$ for all $1 \leq p \leq \infty$. Moreover, given a function $f \in L^1_{loc}(\mathcal{G})$ we may identify it with a measure μ_f by setting

$$\mu_f(g) = \int_{\mathcal{G}} g(x)f(x)dx, \quad g \in C_c(\mathcal{G}). \tag{1.2}$$

Lemma 1.2.1 shows that

$$(\mu_f)_A(g) = \mu_f(g_{A^{-1}}) = \int_{\mathcal{G}} g(Ax)f(x)dx = \int_{\mathcal{G}} g(x)f(A^{-1}(x))dx = \mu_{(f_A)}(g),$$

which means that the extension of the action of $\mathcal{A}$ to Radon measures is consistent with identifying functions as measures.

For some topological vector space Y of functions (measures, distributions) on $\mathcal{G}$ we denote by

$$Y_{\mathcal{A}} := \{f \in Y, f_A = f \text{ for all } A \in \mathcal{A}\}$$

its subspace of invariant elements. If U_A acts continuously on Y for all $A \in \mathcal{A}$ then $Y_{\mathcal{A}} = \cap_{A \in \mathcal{A}} \ker(\mathrm{Id} - U_A)$ is a closed subspace of Y. In particular, the operator U_A acts continuously on the spaces $C_0(\mathcal{G}), C^b(\mathcal{G}), M(\mathcal{G}), L^p(\mathcal{G}), 1 \leq p \leq \infty$ and on $L^1_{loc}(\mathcal{G})$, and, hence, the corresponding subspaces of invariant elements are closed.

Further, if f is a function on $\mathcal{G}$ then we let

$$P_{\mathcal{A}}f(x) := \int_{\mathcal{A}} f(A^{-1}x)dA \tag{1.3}$$

whenever this integral is well-defined (almost everywhere).

Lemma 1.2.2.   *(a) If $P_{\mathcal{A}}f$ is well-defined (a.e.) then $P_{\mathcal{A}}f$ is invariant under $\mathcal{A}$ and $P_{\mathcal{A}}^2 f = P_{\mathcal{A}}f$, i.e., $P_{\mathcal{A}}$ is a projection.*

(b) If f is continuous then also $P_{\mathcal{A}}f$ is continuous. Moreover, $P_{\mathcal{A}}$ maps $C_c(\mathcal{G})$ onto $(C_c)_{\mathcal{A}}(\mathcal{G})$.

(c) The operator $P_{\mathcal{A}}$ is a continuous projection from L^1_{loc} onto its closed subspace $(L^1_{loc})_{\mathcal{A}}$.

(d) If Y is a Banach space of functions (measures, distributions), on which U_A acts isometrically, then

$$P_{\mathcal{A}}f = \int_{\mathcal{A}} f_A dA, \qquad f \in Y, \tag{1.4}$$

to be interpreted as Bochner integral, defines a bounded projection from Y onto its closed subspace $Y_{\mathcal{A}}$ with $\|P_{\mathcal{A}}|Y\| \leq 1$. Moreover, if Y is continuously embedded into $L^1_{loc}(\mathcal{G})$ then the definitions (1.3) and (1.4) are equivalent. In particular, this assertion applies to the spaces $C^b(\mathcal{G}), C_0(\mathcal{G}), L^p(\mathcal{G}), 1 \leq p \leq \infty$ and, except for the continuous embedding into $L^1_{loc}(\mathcal{G})$, also to $M(\mathcal{G})$.

(e) On $L^2(\mathcal{G})$ the operator $P_{\mathcal{A}}$ is an orthogonal projection.

(f) Suppose $\mathbb{S}$ is a locally convex space of test functions, on which $P_{\mathcal{A}}$ acts continuously. Then $P_{\mathcal{A}}$ extends to a continuous projection from $\mathbb{S}'$ onto the closed subspace $(\mathbb{S}')_{\mathcal{A}}$ by

$$(P_{\mathcal{A}}\phi)(f) := \phi(P_{\mathcal{A}}f) \qquad for \ \phi \in \mathbb{S}'(\mathcal{G}), f \in \mathbb{S}(\mathcal{G}). \tag{1.5}$$

Moreover, if $\mathbb{S} \subset L^2(\mathcal{G})$ with dense and continuous embedding (implying $\mathbb{S} \subset \mathbb{S}'$) then the definitions (1.3) and (1.5) coincide on $\mathbb{S}$. Further, if $\mathbb{S}$ is a Banach space continuously embedded into L^1_{loc}, on which $\mathcal{A}$ acts continuously, then the definitions (1.4) and (1.5) are equivalent on $\mathbb{S}'$, in particular on $M(\mathcal{G}) = C_0'(\mathcal{G})$.

(g) Let $\mathbb{S}$ as in (f). Then $\widetilde{P}_{\mathcal{A}} : (\mathbb{S}_{\mathcal{A}})' \to (\mathbb{S}')_{\mathcal{A}}$, $(\widetilde{P}_{\mathcal{A}}\phi)(f) := \phi(P_{\mathcal{A}}f)$ for $\phi \in (\mathbb{S}_{\mathcal{A}})', f \in \mathbb{S}$ defines a homeomorphism. If $\mathbb{S}$ is a Banach space, on which $\mathcal{A}$ acts isometrically, then $\widetilde{P}_{\mathcal{A}}$ is an isometry. Hence, we may unambiguously write $\mathbb{S}'_{\mathcal{A}}$.

Proof: (a) The relation $P_{\mathcal{A}}f(Ax) = P_{\mathcal{A}}f(x)$ for all $A \in \mathcal{A}$ follows from the left invariance of the Haar measure of $\mathcal{A}$ and $P_{\mathcal{A}}^2 f = P_{\mathcal{A}}f$ is an immediate consequence.

(b) By Lebesgue's dominated convergence theorem and compactness of $\mathcal{A}$ we may interchange limits and the integral over $\mathcal{A}$ for continuous functions. This implies the first assertion. Moreover, if f has compact support on K, say, then supp $P_{\mathcal{A}}f \subset \mathcal{A}(K)$, and $\mathcal{A}(K)$ is again compact by compactness of $\mathcal{A}$ and K.

(c) If $f \in L^1_{loc}$ then for any compact set $K \subset \mathcal{G}$ we obtain by Fubini's theorem and Lemma 1.2.1

$$p_K(P_{\mathcal{A}}f) = \int_K \left| \int_{\mathcal{A}} f(A^{-1}x)dA \right| dx \leq \int_{\mathcal{A}} \int_K |f(A^{-1}x)|dxdA = \int_{\mathcal{A}} \int_{\mathcal{A}(K)} |f(x)|dxdA$$

$$\leq \int_{\mathcal{A}} \int_{\mathcal{A}(K)} |f(x)|dxdA = p_{\mathcal{A}(K)}(f).$$

Since by compactness of $\mathcal{A}$ also $\mathcal{A}(K)$ is compact, this shows that $P_{\mathcal{A}}f \in L^1_{loc}$ and the operator $P_{\mathcal{A}} : L^1_{loc} \to L^1_{loc}$ is continuous. As in (a) ones proves that $P_{\mathcal{A}}$ is a projection onto $(L^1_{loc})_{\mathcal{A}}$.

(d) Since $U_{\mathcal{A}}$ is an isometry on Y we have $\int_{\mathcal{A}} \|f_A|Y\|dA = \int_{\mathcal{A}} dA\|f|Y\| = \|f|Y\|$. Hence, the Bochner integral is well-defined, $P_{\mathcal{A}}f$ is an element of Y and it holds $\|P_{\mathcal{A}}|Y\| \leq 1$, see [Yos80, Theorem 1, Chapter V.5]. By left invariance of the Haar measure we further deduce that $P_{\mathcal{A}}f$ is invariant under $\mathcal{A}$ and $P_{\mathcal{A}}^2 = P_{\mathcal{A}}$.

Now assume that Y is continuously embedded into $L^1_{loc}(\mathcal{G})$ and let K be some compact subset of $\mathcal{G}$. For some moment we denote the operator in (1.4) by $\widehat{P}_{\mathcal{A}}$. Since $p_{\mathcal{A}(K)}(f) \leq C_{\mathcal{A}(K)}\|f|Y\|$ for some constant $C_{\mathcal{A}(K)}$ the restriction of $\widehat{P}_{\mathcal{A}}f$ to $\mathcal{A}(K)$ is given by $\int_{\mathcal{A}} (f|_{\mathcal{A}(K)})_A dA$, to be interpreted as a Bochner integral on $L^1(\mathcal{A}(K))$. Since $(L^1(\mathcal{A}(K)))' = L^\infty(\mathcal{A}(K))$ we obtain for all $f \in Y$ and all $g \in L^\infty(\mathcal{A}(K))$

$$\int_{\mathcal{A}(K)} (P_{\mathcal{A}}f)(x)g(x)dx = \int_{\mathcal{A}} \int_{\mathcal{A}(K)} f(A^{-1}x)g(x)dxdA = \int_{\mathcal{A}(K)} \int_{\mathcal{A}} f(A^{-1}x)dAg(x)dx$$

by Fubini's theorem. Since K is arbitrary, we conclude that $\widehat{P}_{\mathcal{A}} = P_{\mathcal{A}}$ on Y.

Further, we already remarked that $U_{\mathcal{A}}$ is an isometry on $L^p(\mathcal{G}), 1 \leq p \leq \infty$, by Lemma 1.2.1. Also for the spaces $C_0(\mathcal{G})$ and $C^b(\mathcal{G})$, this is checked easily.

(e) Assume $f, g \in L^2(\mathcal{G})$. An application of Fubini's theorem together with Lemma

1.2.1 yields

$$\langle P_{\mathcal{A}}f, g\rangle_{L^2(\mathcal{G})} = \int_{\mathcal{G}}\int_{\mathcal{A}} F(A^{-1}x)dA\,\overline{G(x)}dx = \int_{\mathcal{A}}\int_{\mathcal{G}} F(A^{-1}x)\overline{G(x)}dx\,dA$$

$$= \int_{\mathcal{A}}\int_{\mathcal{G}} F(x)\overline{G(A(x))}dx\,dA = \int_{\mathcal{G}} F(x)\overline{\int_{\mathcal{A}} G(A^{-1}x)dA}dx = \langle f, P_{\mathcal{A}}g\rangle_{L^2(\mathcal{G})}.$$

Hereby, also the invariance of the Haar measure of $\mathcal{A}$ under the transform $A \mapsto A^{-1}$ was used, which is due to the compactness of $\mathcal{A}$. Hence, $P_{\mathcal{A}}$ is self-adjoint on $L^2(\mathcal{G})$. The property $P_{\mathcal{A}}^2 = P_{\mathcal{A}}$ was already shown in (a).

(f) By definition, the extension of $P_{\mathcal{A}}$ to $\mathbb{S}'$ coincides with the adjoint of $P_{\mathcal{A}} : \mathbb{S} \to \mathbb{S}$ and, thus, by continuity of $P_{\mathcal{A}}$ on $\mathbb{S}$ it is a continuous operator from $\mathbb{S}'$ into itself with respect to the $\sigma(\mathbb{S}',\mathbb{S})$-topology [Edw65, Corollary 8.6.6], i.e., the standard topology on $\mathbb{S}'$. If $\mathbb{S}$ is a Banach space then also $\mathbb{S}'$ is a Banach space. Again since $P_{\mathcal{A}} : \mathbb{S}' \to \mathbb{S}'$ is an adjoint operator it is continuous also with respect to the norm topology. That $P_{\mathcal{A}}$ is a projection onto $(\mathbb{S}')_{\mathcal{A}}$ is proven as in (a).

If $\mathbb{S} \subset L^2(\mathcal{G})$ with dense and continuous embedding then one identifies elements $f \in \mathbb{S}$ with distributions by $f(g) = \int_{\mathcal{G}} g(x)f(x)dx$. By (e) the definitions (1.3) and (1.5) of $P_{\mathcal{A}}f$ are equivalent.

If $\mathbb{S}$ is a Banach space then also $\mathbb{S}'$ is a Banach space and by (d) the operator in (1.4) is well-defined. Whenever $\mathbb{S}$ is continuously embedded into L^1_{loc} it coincides with the one in (1.3). Since Bochner integrals can be interchanged with bounded operators we have $\phi(P_{\mathcal{A}}f) = \int_{\mathcal{A}} \phi(f_A)dA = \int_{\mathcal{A}} \phi_{A^{-1}}(f)dA$. As strong convergence implies weak-$*$ convergence (with the same limits) we realize that the definitions (1.4) and (1.5) coincide.

(g) Assuming $\tilde{P}_{\mathcal{A}}\phi_1 = \tilde{P}_{\mathcal{A}}\phi_2$ for $\phi_1, \phi_2 \in (\mathbb{S}_{\mathcal{A}})'$ means $\phi_1(P_{\mathcal{A}}f) = \phi_2(P_{\mathcal{A}}f)$ for all $f \in \mathbb{S}$. Since $P_{\mathcal{A}}$ is surjective from $\mathbb{S}$ onto $\mathbb{S}_{\mathcal{A}}$ we conclude that $\tilde{P}_{\mathcal{A}}$ is injective. In order to show surjectivity we suppose that $\tilde{\phi} \in (\mathbb{S}')_{\mathcal{A}}$ and denote by $\phi \in (\mathbb{S}_{\mathcal{A}})'$ the restriction of $\tilde{\phi}$ to $\mathbb{S}_{\mathcal{A}}$. This yields

$$(\tilde{P}_{\mathcal{A}}\phi)(f) = \phi(P_{\mathcal{A}}f) = \tilde{\phi}(P_{\mathcal{A}}f) = (P_{\mathcal{A}}\tilde{\phi})(f) = \tilde{\phi}(f)$$

for all $f \in \mathbb{S}$ by definition of $P_{\mathcal{A}}$ and invariance of $\tilde{\phi}$. Hence, $\tilde{P}_{\mathcal{A}}$ is surjective. Clearly, the operator $\tilde{P}_{\mathcal{A}} : (\mathbb{S}_{\mathcal{A}})' \to \mathbb{S}'$ is the adjoint operator of $P_{\mathcal{A}} : \mathbb{S} \to \mathbb{S}_{\mathcal{A}}$ and thus it is continuous with respect to the $\sigma((\mathbb{S}_{\mathcal{A}})', \mathbb{S}_{\mathcal{A}})$ topology on $(\mathbb{S}_{\mathcal{A}})'$ and the $\sigma(\mathbb{S}', \mathbb{S})$ topology on $\mathbb{S}'$ [Edw65, Corollary 8.6.6], i.e., the standard topologies. By Corollary 2 on page 135 in [Sch71] the $\sigma((\mathbb{S}_{\mathcal{A}})', \mathbb{S}_{\mathcal{A}})$ topology coincides with the one induced from $\sigma(\mathbb{S}', \mathbb{S})$. Hence, the operator $\tilde{P}_{\mathcal{A}} : (\mathbb{S}_{\mathcal{A}})' \to (\mathbb{S}')_{\mathcal{A}}$ is continuous. Its inverse operator is the restriction operator $(\mathbb{S}')_{\mathcal{A}} \to (\mathbb{S}_{\mathcal{A}})'$, $\phi \mapsto \phi|_{\mathbb{S}_{\mathcal{A}}}$. Again, since the $\sigma((\mathbb{S}_{\mathcal{A}})', \mathbb{S}_{\mathcal{A}})$ topology and the one induced from $\sigma(\mathbb{S}', \mathbb{S})$ coincide also the inverse operator is continuous. Thus we proved the first claim. Moreover, if $\mathbb{S}$ is a Banach space on which $\mathcal{A}$ acts isometrically then we obtain

$$\|P_{\mathcal{A}}\phi|(\mathbb{S}')_{\mathcal{A}}\| = \sup_{f \in \mathbb{S}, \|f\|=1} |\phi(P_{\mathcal{A}}f)| = \sup_{f \in \mathbb{S}_{\mathcal{A}}, \|f\|=1} |\phi(f)| = \|\phi|(\mathbb{S}_{\mathcal{A}})'\|.$$

Hereby, we used that $P_{\mathcal{A}}$ is projection with $\|P_{\mathcal{A}}|\mathbb{S}\| \leq 1$. This concludes the proof. $\square$

In particular, the previous Lemma states that for all reasonable spaces the different definitions of $P_{\mathcal{A}}$ are equivalent, provided more than one definition applies.

Now for $x \in \mathcal{G}$ let $\mathcal{A}x := \{Ax, A \in \mathcal{A}\}$ denote the orbit of x under $\mathcal{A}$ and define

$$\mathcal{K} := \mathcal{G}^{\mathcal{A}} := \{\mathcal{A}x, x \in \mathcal{G}\}$$

to be the collection of all such orbits. We endow $\mathcal{K}$ with the quotient topology: Denote by $\iota : \mathcal{G} \to \mathcal{K}, x \mapsto \mathcal{A}x$ the canonical projection. A subset U of $\mathcal{K}$ is open iff $\iota^{-1}(U)$ is open in $\mathcal{G}$. Clearly, $\mathcal{K}$ becomes a locally compact space in this way. A function on $\mathcal{G}$ that is invariant under $\mathcal{A}$ can be interpreted as a function on $\mathcal{K}$. Lemma 1.2.2(g) shows that also invariant measures (distributions) on $\mathcal{G}$ can be interpreted as measures (distributions) on $\mathcal{K}$. In particular, $M_{\mathcal{A}}(\mathcal{G})$ is isometrically isomorphic to $M(\mathcal{K})$. We will usually denote a function (measure) on $\mathcal{K}$ and its corresponding invariant one on $\mathcal{G}$ with the same symbol. For the projection of the Haar measure μ onto $\mathcal{K}$, however, we use the extra symbol m, i.e.,

$$\int_{\mathcal{K}} f(\mathcal{A}x)dm(\mathcal{A}x) = \int_{\mathcal{G}} f(x)d\mu(x) \quad \text{for all } f \in C_c(\mathcal{K}). \tag{1.6}$$

Then $L_{\mathcal{A}}^p(\mathcal{G})$ is isometrically isomorphic to $L^p(\mathcal{K}, m) = L^p(\mathcal{K})$ for $1 \leq p \leq \infty$.

For the rest of this section we consider convolutions. Let us first define the left and right translation operators for functions on $\mathcal{G}$ by

$$L_x f(y) := f(x^{-1}y) \qquad \text{and} \qquad R_x f(y) := f(yx).$$

For $f, g \in L_{loc}^1(\mathcal{G})$ the convolution is given by

$$f * g(x) := \int_{\mathcal{G}} f(y)g(y^{-1}x)d\mu(y) = \int_{\mathcal{G}} f(y)L_y g(x)d\mu(y) \tag{1.7}$$

whenever this integral is well-defined for almost all $x \in \mathcal{G}$. It may also be expressed by [Fol95, p.51]

$$f * g(x) = \int_{\mathcal{G}} g(y)f(xy^{-1})\Delta(y^{-1})d\mu(y) = \int_{\mathcal{G}} g(y)R_{y^{-1}}f(x)\Delta(y^{-1})d\mu(y). \tag{1.8}$$

The convolution is commutative if and only if the group $\mathcal{G}$ is commutative. Young's theorem states that

$$\|f * g|L^p(\mathcal{G})\| \leq \|f|L^1(\mathcal{G})\| \|g|L^p(\mathcal{G})\| \quad \text{for all } f \in L^1(\mathcal{G}), g \in L^p(\mathcal{G}), 1 \leq p \leq \infty,$$

see [Fol95, Proposition 2.39a]. In particular, $L^1(\mathcal{G})$ is a Banach algebra and $L^p(\mathcal{G})$ is a left-L^1-Banach module.

We may extend the convolution to $M(\mathcal{G})$. Indeed for $\tau, \rho \in M(\mathcal{G})$ and $h \in C_0(\mathcal{G})$ we let

$$\tau * \rho(h) := \int_{\mathcal{G} \times \mathcal{G}} h(xy) d(\tau \otimes \rho)(x, y). \tag{1.9}$$

Denoting the Dirac measure at $x \in \mathcal{G}$ by $\epsilon_x(h) = h(x)$, $h \in C^b(\mathcal{G})$, we note that

$$\epsilon_x * \epsilon_y = \epsilon_{xy} \quad \text{and} \quad \epsilon_x * h(y) = \epsilon_{x^{-1}} * \epsilon_y(h) = L_x h(y), \quad x, y \in \mathcal{G}. \tag{1.10}$$

Together with the involution $\tau^*(E) = \overline{\tau(E^{-1})}$, $E \subset \mathcal{G}$ a measurable subset, $(M(\mathcal{G}), *)$ becomes a Banach-$*$-algebra with unit ϵ_e. Identifying a function $f \in L^1(\mathcal{G})$ with the measure μ_f as in (1.2), it holds $\mu_{f*g} = \mu_f * \mu_g$. In this sense $L^1(\mathcal{G})$ is a $*$-subalgebra of $M(\mathcal{G})$, in fact even a two-sided ideal with inherited involution $f^*(x) = \Delta(x^{-1})\overline{f(x^{-1})}$. The convolution of $\tau \in M(\mathcal{G})$ with $f \in L^1(\mathcal{G})$ can be written as

$$\tau * f(x) = \int_{\mathcal{G}} f(y^{-1}x) d\tau(y).$$

The convolution may be further extended. Let $\mathbb{S}$ be a locally convex vector space of test functions on $\mathcal{G}$. For $f, g \in \mathbb{S}$ we define the tensor product as the function $(f \otimes g)(x, y) := f(x)g(y)$ on $\mathcal{G} \times \mathcal{G}$. For two continuous functionals (distributions) $\phi, \psi \in \mathbb{S}'$ we let $(\phi \otimes \psi)(f \otimes g) := \phi(f)\psi(g)$. By linearity this defines a unique functional $\phi \otimes \psi$ on $\mathbb{S}_0(\mathcal{G} \times \mathcal{G}) := \text{span}\{f \otimes g, f, g \in \mathbb{S}\}$. Depending on ϕ and ψ it might be possible to extend the tensor product $\phi \otimes \psi$ uniquely to a larger function space $\mathbb{S}_{\phi,\psi} \supset \mathbb{S}_0(\mathcal{G} \times \mathcal{G})$ on $\mathcal{G} \times \mathcal{G}$ such that

$$(\phi \otimes \psi)(h) = \phi^{(x)}(\psi^{(y)}(h(x, y))) = \psi^{(y)}(\phi^{(x)}(h(x, y))) \quad \text{for all } h \in \mathbb{S}_{\phi,\psi},$$

where $\phi^{(x)}$ denotes the application of ϕ to a function depending on the x-variable. In other words, the application of ϕ and ψ is independent of the order. Now, for a function $f \in \mathbb{S}$ we define $Tf(x, y) := f(xy)$. Whenever $T\mathbb{S} \subset \mathbb{S}_{\phi,\psi}$ we may define the convolution by

$$(\phi * \psi)(f) = (\phi \otimes \psi)(Tf), \quad \text{for all } f \in \mathbb{S}. \tag{1.11}$$

Under the described assumption $\phi * \psi$ is again an element of $\mathbb{S}'$.

The continuous functionals on $\mathbb{S} = C_c(\mathcal{G})$ can be identified with the Radon measures and the tensor product of two Radon measures is their product measure. In particular, for measures in $M(\mathcal{G}) = (C_0(\mathcal{G}))'$ definition (1.11) coincides with (1.9). Moreover, since $L^1_{loc}(\mathcal{G})$ can be identified as a space of Radon measures definition (1.11) applies to the convolution of two locally integrable functions f, g, whenever $f * g$ is well-defined. For more details on the convolution of Radon measures, we refer to [Edw65, Chapter 4.19].

We note that (1.11) applies also to the convolution of distributions in $\mathcal{S}'(\mathbb{R}^d)$, whenever it is defined [Hör83].

Now, we show that the convolution of two invariant functions (measures, distributions) is again invariant under $\mathcal{A}$.

Lemma 1.2.3. *(a) Suppose* $\mathbb{S}$ *is a locally convex vector space of test functions on* $\mathcal{G}$, *on which* $\mathcal{A}$ *acts continuously. Further assume that* $\phi, \psi \in \mathbb{S}'$ *are distributions whose convolution is well-defined. Then it holds*

$$(\phi_A) * (\psi_A) = (\phi * \psi)_A \qquad \text{for all } A \in \mathcal{A}.$$

In particular, the convolution of two invariant distribution is again invariant.

(b) Consequently, $M_{\mathcal{A}}(\mathcal{G})$ *is a closed* *-subalgebra of* $M(\mathcal{G})$ *and* $L^1_{\mathcal{A}}(\mathcal{G})$ *is a closed* *-subalgebra of* $L^1(\mathcal{G})$.

Proof: (a) Let $A \in \mathcal{A}$. Since $\mathcal{A}$ acts continuously on $\mathbb{S}$ the elements ϕ_A, ψ_A are continuous functionals on $\mathbb{S}$. For all $f \in \mathbb{S}$ we obtain

$$(\phi_A) * (\psi_A)(f) = \phi^{(x)}(\psi^{(y)}(f(A(x)A(y)))) = \phi^{(x)}(\psi^{(y)}(f_{A^{-1}}(xy)))$$
$$= (\phi * \psi)(f_{A^{-1}}) = (\phi * \psi)_A(f).$$

(b) The assertion follows from (a) by taking $\mathbb{S} = C_0(\mathcal{G})$. $\qquad\square$

Although the space of orbits $\mathcal{K}$ has no group structure (in general), the previous lemma states that $L^1(\mathcal{K}) \cong L^1_{\mathcal{A}}(\mathcal{G})$ still possesses a convolution structure. It is possible to rewrite the convolution of invariant functions as follows. We define a **generalized translation** by

$$\mathcal{T}_y f(x) = \int_{\mathcal{A}} f(A(y)x)dA, \qquad (1.12)$$

for $f \in (L^1_{loc})_{\mathcal{A}}$. Associated to $\mathcal{T}$ are the **generalized left translation** $\mathcal{L}$ and the **generalized right translation** $\mathcal{R}$ defined by

$$\mathcal{L}_y f(x) := \mathcal{T}_{y^{-1}} f(x) = \int_{\mathcal{A}} f(A(y^{-1})x)dA, \qquad (1.13)$$

$$\mathcal{R}_y f(x) := \mathcal{T}_x f(y) = \int_{\mathcal{A}} f(A(x)y)dA. \qquad (1.14)$$

Let us furthermore introduce the "invariant Dirac measures"

$$\epsilon_{Ax}(h) := (P_{\mathcal{A}}h)(x) = \int_{\mathcal{A}} h(Ax)dA, \quad h \in C^b(\mathcal{G}). \qquad (1.15)$$

Clearly, if h is invariant under $\mathcal{A}$ then $\epsilon_{Ax}(h) = h(x)$ for all $x \in \mathcal{G}$.

Lemma 1.2.4. *(a) Suppose $f \in (L^1_{loc})_{\mathcal{A}}(\mathcal{G})$. Then the function $(x,y) \mapsto \mathcal{T}_y f(x)$ is invariant in $x \in \mathcal{G}$ and in $y \in \mathcal{G}$ and, hence, we may unambiguously write $\mathcal{T}_{\mathcal{A}y} f(\mathcal{A}x)$ for $\mathcal{A}x, \mathcal{A}y \in \mathcal{K}$. Consequently, the same holds for $\mathcal{L}_y f(x)$ and $\mathcal{R}_y f(x)$. Moreover, we have*

$$\mathcal{T}_y f(x) = \int_{\mathcal{A}} f(yA(x))dA$$

and with the usual left translation L_y on $\mathcal{G}$

$$\mathcal{L}_y f(x) = (P_{\mathcal{A}} L_y f)(x) = \epsilon_{\mathcal{A}y} * f(x) = \epsilon_{\mathcal{A}y^{-1}} * \epsilon_{\mathcal{A}x}(f).$$

(b) Let Y denote one of the spaces $C_{\mathcal{A}}(\mathcal{G}), C^b_{\mathcal{A}}(\mathcal{G}), (C_0)_{\mathcal{A}}(\mathcal{G}), (C_c)_{\mathcal{A}}(\mathcal{G})$ or $L^p_{\mathcal{A}}(\mathcal{G})$, $1 \le p \le \infty$. Then $\mathcal{L}_x f \in Y$ holds for all $f \in Y$ and $x \in \mathcal{G}$. Moreover, we have (except for the spaces $C(\mathcal{G})$ and $C_c(\mathcal{G})$ where it makes no sense)

$$\|\mathcal{L}_x f | Y\| \le \|f | Y\| \qquad \text{for all } x \in \mathcal{G}, f \in Y.$$

(c) Suppose $f \in L^1(\mathcal{K})$. For all $\mathcal{A}x \in \mathcal{K}$ it holds

$$\int_{\mathcal{K}} \mathcal{L}_{\mathcal{A}x} f(\mathcal{A}y) dm(\mathcal{A}y) = \int_{\mathcal{K}} f(\mathcal{A}y) dm(\mathcal{A}y).$$

(d) Suppose f, g are $\mathcal{A}$-invariant functions, whose convolution is well-defined. Then it holds

$$f * g(x) = \int_{\mathcal{K}} f(\mathcal{A}y) \mathcal{L}_{\mathcal{A}y} g(\mathcal{A}x) dm(\mathcal{A}y). \tag{1.16}$$

(e) Suppose $\tau, \rho \in M_{\mathcal{A}}(\mathcal{G}) \cong M(\mathcal{K})$ and $h \in C^b_{\mathcal{A}}(\mathcal{G}) \cong C^b(\mathcal{K})$. Then we have

$$\tau * \rho(h) = \int_{\mathcal{K}} \int_{\mathcal{K}} \mathcal{T}_{\mathcal{A}x} h(\mathcal{A}y) d\tau(\mathcal{A}x) d\rho(\mathcal{A}y).$$

Proof: (a) Suppose $B, C \in \mathcal{A}$. Recall that the Haar measure of $\mathcal{A}$ is left and right invariant by compactness of $\mathcal{A}$. Invariance of f thus yields

$$\mathcal{T}_{Cy} f(Bx) = \int_{\mathcal{A}} f(AC(y)B(x))dA = \int_{\mathcal{A}} f(B^{-1}AC(y)x)dA = \int_{\mathcal{A}} f(A(y)x)dA$$
$$= \mathcal{T}_x f(y).$$

By compactness of $\mathcal{A}$ it holds $dA = dA^{-1}$. Thus by invariance of f, we have

$$\mathcal{T}_y f(x) = \int_{\mathcal{A}} f(A(y)x)dA = \int_{\mathcal{A}} f(yA^{-1}(x))dA = \int_{\mathcal{A}} f(yA(x))dA.$$

The relation $\mathcal{L}_y = P_{\mathcal{A}} L_y$ follows as immediate consequence. Moreover, it holds

$$\epsilon_{\mathcal{A}y} * f(x) = \int_{\mathcal{G}} f(z^{-1}x) d\epsilon_{\mathcal{A}y}(z) = \int_{\mathcal{A}} f(A(y^{-1})x) dA = \mathcal{L}_y f(x)$$

and

$$\epsilon_{\mathcal{A}y^{-1}} * \epsilon_{\mathcal{A}x}(f) = \int_{\mathcal{G}} \int_{\mathcal{G}} f(wz) d\epsilon_{\mathcal{A}y^{-1}}(w) d\epsilon_{\mathcal{A}x}(z) = \int_{\mathcal{A}} \int_{\mathcal{A}} f(A(y^{-1})B(x)) dA dB$$

$$= \int_{\mathcal{A}} \mathcal{L}_y f(x) dB = \mathcal{L}_y f(x).$$

(b) As $\mathcal{L}_x = P_{\mathcal{A}} L_x$ and since L_x is an isometry on each of the spaces $L^p(\mathcal{G}), C_0(\mathcal{G})$, $C^b(\mathcal{G})$ and leaves $C(\mathcal{G}), C_c(\mathcal{G})$ invariant the assertion follows from Lemma 1.2.2(b) and (d).

(c) Using Fubini's theorem and the left invariance of the Haar measure of $\mathcal{G}$ we obtain

$$\int_{\mathcal{K}} \mathcal{L}_{\mathcal{A}x} f(\mathcal{A}y) dm(\mathcal{A}y) = \int_{\mathcal{G}} \int_{\mathcal{A}} f(A(x^{-1})y) dA dy = \int_{\mathcal{A}} \int_{\mathcal{G}} f(A(x^{-1})y) dy dA$$

$$= \int_{\mathcal{A}} \int_{\mathcal{G}} f(y) dy dA = \int_{\mathcal{K}} f(\mathcal{A}y) dm(\mathcal{A}y).$$

(d) By Lemma 1.2.3 the function $A \mapsto f * g(A(x))$ is constant (a.e.) for almost all $x \in \mathcal{G}$. Application of Fubini's theorem yields

$$(f * g)(x) = \int_{\mathcal{A}} (f * g)(Ax) dA = \int_{\mathcal{G}} f(y) \int_{\mathcal{A}} g(y^{-1}A(x)) dA dx$$

$$= \int_{\mathcal{K}} f(\mathcal{A}y) \mathcal{L}_{\mathcal{A}y} g(\mathcal{A}x) dm(y).$$

The assertion in (e) is deduced similarly. $\qquad\qquad \square$

The operator $\mathcal{L}$ is a natural generalization of a left translation. First we apply the usual left translation and since the resulting function is no longer invariant we project it back onto the invariant functions using $P_{\mathcal{A}}$.

1.3 Harmonic Analysis of Radial Functions

In this section we treat our main example of functions that are invariant under symmetry groups, that is, we specialize to radially symmetric functions on $\mathbb{R}^d$. Although the corresponding results seem to be quite well-known we give detailed proofs. We do so because we are not aware of a reference that provides a complete and easy introduction.

Let us first give the precise definition of radial symmetry. We denote by $O(d)$ the group of all orthogonal $d \times d$ matrices and $SO(d)$ its subgroup of matrices with determinant 1.

Definition 1.3.1. A function (measure, distribution) on $\mathbb{R}^d$, which is invariant under $O(d)$, is called **radial**. Equivalently, a function f is radial iff there exists a function f on $\mathbb{R}_+ := [0, \infty)$ such that $f(x) = \mathsf{f}(|x|)$.

Since the case $d = 1$ needs special treatment in some of the following statements, and since we are mainly interested in the multidimensional case anyway, we assume $d \geq 2$ throughout this section. Invariance under $O(d)$ is then equivalent to invariance under $SO(d)$.

Let us now specialize the results of the previous section to our situation. Clearly, the space of orbits $\mathcal{K} = (\mathbb{R}^d)^{SO(d)}$ consists of all spheres centered at the origin and, hence, is isomorphic to $\mathbb{R}_+$. When $f, g, h, \ldots$ are radial functions on $\mathbb{R}^d$ we agree upon denoting the corresponding functions on $\mathbb{R}_+$ by $\mathsf{f}, \mathsf{g}, \mathsf{h}, \ldots$ throughout this thesis.

Instead of the subscript $SO(d)$ we will usually use the more intuitive subscript *rad*, in particular, $L_{rad}^p(\mathbb{R}^d)$ denotes the closed subspace of $L^p(\mathbb{R}^d)$ of radial L^p-functions. By polar decomposition we obtain for $f \in L_{rad}^1(\mathbb{R}^d)$

$$\int_{\mathbb{R}^d} f(x)dx = \int_0^\infty r^{d-1} \int_{S^{d-1}} f(r\xi)dS(\xi)dr = |S^{d-1}| \int_0^\infty \mathsf{f}(r)r^{d-1}dr.$$

Hereby dS denotes the surface measure of the sphere S^{d-1} and $|S^{d-1}| = \frac{2\pi^{d/2}}{\Gamma(d/2)}$ its area, see also Appendix A. Hence, the corresponding measure m on $\mathcal{K} \cong \mathbb{R}_+$ defined in (1.6) is given by

$$d\mu_d(r) := |S^{d-1}|r^{d-1}dr. \tag{1.17}$$

Thus, $L_{rad}^p(\mathbb{R}^d)$ is isometrically isomorphic to $L^p(\mathbb{R}_+, \mu_d)$ for all $1 \leq p \leq \infty$.

By Lemma 1.2.2(g) also radial measures (resp. distributions) on $\mathbb{R}^d$ can be considered as measures (resp. distributions) on $\mathbb{R}_+$. In particular, $M_{rad}(\mathbb{R}^d)$ is isomorphic to $M(\mathbb{R}_+)$ and $\mathcal{S}'_{rad}(\mathbb{R}^d)$ is isomorphic to $\mathcal{S}'(\mathbb{R}_+)$, where $\mathcal{S}(\mathbb{R}_+) := \{\mathsf{f}|_{\mathbb{R}_+}, \mathsf{f} \in \mathcal{S}(\mathbb{R}), \mathsf{f} \text{ is even}\}$.

The operator $P_{rad} := P_{SO(d)}$ defined in (1.3) may be written for functions in $L_{loc}^1(\mathbb{R}^d)$ as

$$P_{\mathrm{rad}}f(x) = \frac{1}{|S^{d-1}|} \int_{S^{d-1}} f(|x|\xi)dS(\xi). \tag{1.18}$$

As stated in Lemma 1.2.2 P_{rad} extends to $M(\mathbb{R}^d)$ and to $\mathcal{S}'(\mathbb{R}^d)$ by (1.4) and (1.5). Let us now study the convolution (1.7). On $\mathbb{R}^d$ it reads

$$f * g(x) = \int_{\mathbb{R}^d} f(y)g(x-y)dy. \tag{1.19}$$

Since $\mathbb{R}^d$ is commutative also the convolution (1.19) is commutative. Together with the involution on $\mathbb{R}^d$ given by

$$f^*(x) := \overline{f(-x)} \tag{1.20}$$

$L^1(\mathbb{R}^d)$ becomes a Banach-$*$-algebra. Furthermore, we have an extension of the convolution to $M(\mathbb{R}^d)$ as in (1.9) and to suitable distributions in $\mathcal{S}'(\mathbb{R}^d)$ as in (1.11), see also [Hör83]. We remark, however, that the convolution is not well-defined on all of $\mathcal{S}'(\mathbb{R}^d)$. Lemma 1.2.3 immediately yields the following result.

Lemma 1.3.1. *The convolution preserves radiality, i.e., if $f, g \in \mathcal{S}'(\mathbb{R}^d)$ are radial functions (measures, distributions) then $f * g$ is again radial whenever the convolution exists.*

Our next aim is to express the convolution of radial functions by means of their corresponding functions on $\mathbb{R}_+$. In view of Lemma 1.2.4 the generalized translation (1.13) is needed for this task. In our special case it reads

$$\tau_y f(x) := \int_{SO(d)} f(x - Ay) dA, \tag{1.21}$$

where dA denotes the normalized Haar measure on $SO(d)$. If f is radial then $\tau_y f(x)$ is radial in x and y by Lemma 1.2.4. We denote the corresponding operator on $L^1_{loc}(\mathbb{R}_+)$ also by τ, i.e., $\tau_s f(r) = \tau_y f(x)$ with $s = |y|, r = |x|$. In view of formula (A.3) for the integration on $SO(d)$ we have

$$\tau_s f(r) = \frac{1}{|S^{d-1}|} \int_{S^{d-1}} f(|s\eta - r\xi|) dS(\xi), \quad r, s \in \mathbb{R}_+, \eta \in S^{d-1}, \tag{1.22}$$

where the expression on the right hand side is independent of the choice of $\eta \in S^{d-1}$. In light of Lemma 1.2.4(d) we define a corresponding convolution on $\mathbb{R}_+$ by

$$f \star g(s) := \int_0^\infty f(r) \tau_s g(r) d\mu_d(r). \tag{1.23}$$

We immediately obtain the following result.

Lemma 1.3.2. *Suppose f, g are radial functions on $\mathbb{R}^d$ with corresponding functions f, g on $\mathbb{R}_+$. If their convolution is well-defined then it holds*

$$f * g(x) = f \star g(|x|).$$

Consequently, $L^1(\mathbb{R}_+, \mu_d)$ becomes a commutative Banach algebra with the convolution (1.23).

We remark further that the commutativity of $\star$ written down explicitly yields

$$\int_0^\infty \mathsf{f}(r)\tau_s\mathsf{g}(r)d\mu_d(r) \;=\; \int_0^\infty \tau_s\mathsf{f}(r)\mathsf{g}(r)d\mu_d(r), \tag{1.24}$$

which appears to be quite useful at certain places.

In order to study properties of the generalized translation τ_s we introduce the "radial Dirac measures" $\epsilon_{(r)} = \epsilon_{SO(d)x}$, $r = |x|$, see (1.15), i.e.,

$$\epsilon_{(r)}(f) \;=\; (P_{rad}f)(x) \;=\; \frac{1}{|S^{d-1}|}\int_{S^{d-1}} f(r\xi)dS(\xi), \quad f \in C^b(\mathbb{R}^d).$$

Furthermore, the usual translation operator on $\mathbb{R}^d$ will be denoted by

$$T_x F(y) := F(x-y) \tag{1.25}$$

throughout this thesis.

Lemma 1.3.3. *Let $g \in (L^1_{loc})_{rad}(\mathbb{R}^d)$ with corresponding function $\mathsf{g} \in L^1_{loc}(\mathbb{R}_+)$.*

(a) For $x,y \in \mathbb{R}^d$ with $r = |x|$ and $s = |y|$ we have

$$\tau_s\mathsf{g}(r) \;=\; (P_{rad}T_y g)(x) \tag{1.26}$$

and

$$\tau_s\mathsf{g}(r) \;=\; \epsilon_{(s)} * \epsilon_{(r)}(g) \;=\; \epsilon_{(s)} * g(x). \tag{1.27}$$

(b) For all $r,s \in \mathbb{R}_+$ we have

$$\tau_s\mathsf{g}(r) \;=\; \tau_r\mathsf{g}(s), \qquad \tau_s 1 = 1 \qquad and \qquad \tau_0\mathsf{g}(r) \;=\; \mathsf{g}(r).$$

Moreover, if g is positive then so is $\tau_s\mathsf{g}$ for all $s \in \mathbb{R}_+$.

(c) The number $\tau_r\mathsf{g}(s)$ depends only on the values of g in the interval $[|s-r|, s+r]$. Consequently, if $\mathrm{supp}\,\mathsf{g} \subset [a,b]$ then $\mathrm{supp}\,\tau_s\mathsf{g} \subset [\min\{0, a-s\}, b+s]$.

(d) For all $\mathsf{f} \in L^1(\mathbb{R}_+, \mu_d)$ and $s \in \mathbb{R}_+$ we have

$$\int_0^\infty \tau_s\mathsf{f}(r)d\mu_d(r) \;=\; \int_0^\infty \mathsf{f}(r)d\mu_d(r).$$

(e) If Y is one of the spaces $L^p(\mathbb{R}_+, \mu_d), 1 \leq p \leq \infty$, $C(\mathbb{R}_+)$, $C_0(\mathbb{R}_+), C^b(\mathbb{R}_+)$ or $C_c(\mathbb{R}_+)$ then $\tau_s\mathsf{f} \in Y$ for all $\mathsf{f} \in Y$ and it holds (except for $C(\mathbb{R}_+)$ and $C_c(\mathbb{R}_+)$)

$$\|\tau_s\mathsf{f}|Y\| \;\leq\; \|\mathsf{f}|Y\|.$$

(f) *The generalized translation can also be expressed in the form*

$$\tau_s g(r) = \frac{|S^{d-2}|}{|S^{d-1}|} \int_{-1}^{1} g(\sqrt{s^2 - 2rst + r^2})(1 - t^2)^{\frac{d-3}{2}} dt$$

$$= \frac{|S^{d-2}|}{|S^{d-1}|} \int_{0}^{\pi} g(\sqrt{s^2 - 2rs\cos\theta + r^2})\sin^{d-2}(\theta)d\theta. \qquad (1.28)$$

Proof: (a) The claim follows from Lemma 1.2.4(a).

(b) The first assertion is a consequence of the commutativity of the convolution on $\mathbb{R}^d$ and the other stated properties are easy to see.

(c) The argument $|s\eta - r\xi|$ of $\mathbf{g}$ in (1.22) varies between $|r - s|$ and $r + s$ since η is a fixed unit vector and ξ varies through all of S^{d-1}. The second assertion is an immediate consequence.

The assertions (d) and (e) are special cases of Lemma 1.2.4(c) and (b), respectively.

(f) We use the integration rule outlined in Appendix A. So we choose a fixed vector $\eta \in S^{d-1}$ and let $S_\eta := \{x \in S^{d-1}, x \cdot \eta = 0\} \cong S^{d-2}$. By formula (A.1) we obtain

$$\tau_s g(r) = \frac{1}{|S^{d-1}|} \int_{S^{d-1}} g(|s\eta - r\xi|)dS^{d-1}(\xi)$$

$$= \frac{1}{|S^{d-1}|} \int_{S_\eta} \int_{-1}^{1} g(|(s - rt)\eta - r\sqrt{1 - t^2}\zeta|)(1 - t^2)^{\frac{d-3}{2}} dt dS^{d-2}(\zeta). \qquad (1.29)$$

Since η is orthogonal to $\zeta \in S_\eta$ we obtain for the argument of g by the theorem of Pythagoras

$$|(s - rt)\eta - r\sqrt{1 - t^2}\zeta| = \sqrt{(s - rt)^2 + r^2(1 - t^2)} = \sqrt{s^2 - 2rst + r^2},$$

and, hence, the integrand in (1.29) is independent of ζ. Thus, we obtain

$$\tau_s g(r) = \frac{|S^{d-2}|}{|S^{d-1}|} \int_{-1}^{1} g(\sqrt{s^2 - 2rst + r^2})(1 - t^2)^{\frac{d-3}{2}} dt.$$

The transformation $t = \cos\theta$ yields (1.28). □

Remark 1.3.1. (a) By equation (1.26) the generalized translation is the most intuitive way to define a translation that preserves radiality. Indeed first one applies a usual translation to a radial function and since the result is no longer radial (except in trivial cases) one projects back onto the radial functions.

(b) Although the generalized translation τ_s shares some common properties with the usual translation T_x we have in general

$$\tau_s \tau_r \neq \tau_{s+r}, \quad r, s \in \mathbb{R}_+ \setminus \{0\}.$$

(c) We may extend the generalized translation also to radial measures and distributions. Indeed, the usual translation is extended by $(T_x\phi)(f) = \phi(T_{-x}f)$ for $\phi \in \mathcal{S}'(\mathbb{R}^d)$, $f \in \mathcal{S}(\mathbb{R}^d)$. By Lemma 1.2.2(d) and (f) also P_{rad} extends to $\mathcal{S}'(\mathbb{R}^d)$ and this suggests to define $\tau_s\phi := P_{rad}T_x\phi$, $r = |x|$ for $\phi \in \mathcal{S}'_{rad}(\mathbb{R}^d)$. For $\rho \in M(\mathbb{R}_+) \cong M_{rad}(\mathbb{R}^d)$ also the formula

$$\tau_s\rho = \epsilon_{(s)} * \rho \tag{1.30}$$

applies and the convolution of ρ with $\sigma \in M(\mathbb{R}_+)$ is defined by

$$\rho \star \sigma(\mathsf{h}) = \int_0^\infty \int_0^\infty \tau_s\mathsf{h}(r)d\rho(r)d\sigma(s), \quad \mathsf{h} \in C_0(\mathbb{R}_+).$$

Let us now consider the Fourier transform. For $f \in L^1(\mathbb{R}^d)$ it is defined by

$$\mathcal{F}(f)(\xi) := \hat{f}(\xi) := \int_{\mathbb{R}^d} f(x)e^{-2\pi i x \cdot \xi}dx, \quad \xi \in \mathbb{R}^d.$$

The Plancherel theorem states that $\mathcal{F}$ extends to a unitary transform on $L^2(\mathbb{R}^d)$, in particular

$$\langle \hat{f}, \hat{g} \rangle_{L^2(\mathbb{R}^d)} = \langle f, g \rangle_{L^2(\mathbb{R}^d)} \quad \text{for all } f, g \in L^2(\mathbb{R}^d). \tag{1.31}$$

Furthermore, it is well-known that $\mathcal{F}$ is a topological isomorphism of the Schwartz space $\mathcal{S}(\mathbb{R}^d)$ into itself. This allows to extend the Fourier transform also to $\mathcal{S}'(\mathbb{R}^d)$ by $\hat{\phi}(f) := \phi(\hat{f})$ for $\phi \in \mathcal{S}'(\mathbb{R}^d), f \in \mathcal{S}(\mathbb{R}^d)$. The next lemma states that the Fourier transform preserves radiality.

Lemma 1.3.4. *(a) Suppose $f \in L^1(\mathbb{R}^d)$. Then it holds*

$$\widehat{(f_A)}(\xi) = (\hat{f})_A(\xi) \quad \text{for all } A \in O(d). \tag{1.32}$$

Moreover, (1.32) extends to $\mathcal{S}'(\mathbb{R}^d)$, i.e., $\widehat{(\phi_A)} = (\hat{\phi})_A$ for all $\phi \in \mathcal{S}'(\mathbb{R}^d), A \in O(d)$.

(b) Consequently, if $\phi \in \mathcal{S}'(\mathbb{R}^d)$ is radial then also its Fourier transform $\hat{\phi}$ is radial.

Proof: By invariance of the Lebesgue measure under the action of $O(d)$ we have

$$\widehat{(f_A)}(\xi) = \int_{\mathbb{R}^d} f(A^{-1}x)e^{-2\pi i \xi \cdot x}dx = \int_{\mathbb{R}^d} f(x)e^{-2\pi i \xi \cdot Ax}dx$$
$$= \int_{\mathbb{R}^d} f(x)e^{-2\pi i A^{-1}\xi \cdot x}dx = (\hat{f})_A(\xi) \tag{1.33}$$

for all $A \in O(d)$. Relation (1.32) applies in particular to Schwartz functions. Therefore, we obtain for $\phi \in \mathcal{S}'(\mathbb{R}^d)$ and all $f \in \mathcal{S}(\mathbb{R}^d), A \in O(d)$

$$\langle \widehat{(\phi_A)}, f \rangle = \langle \phi_A, \hat{f} \rangle = \langle \phi, \hat{f}_{A^{-1}} \rangle = \langle \hat{\phi}, (\hat{f})_{A^{-1}} \rangle = \langle (\hat{\phi})_A, \hat{f} \rangle.$$

This proves (a), and (b) follows as an immediate consequence. $\qquad\square$

We now aim at expressing the Fourier transform of a radial function f by some integral over $\mathbb{R}_+$ involving the corresponding function f. To do this we need to introduce a special function.

Definition 1.3.2. For $r \in \mathbb{R}_+, \eta \in S^{d-1}$ we define the spherical Bessel function by

$$\mathcal{B}_d(r) := \frac{1}{|S^{d-1}|} \int_{S^{d-1}} e^{-2\pi i r \eta \cdot \zeta} dS(\zeta). \qquad (1.34)$$

The substitution $\eta \to A\eta, A \in SO(d)$, shows that (1.34) is independent of the choice of η. Further, denoting $e_\xi(x) := e^{-2\pi i \xi \cdot x}$ for some $\xi \in S^{d-1}$ we observe that

$$\mathcal{B}_d(r) = P_{rad}(e_\xi)(r\eta) = \mathcal{F}(\epsilon_{(r)})(\eta), \quad r \in \mathbb{R}_+, \qquad (1.35)$$

independently of the choice of $\eta, \xi \in S^{d-1}$.

Definition 1.3.3. For $f \in L^1(\mathbb{R}_+, \mu_d)$ the **Hankel transform** is defined by

$$H_d(f)(\lambda) := \hat{f}(\lambda) := \int_0^\infty f(r) \mathcal{B}_d(\lambda r) d\mu_d(r). \qquad (1.36)$$

The Hankel transform – sometimes also called Fourier-Bessel transform – is essentially the Fourier transform of a radial function as stated in the next lemma.

Lemma 1.3.5. *Suppose* $f \in L^1_{rad}(\mathbb{R}^d)$ *with corresponding* f $\in L^1(\mathbb{R}_+, \mu_d)$. *Then it holds*

$$\hat{f}(\xi) = H_d(f)(|\xi|), \qquad \text{for all } \xi \in \mathbb{R}^d. \qquad (1.37)$$

Proof: A change to polar coordinates in the Fourier integral yields

$$\hat{f}(\xi) = \int_{\mathbb{R}^d} f(x) e^{-2\pi i \xi \cdot x} dx = \int_0^\infty r^{d-1} \int_{S^{d-1}} f(r\zeta) e^{-2\pi i r \xi \cdot \zeta} dS(\zeta) dr$$

$$= \int_0^\infty f(r) r^{d-1} \int_{S^{d-1}} e^{-2\pi i r \lambda \eta \cdot \zeta} dS(\zeta) dr \qquad (1.38)$$

where we have written $\xi = \lambda \eta$ with $\lambda = |\xi| \in \mathbb{R}_+, \eta \in S^{d-1}$. The inner integral coincides with the definition of $\mathcal{B}_d(\lambda r)$ (up to the normalization constant $|S^{d-1}|$) and hence we identify (1.38) with the Hankel transform (1.36) of f. $\qquad\square$

By relation (1.37) the Hankel transform inherits many properties from the Fourier transform, which we list in the following Corollary.

Corollary 1.3.6. *(a) The Hankel transform extends to a unitary isomorphism of $L^2(\mathbb{R}_+, \mu_d)$ into itself.*

(b) The Hankel transform is self-inverse, i.e., $(H_d)^2 = \mathrm{Id}$ on $L^2(\mathbb{R}_+, \mu_d)$.

(c) For $\mathsf{f}, \mathsf{g} \in L^1(\mathbb{R}_+, \mu_d)$ the convolution theorem

$$(\mathsf{f} \star \mathsf{g})\hat{\ } = \hat{\mathsf{f}}\hat{\mathsf{g}}$$

holds, where $\star$ denotes the convolution on $\mathbb{R}_+$ defined in (1.23).

(d) The Hankel transform is a topological isomorphism of $\mathcal{S}(\mathbb{R}_+)$ and extends to a topological isomorphism of $\mathcal{S}'(\mathbb{R}_+)$ by

$$\hat{\phi}(\mathsf{f}) := \phi(\hat{\mathsf{f}}), \quad \phi \in \mathcal{S}'(\mathbb{R}_+), \mathsf{f} \in \mathcal{S}(\mathbb{R}_+).$$

The name of the function $\mathcal{B}_d$ in (1.34) is due to its relation (see below) to the Bessel function J_ν of the first kind defined by

$$J_\nu(t) = \sum_{k=0}^\infty \frac{(-1)^k}{k!\Gamma(k+\nu+1)} \left(\frac{t}{2}\right)^{2k+\nu}.$$

Let us collect some of the properties of $\mathcal{B}_d$ in the next lemma.

Lemma 1.3.7. *(a) It holds $\mathcal{B}_d(0) = 1$ and $|\mathcal{B}_d(t)| \leq 1$ for all $t \in \mathbb{R}_+$.*

(b) The spherical Bessel function has the representations

$$\mathcal{B}_d(t) = \frac{|S^{d-2}|}{|S^{d-1}|} \int_{-1}^1 e^{-2\pi i t u}(1-u^2)^{\frac{d-3}{2}} du \tag{1.39}$$

$$= j_\nu(2\pi t), \quad \nu = \frac{d-2}{2} \tag{1.40}$$

where

$$j_\nu(x) := \Gamma(\nu+1)(2/x)^\nu J_\nu(x) = \Gamma(\nu+1) \sum_{k=0}^\infty \frac{(-1)^k (x/2)^{2k}}{k!\Gamma(k+\nu+1)}. \tag{1.41}$$

We have in particular

$$\mathcal{B}_1(t) = \cos(2\pi t), \quad \mathcal{B}_2(t) = J_0(2\pi t) \quad and \quad \mathcal{B}_3(t) = \frac{\sin(2\pi t)}{2\pi t}.$$

Proof: (a) Directly from the definition it follows $\mathcal{B}_d(0) = 1$ and one easily deduces $|\mathcal{B}_d(t)| \leq 1$ for all $t \in \mathbb{R}_+$.

(b) Using the integration formula (A.1) we obtain

$$
\begin{aligned}
\mathcal{B}_d(t) &= \frac{1}{|S^{d-1}|} \int_{S^{d-1}} e^{-2\pi i t \eta \cdot \zeta} dS(\zeta) \\
&= \frac{1}{|S^{d-1}|} \int_{S_\eta} \int_{-1}^{1} e^{-2\pi i t \eta \cdot (u\eta + \sqrt{1-u^2}\xi)} (1-u^2)^{\frac{d-3}{2}} du\, dS^{d-2}(\xi) \\
&= \frac{|S^{d-2}|}{|S^{d-1}|} \int_{-1}^{1} e^{-2\pi i t u} (1-u^2)^{\frac{d-3}{2}} du.
\end{aligned}
$$

According to formula (4.7.5) in [AAR99] the Bessel function of the first kind has the integral representation

$$
J_\nu(x) = \frac{1}{\sqrt{\pi}\Gamma(\nu + 1/2)} \left(\frac{x}{2}\right)^\nu \int_{-1}^{1} e^{ixu} (1-u^2)^{\nu - 1/2} du
$$

which in turn implies the representation

$$
j_\nu(x) = \frac{\Gamma(\nu + 1)}{\sqrt{\pi}\Gamma(\nu + 1/2)} \int_{-1}^{1} e^{-iux} (1-u^2)^{\nu - 1/2} du.
$$

Setting $\nu = \frac{d-2}{2}$ and using $|S^{d-1}| = \frac{2\pi^{d/2}}{\Gamma(d/2)}$ we obtain $\mathcal{B}_d(t) = j_\nu(2\pi t)$. Let us remark that one can show the identity (1.40) also directly by expanding the exponential function into its power series and interchanging the integral and the series. Then the resulting integrals are essentially Beta functions and after some calculations involving identities of the Gamma function we arrive at the power series (1.41). The explicit expressions of $\mathcal{B}_d$ for $d = 1, 2, 3$ are either derived directly from the integral representation (1.39) or by its relation to the Bessel function (1.40). □

A crucial result in the harmonic analysis of radial functions is the product formula for $\mathcal{B}_d$.

Theorem 1.3.8. *For $r, s \in \mathbb{R}_+$ it holds*

$$
\mathcal{B}_d(r)\mathcal{B}_d(s) = \tau_r \mathcal{B}_d(s). \tag{1.42}
$$

Proof: Recall that $\mathcal{B}_d(r)$ may be interpreted as a Fourier transform of the "radial Dirac measure" $\epsilon_{(r)}$, i.e., $\mathcal{B}_d(r) = \mathcal{F}(\epsilon_{(r)})(\eta)$ where $\eta \in S^{d-1}$ is arbitrary, see (1.35). The convolution theorem yields

$$
\mathcal{B}_d(r)\mathcal{B}_d(s) = \mathcal{F}(\epsilon_{(r)})(\eta)\mathcal{F}(\epsilon_{(s)})(\eta) = \mathcal{F}(\epsilon_{(r)} * \epsilon_{(r)})(\eta) = \epsilon_r * \epsilon_s(e^{-2\pi i \eta \cdot})
$$

Since for a radial measure ρ it holds $\rho(f) = \rho(P_{\mathrm{rad}}f)$ by Lemma 1.2.2(f) and $\epsilon_{(r)} * \epsilon_{(s)}$ is indeed radial we further deduce

$$
\mathcal{B}_d(r)\mathcal{B}_d(s) = \epsilon_{(r)} * \epsilon_{(s)}(P_{\mathrm{rad}}(e^{-2\pi i \eta \cdot})) = \epsilon_{(r)} * \epsilon_{(s)}(\mathcal{B}_d) = \tau_r \mathcal{B}_d(s). \quad \square
$$

Let $\mathcal{B}_d^\lambda(t) := \mathcal{B}_d(\lambda t)$. Bearing in mind the explicit formula (1.21) for τ_s we obtain as a generalization of (1.42)

$$\mathcal{B}_d^\lambda(r)\mathcal{B}_d^\lambda(s) = \tau_r\mathcal{B}_d^\lambda(s). \tag{1.43}$$

Let us further introduce the "modulation" operator

$$(m_\lambda\mathsf{f})(r) := \mathcal{B}_d(\lambda r)\mathsf{f}(r).$$

As a consequence of the product formula we obtain the following useful result.

Corollary 1.3.9. *Let* $s, \lambda \in \mathbb{R}_+$ *and* $\mathsf{f} \in L^1(\mathbb{R}_+, \mu_d) \cup L^2(\mathbb{R}_+, \mu_d)$. *Then we have*

$$(\tau_s\mathsf{f})\hat{\ } = m_s\hat{\mathsf{f}} \quad and \quad (m_\lambda\mathsf{f})\hat{\ } = \tau_\lambda\hat{\mathsf{f}}.$$

Proof: The product formula (1.43) together with (1.24) yields for $\mathsf{f} \in L^1(\mathbb{R}_+, \mu_d)$

$$(\tau_s\mathsf{f})\hat{\ }(\lambda) = \int_0^\infty \tau_s\mathsf{f}(r)\mathcal{B}_d(r\lambda)d\mu_d(r) = \int_0^\infty \mathsf{f}(r)\tau_s\mathcal{B}_d^\lambda(r)d\mu_d(r)$$

$$= \mathcal{B}_d(\lambda s)\int_0^\infty \mathsf{f}(r)\mathcal{B}_d(r\lambda)d\mu_d(r) = \mathcal{B}_d(\lambda s)\hat{\mathsf{f}}(\lambda) = (m_s\hat{\mathsf{f}})(\lambda).$$

Similarly, we obtain $(m_\lambda\mathsf{f})\hat{\ } = \tau_\lambda\hat{\mathsf{f}}$. By density arguments both relations extend to $L^2(\mathbb{R}_+, \mu_d)$. $\qquad\square$

We end this section by considering the special case $d = 3$, which will become important in Chapter 3. The Hankel transform is then given by

$$\hat{\mathsf{f}}(\lambda) = 4\pi \int_0^\infty \mathsf{f}(r)\frac{\sin(2\pi r\lambda)}{2\pi r\lambda}r^2 dr = \frac{2}{\lambda}\int_0^\infty \mathsf{f}(r)\sin(2\pi r\lambda)r dr. \tag{1.44}$$

Let us suppose f is continued to an even function on $\mathbb{R}$. Then a simple calculation using the Euler formula for the sine function yields

$$\hat{\mathsf{f}}(\lambda) = \frac{i}{\lambda}\mathcal{F}(\cdot\mathsf{f})(\lambda) = -\frac{\mathcal{F}(\mathsf{f})'(\lambda)}{2\pi\lambda}, \tag{1.45}$$

where $\mathcal{F}$ denotes the Fourier transform on $\mathbb{R}$ and $\cdot$ denotes the independent variable, i.e., $(\cdot\mathsf{f})(r) = r\mathsf{f}(r)$. The generalized translation can also be expressed differently. Since the Hankel transform is self-inverse it holds by Corollary 1.3.9

$$\tau_s\mathsf{f}(r) = \widehat{(\tau_s\mathsf{f})}(r) = \widehat{(\mathcal{B}_3^s\hat{\mathsf{f}})}(r) = 4\pi \int_0^\infty \hat{\mathsf{f}}(\lambda)\frac{\sin(2\pi s\lambda)}{2\pi s\lambda}\frac{\sin(2\pi r\lambda)}{2\pi r\lambda}\lambda^2 d\lambda$$

$$= \frac{1}{\pi rs}\int_0^\infty \hat{\mathsf{f}}(\lambda)\sin(2\pi s\lambda)\sin(2\pi r\lambda)d\lambda.$$

Again we continue $\hat{f}$ to an even function on $\mathbb{R}$. Applying again Euler's formula yields after a short calculation

$$\tau_s f(r) = \frac{1}{4\pi rs} \left(\mathcal{F}(\hat{f})(r - s) - \mathcal{F}(\hat{f})(r + s) \right), \qquad (1.46)$$

where $\mathcal{F}$ denotes the Fourier transform on $\mathbb{R}$ and $\hat{f}$ is the (evenly continued) Hankel transform of f. We remark that, in general, (1.46) should be understood in the distributional sense.

1.4 Hypergroups

In the two previous sections we encountered the situation where we have a locally compact space $\mathcal{K}$ without a group structure but whose measure algebra $M(\mathcal{K})$ still allows a convolution. Motivated by this observation Jewett [Jew75], Dunkl [Dun73] and Spector [Spe75] independently introduced in the 1970's the notion of a hypergroup. Its axioms are designed such that it generalizes a locally compact group and such that still a rich harmonic analysis can be developed.

Our motivation to discuss hypergroups here stems from the fact that invariant functions on groups – and as a particular case radial functions on $\mathbb{R}^d$ – generate examples of hypergroups. We believe that hypergroup theory gives some further insights into the underlying structure. Actually, the investigations done in this thesis started with the idea to find an analogue of the short time Fourier transform for (a suitable class of) commutative hypergroups. (This problem, however, still remains open for the general case, see also Appendix B). Moreover, with hypergroups in mind one might get ideas how to generalize some of the results presented in this thesis. At some points we will comment on how this could possibly be done.

1.4.1 Basic Definitions and Examples

Before giving the precise definition of a hypergroup we have to introduce the Michael topology [Mic51]. Suppose that $\mathcal{K}$ is a nonvoid locally compact Hausdorff space. By $\mathcal{C}(\mathcal{K})$ we denote the set of all non-empty compact subsets of $\mathcal{K}$. For subsets $U, V \subset \mathcal{K}$ we define

$$\mathcal{C}_U(V) := \{C \in \mathcal{C}(\mathcal{K}) : C \cap U \neq \emptyset, C \subset V\}.$$

The Michael topology on $\mathcal{C}(\mathcal{K})$ is defined as the one generated by the subbasis of all $\mathcal{C}_U(V)$ for which U and V are open subsets of $\mathcal{K}$. For the various properties of this topology we refer to [Mic51]. We only list the following, see also [Jew75, Section 2.5], [BH94, p.7], [Fil00].

Theorem 1.4.1. *(a) $\mathcal{C}(\mathcal{K})$ is a locally compact Hausdorff space.*

(b) If $\mathcal{K}$ is compact then $\mathcal{C}(\mathcal{K})$ is also compact.

(c) The mapping $\mathcal{K} \to \mathcal{C}(\mathcal{K})$, $x \mapsto \{x\}$ is a homeomorphism onto a closed subset of $\mathcal{C}(\mathcal{K})$.

For $x \in \mathcal{K}$ we denote the Dirac measure in x by ϵ_x. Denote by $M^1(\mathcal{K})$ the set of probability measures on $\mathcal{K}$. We assume now that we have given a mapping

$$\mathcal{K} \times \mathcal{K} \to M^1(\mathcal{K}), \quad (x,y) \mapsto \epsilon_x * \epsilon_y$$

which is continuous with respect to the weak-$*$ topology on $M(\mathcal{K})$. We extend this mapping onto $M(\mathcal{K})$ by setting

$$\tau * \rho(f) = \int_{\mathcal{K} \times \mathcal{K}} \epsilon_x * \epsilon_y(f) d(\tau \otimes \rho)(x,y), \quad \tau, \rho \in M(\mathcal{K}), f \in C_0(\mathcal{K}). \qquad (1.47)$$

The function $(x,y) \mapsto \epsilon_x * \epsilon_y(f)$ is continuous by the weak-$*$ continuity assumption. Since $\epsilon_x * \epsilon_y \in M^1(\mathcal{K})$ it holds $|\epsilon_x * \epsilon_y(f)| \leq \|f\|_\infty$ and hence,

$$|\tau * \rho(f)| \leq \|f\|_\infty \|\tau\| \|\rho\|. \qquad (1.48)$$

Thus the bilinear mapping $*$ in (1.47), which we call convolution, is well-defined. Furthermore, (1.48) implies

$$\|\tau * \rho\| \leq \|\tau\| \|\rho\|. \qquad (1.49)$$

Let further $\tilde{\ } : \mathcal{K} \to \mathcal{K}$ denote a homeomorphic involution on $\mathcal{K}$ that is extended onto $M(\mathcal{K})$ by setting

$$\tilde{\rho}(E) := \rho(\tilde{E}), \quad \rho \in M(\mathcal{K}),$$

where E is a Borel set and $\tilde{E} := \{\tilde{x}, x \in E\}$. Now we are able to state the definition of a hypergroup.

Definition 1.4.1. Let $\mathcal{K}$ be a locally compact space and suppose $*$ and $\tilde{\ }$ are as above. We call $(\mathcal{K}, *, \tilde{\ })$ a hypergroup if the following axioms are satisfied.

(H1) (Associativity) It holds $(\epsilon_x * \epsilon_y) * \epsilon_z = \epsilon_x * (\epsilon_y * \epsilon_z)$ for all $x,y,z \in \mathcal{K}$.

(H2) The support $\operatorname{supp} \epsilon_x * \epsilon_y$ is compact for all $x,y \in \mathcal{K}$.

(H3) It holds $(\epsilon_x * \epsilon_y)\tilde{\ } = \epsilon_{\tilde{y}} * \epsilon_{\tilde{x}}$.

(H4) There exists a (necessarily unique) neutral element $e \in \mathcal{K}$ such that $\epsilon_x * \epsilon_e = \epsilon_e * \epsilon_x = \epsilon_x$ for all $x \in \mathcal{K}$.

(H5) We have $e \in \operatorname{supp} \epsilon_x * \epsilon_y$ if and only if $x = \tilde{y}$.

(H6) The mapping $\mathcal{K} \times \mathcal{K} \to \mathcal{C}(\mathcal{K}), (x,y) \mapsto \operatorname{supp} \epsilon_x * \epsilon_y$ is continuous with respect to the Michael topology on $\mathcal{C}(\mathcal{K})$.

If, in addition, $\epsilon_x * \epsilon_y = \epsilon_y * \epsilon_x$ for all $x, y \in \mathcal{K}$ then the hypergroup $\mathcal{K}$ is said to be commutative.

For short we usually write $\mathcal{K}$ instead of $(\mathcal{K}, *, \tilde{\ })$. Observe that inequality (1.49) together with the associativity axiom (H1) implies that $M(\mathcal{K})$ is a Banach algebra, which is commutative iff $\mathcal{K}$ is commutative.

Remark 1.4.1. (a) Suppose $\mathcal{K} = \mathcal{G}$ is a locally compact group. Let $*$ be the group convolution on $M(\mathcal{K})$, i.e., $\epsilon_x * \epsilon_y = \epsilon_{xy}$, see (1.10) and set $\tilde{x} = x^{-1}$. Then $\mathcal{G}$ is hypergroup. Indeed associativity (H1) follows from the associativity of the group convolution and $\operatorname{supp} \epsilon_x * \epsilon_y = \{xy\}$ is clearly compact which verifies (H2). The property (H3) follows from the rule $(xy)^{-1} = y^{-1}x^{-1}$ and the neutral element e is of course given by the neutral element of the group justifying (H4). Since $e \in \operatorname{supp} \epsilon_x * \epsilon_y = \{xy\}$ if and only if $x = y^{-1}$ we obtain (H5) and (H6) follows with Theorem 1.4.1(c).

Thus hypergroups generalize locally compact groups, which also explains the terminology.

(b) The involution $\tilde{\ }$ plays the role of a substitute for the group inverse. Property (H3) ensures that $\tilde{\ }$ behaves similar as the inverse.

(c) In contrast to the group case, the convolution of Dirac measures is no longer a Dirac measure in general. The support of the resulting measure may be spread. Roughly speaking, instead of assigning a single element to the multiplication of two elements one assigns a "continuous convex combination" of elements, which makes it necessary to work with measures on $\mathcal{K}$ instead of working on $\mathcal{K}$ directly.

(d) One may further generalize hypergroups with the aim of dropping the assumption that $\epsilon_x * \epsilon_y$ is a positive measure. In order to be able to work with Banach algebra techniques one needs then the requirement that

$$\|\epsilon_x * \epsilon_y\| \leq C \text{ for all } x, y \in \mathcal{K}$$

for some constant $C > 0$. This leads to the notion of signed hypergroups, see e.g. [Rös95a, Ros95b]. Another generalization, which goes even beyond signed hypergroup, was done in [LOR02, OW96], at least in the discrete setting.

Let us first show that for a locally compact group $\mathcal{G}$ and a compact automorphism group $\mathcal{A}$ of $\mathcal{G}$ the set of orbits $\mathcal{K} := \mathcal{G}^{\mathcal{A}}$ introduced in the previous section indeed carries the structure of a hypergroup. For the convolution of Dirac measures we set

$$\epsilon_{\mathcal{A}x} * \epsilon_{\mathcal{A}y}(F) = T_x F(y), \quad \mathcal{A}x, \mathcal{A}y \in \mathcal{K}, F \in C_0(\mathcal{K}) \cong C_0^{\mathcal{A}}(\mathcal{G}),$$

where T_x was defined in (1.12). Thus the convolution on $\mathcal{K}$ is the one inherited from $\mathcal{G}$. The involution on $\mathcal{K}$ is defined by $\widetilde{\mathcal{A}x} := \mathcal{A}x^{-1}$. In order to show that $(\mathcal{K}, *, \tilde{\ })$

is a hypergroup we need to state an auxiliary result first. If A, B are subsets of a hypergroup $\mathcal{K}$ then the set $A * B$ is defined by

$$A * B := \bigcup_{x \in A, y \in B} \mathrm{supp}(\epsilon_x * \epsilon_y).$$

Note that $A * B = A \cdot B$ if $\mathcal{K} = \mathcal{G}$ is a group.

Lemma 1.4.2. *[Jew75, Lemma 3.2]*

(a) *If A and B are compact subsets of $\mathcal{K}$ then also $A * B$ is compact.*

(b) *The convolution of sets is a continuous operation in $\mathcal{C}(\mathcal{K})$.*

With this preparation we are ready to show the following.

Theorem 1.4.3. *[Jew75, Theorem 8.3A] Suppose $\mathcal{G}$ is a locally compact group with compact automorphism group $\mathcal{A}$. Then $\mathcal{K} = \mathcal{G}^{\mathcal{A}}$ with the operations $*$ and $\tilde{\ }$ defined above forms a hypergroup with identity element $\mathcal{A}e = \{e\}$.*

Proof: First observe that $M(\mathcal{K})$ is continuously embedded into $M(\mathcal{G})$ by identifying a measure on $\mathcal{K}$ with its corresponding invariant one on $\mathcal{G}$. The Dirac measures $\epsilon_{\mathcal{A}x}$ are clearly probability measures when interpreted as measures on $\mathcal{G}$ and thus also their convolution is a probability measure. Since the quotient topology inherited from $\mathcal{G}$ is taken on $\mathcal{G}^{\mathcal{A}}$ it is simple to verify that the mapping $\mathcal{G}^{\mathcal{A}} \times \mathcal{G}^{\mathcal{A}} \to M^1(\mathcal{G}^{\mathcal{A}})$, $(\mathcal{A}x, \mathcal{A}y) \mapsto \epsilon_{\mathcal{A}x} * \epsilon_{\mathcal{A}y}$ is weak-$*$ continuous.

Associativity (H1) follows from the associativity of $M(\mathcal{G})$. Since $\mathcal{A}$ is compact, the sets $\mathcal{A}x, x \in \mathcal{G}$ are compact in $\mathcal{G}$ and thus by Lemma 1.4.2 $\mathcal{A}x * \mathcal{A}y$ (interpreted as subset of $\mathcal{G}$) is compact. By construction of the topology on $\mathcal{G}^{\mathcal{A}}$ the set $\mathrm{supp}\,\epsilon_{\mathcal{A}x} * \epsilon_{\mathcal{A}y}$ is also compact as subset of $\mathcal{K}$. Property (H3) follows from

$$(\epsilon_{\mathcal{A}x} * \epsilon_{\mathcal{A}y})^{\tilde{\ }}(f) = \int_A f((xA(y))^{-1})dA = \int_A f(A(y^{-1})x^{-1})dA = \epsilon_{\widetilde{\mathcal{A}y}} * \epsilon_{\widetilde{\mathcal{A}x}}(f)$$

for all $f \in C_0(\mathcal{K}) \cong C_0^{\mathcal{A}}(\mathcal{G})$. Hereby, we used Lemma 1.2.4(a) in the fourth equation. Property (H4) is verified with the neutral element $\mathcal{A}e \in \mathcal{K}$. Furthermore, $\mathcal{A}e \in \mathrm{supp}\,\epsilon_{\mathcal{A}x} * \epsilon_{\mathcal{A}y}$ is easily seen to be equivalent to $\mathcal{A}x = \mathcal{A}y^{-1}$, which shows (H5). For axiom (H6) we first note that $\mathcal{G}^{\mathcal{A}}$ can be identified with a closed subset of $\mathcal{C}(\mathcal{G})$ and the relative topology on $\mathcal{G}^{\mathcal{A}}$ inherited from $\mathcal{C}(\mathcal{G})$ coincides with the quotient topology inherited from $\mathcal{G}$, see Theorem 8.1A in [Jew75]. Thus the inverse ι^{-1} of the canonical projection $\iota : \mathcal{G} \to \mathcal{G}^{\mathcal{A}}, x \mapsto \mathcal{A}x$ is a continuous mapping $\iota^{-1} : \mathcal{G}^{\mathcal{A}} \to \mathcal{C}(\mathcal{G})$. By Lemma 1.4.2(b) the mapping

$$\hat{\imath} \circ * \circ (\iota^{-1} \times \iota^{-1}) : \mathcal{G}^{\mathcal{A}} \times \mathcal{G}^{\mathcal{A}} \to \mathcal{C}(\mathcal{G}^{\mathcal{A}}), \quad (\mathcal{A}x, \mathcal{A}y) \mapsto \mathrm{supp}\,\epsilon_{\mathcal{A}x} * \epsilon_{\mathcal{A}y}$$

is hence continuous as concatenation of continuous mappings. Hereby $*$ has to be understood as convolution of sets in $\mathcal{C}(\mathcal{G})$. Further note that $\hat{\iota} : \mathcal{C}(\mathcal{G}) \to \mathcal{C}(\mathcal{G}^{\mathcal{A}})$, $\hat{\iota}(C) = \iota(C), C \in \mathcal{C}(\mathcal{G})$ is continuous since $\iota : \mathcal{G} \to \mathcal{G}^{\mathcal{A}}$ is continuous. (In general, if $f : X \to Y$ is a continuous mapping, then $\hat{f} : \mathcal{C}(X) \to \mathcal{C}(Y), \hat{f}(C) = f(C), C \in \mathcal{C}(X)$ is also continuous, since $f^{-1}(\mathcal{C}_U(V)) = \mathcal{C}_{f^{-1}(U)}(f^{-1}(V))$ for open sets $U, V \subset Y$.) $\square$

Hypergroups of the type described above are usually referred to as **orbit hypergroups**. In particular, taking $\mathcal{G} = \mathbb{R}^d$ and $\mathcal{A} = SO(d)$ we get $\mathcal{K} \cong \mathbb{R}_+$ and the corresponding convolution of measures is related to the generalized translation τ of Section 1.3 by (1.27), whereas the involution is the identity since the orbits under $SO(d)$ of x and $-x$ coincide. This hypergroup is called the **Bessel-Kingman hypergroup** of index $\nu = \frac{d-2}{2}$. Here, the index ν refers to the corresponding Bessel functions J_ν resp. j_ν in Theorem 1.3.7.

Let us give some further examples of hypergroups.

Example 1.4.1. (a) As a generalization of the situation in Section 1.3 the definition of the **Bessel-Kingman hypergroup** also makes sense for arbitrary real-valued $\nu > -1/2$. We take $\mathcal{K} = \mathbb{R}_+$ and define for $r, s \in \mathbb{R}_+$

$$\epsilon_r * \epsilon_s(\mathsf{f}) = \frac{\Gamma(\nu + 1)}{\sqrt{\pi}\Gamma(\nu)} \int_0^\pi \mathsf{f}(\sqrt{r^2 + 2rs\cos\theta + s^2}) \sin^{2\nu}\theta \, d\theta, \quad \mathsf{f} \in C_0(\mathbb{R}_+). \quad (1.50)$$

The involution on $\mathcal{K}$ is defined by $\tilde{r} := r$. One can show that $(\mathbb{R}_+, *, \tilde{\ })$ is a hypergroup. Note that if $\nu = \frac{d-2}{2}, d \in \mathbb{N}$, it holds

$$\frac{\Gamma(\nu + 1)}{\sqrt{\pi}\Gamma(\nu)} = \frac{|S^{d-2}|}{|S^{d-1}|}$$

and by Lemma 1.3.3(f) the convolution in (1.50) coincides with the one inherited from the convolution of radial measures on $\mathbb{R}^d$, see Lemma 1.3.3(a). Motivated by this geometrical interpretation if ν is a half-integer, one could interpret functions on the Bessel-Kingman hypergroup of index ν as radial functions in (real-valued) dimension $d = 2\nu + 2$. In this sense, the Bessel-Kingman hypergroup interpolates between dimensions. We remark that the name of this hypergroup is attached to J.F.C. Kingman who studied random walks with radial symmetry in [Kin65].

(b) Assume that $\mathcal{G}$ is a locally compact group and H a compact subgroup of $\mathcal{G}$ with normalized Haar measure ν. Let $\mathcal{G}//H$ be the collection of **double cosets** $\{HxH, x \in \mathcal{G}\}$. It becomes a locally compact space with the quotient topology inherited from $\mathcal{G}$. The space $\mathcal{G}//H$ with

$$\epsilon_{HxH} * \epsilon_{HyH}(f) = \int_H f(HxtyH)d\nu(t), \quad f \in C_0(\mathcal{G}//H), \quad (1.51)$$

and involution $(HxH)^{\check{}} = Hx^{-1}H$ is a hypergroup with identity $H = HeH$, see [Jew75, Theorem 8.2B] or [BH94, Theorem 1.1.9]. Hypergroups of this type are called **double coset hypergroups**. We remark that $M(\mathcal{G}//H)$ can be viewed as a certain subalgebra of $M(\mathcal{G})$. For a function f on $\mathcal{G}$ let $f_h(x) = f(hx)$ and $f^h(x) = f(xh)$ for $x \in \mathcal{G}, h \in H$. The function f is said to be H-biinvariant if $f_h^k = f$ for all $h, k \in H$ and a measure $\tau \in M(\mathcal{G})$ is called H-biinvariant if $\tau(g_h^k) = \tau(g)$ for all $k, h \in H, g \in C_0(\mathcal{G})$. A simple computation shows that the convolution of two H-biinvariant measures is again H-biinvariant and the closed subalgebra of $M(\mathcal{G})$ of H-biinvariant measures may indeed be identified with $M(\mathcal{G}//H)$. It is a remarkable fact that in many situations the algebra $M(\mathcal{G}//H)$ is commutative although $\mathcal{G}$ is not. In this case $(\mathcal{G}, H)$ is called a **Gelfand pair**.

Orbit hypergroups may be seen as a special case of double coset hypergroups. Indeed, for a compact automorphism group $\mathcal{A}$ we let $\mathcal{G} \rtimes \mathcal{A}$ be the semi-direct product group. Then the hypergroups $\mathcal{G}^{\mathcal{A}}$ and $(\mathcal{G} \rtimes \mathcal{A})//\mathcal{A}$ are isomorphic, see also [Jew75, Theorem 8.3B].

Let us now continue with the general discussion of hypergroups. Using the convolution of Dirac measures we may define a translation on a hypergroup $\mathcal{K}$ analogously to the group case (1.10) by

$$T_y f(x) := \epsilon_y * \epsilon_x(f), \quad \text{for } f \in C_0(\mathcal{K}), x, y \in \mathcal{K}. \tag{1.52}$$

Moreover, we also introduce a left and right translation by

$$\mathcal{L}_y f(x) := T_{\check{y}} f(x) = \epsilon_{\check{y}} * \epsilon_x(f)$$
$$\text{and} \quad \mathcal{R}_y f(x) := T_x f(y) = \epsilon_x * \epsilon_y(f), \quad f \in C_0(\mathcal{K}), x, y \in \mathcal{K}. \tag{1.53}$$

Lemma 3.1A in [Jew75] (see also [Fil00, Lemma 1.2.1]) states that these translation operators extend to all continuous functions on $\mathcal{K}$ and the mapping $(x, y) \mapsto T_x f(y)$ is continuous for any continuous function f. Moreover, we have

$$\|T_x f\|_\infty \leq \|f\|_\infty \quad \text{for all } f \in C^b(\mathcal{K})$$

and T_x maps $C^b(\mathcal{K})$ into $C^b(\mathcal{K})$, $C_0(\mathcal{K})$ into $C_0(\mathcal{K})$ and $C_c(\mathcal{K})$ into $C_c(\mathcal{K})$, see [Fil00, Theorem 1.2.2] or [BH94, Proposition 1.2.16].

A key result in harmonic analysis on locally compact groups is the existence of a Haar measure. Here is the analogous definition for hypergroups.

Definition 1.4.2. Suppose $\mathcal{K}$ is a hypergroup. A positive non-zero Radon measure m on $\mathcal{K}$ that satisfies
$$m(T_x f) = m(f) \quad \text{for all } f \in C_c(\mathcal{K})$$
is called a Haar measure.

It is an open question whether a Haar measure exists on every hypergroup. However, for most important cases the existence is settled. In particular, on commutative, discrete and compact hypergroups, and on orbit and double coset hypergroups a Haar measure is known to exist, see [BH94, Theorem 1.3.15] and Theorems 7.1A, 7.2B, 8.1B and 8.2B in [Jew75]. Moreover, in case it exists, the Haar measure is unique up to multiplication by a positive constant [Jew75, Theorem 5.2]. Since by associativity (H1) the right translation $\mathcal{R}_x$ commutes with the translation $\mathcal{T}_y$ for all $x, y \in \mathcal{K}$ one easily verifies that the measure $m_x(f) := m(\mathcal{R}_x f)$ is also a Haar measure. By uniqueness there exists hence a function Δ on $\mathcal{K}$, called the **modular function**, such that

$$\Delta(\tilde{x}) \int_{\mathcal{K}} f(y) dm(y) = \int_{\mathcal{K}} \mathcal{R}_x f(y) dm(y).$$

Similar to the group case the modular function is continuous and satisfies $\Delta(e) = 1$, $\Delta(\tilde{x}) = \Delta(x)^{-1}$ and $\mathcal{T}_x \Delta(y) = \Delta(x)\Delta(y)$ [Jew75, Theorems 5.3B,C]. The hypergroup $\mathcal{K}$ is called **unimodular** if $\Delta \equiv 1$. Commutative and compact hypergroups are unimodular [Jew75, Theorem 7.2C].

Let us give the Haar measure for our main examples.

Example 1.4.2.   (a) Let $(\mathbb{R}_+, *, \tilde{\ })$ be the Bessel-Kingman hypergroup of index $\nu > -1/2$. The Haar measure of this hypergroup is given by

$$d\mu_{(\nu)}(r) := \frac{2\pi^{\nu+1}}{\Gamma(\nu+1)} r^{2\nu+1} dr.$$

If $\nu = \frac{d-2}{2}$ we have $\mu_{(\nu)} = \mu_d$, see (1.17), and a proof of the Haar measure property was already provided in Lemma 1.3.3(d).

(b) By Lemma 1.2.4(c) the Haar measure on a orbit hypergroup $\mathcal{G}^{\mathcal{A}}$ is given by the measure m defined in (1.6). The modular function is the one inherited from $\mathcal{G}$ by $\Delta_{\mathcal{G}^{\mathcal{A}}}(\mathcal{A}x) = \Delta_{\mathcal{G}}(x)$.

(c) Also on a double coset hypergroup $\mathcal{G}//H$ the Haar measure m is given by the projection from the Haar measure μ of $\mathcal{G}$ [Jew75, Theorem 8.2B]

$$\int_{\mathcal{G}//H} f(HxH) dm(HxH) = \int_{\mathcal{G}} f(x) d\mu(x),$$

where functions on $\mathcal{G}//H$ are identified with H-biinvariant functions on $\mathcal{G}$.

We assume from now on that a Haar measure m exists on $\mathcal{K}$. In this case the spaces $L^p(\mathcal{K}) = L^p(\mathcal{K}, m)$, $1 \le p \le \infty$, are well-defined. The translation $\mathcal{T}$ may be extended to the L^p-spaces and we have the following theorem.

Theorem 1.4.4. *[BH94, Theorem 1.3.5], [Fil00, Satz 1.2.7, 1.2.8] Suppose* $1 \leq p \leq \infty$ *and* $f \in L^p(\mathcal{K})$. *Then it holds*

$$\|T_x f\|_p \leq \|f\|_p. \tag{1.54}$$

It is a remarkable fact that in contrast to the group case we may have a strict inequality in (1.54). In fact, this phenomenon occurs for the Bessel-Kingman hypergroups and for many orbit hypergroups.

The convolution of two functions on $\mathcal{K}$ is defined by

$$f * g(x) := \int_{\mathcal{K}} f(y) \mathcal{L}_y g(x) dm(y) \tag{1.55}$$

whenever this expression exists (a.e.). Associated to a measurable function f on $\mathcal{K}$ is the measure fm defined by

$$(fm)(E) := \int_E f(x) dm(x), \qquad E \text{ measurable subset of } \mathcal{K}.$$

In this way, $L^1(\mathcal{K})$ is isometrically embedded into $M(\mathcal{K})$. One can show further [Fil00, p.21] that $(f*g)m = (fm)*(gm)$. Thus, the definition (1.55) is natural. If $f, g \in L^1(\mathcal{K})$ it follows from the Banach-algebra inequality of $M(\mathcal{K})$ that

$$\|f * g\|_1 \leq \|f\|_1 \|g\|_1.$$

Hence, $L^1(\mathcal{K})$ is Banach algebra under convolution, actually an ideal of $M(\mathcal{K})$.

Define an involution on $M(\mathcal{K})$ by

$$\rho^*(E) = \overline{\rho(\tilde{E})}, \quad E \text{ measurable .}$$

On the subalgebra $L^1(\mathcal{K})$ this involution takes the form $f^*(x) = \Delta(\tilde{x})\overline{f(\tilde{x})}$. Equipped with this involution $L^1(\mathcal{K})$ and $M(\mathcal{K})$ become Banach-*-algebras. Further, the convolution extends to other L^p-spaces. In particular, there is a Young theorem on hypergroups.

Theorem 1.4.5. *[Jew75, Theorem 6.2] Let* $f \in L^1(\mathcal{K}), g \in L^p(\mathcal{K})$, $1 \leq p \leq \infty$. *Then it holds*

$$\|f * g\|_p \leq \|f\|_1 \|g\|_p.$$

1.4.2   Harmonic Analysis on Commutative Hypergroups

For commutative hypergroups $\mathcal{K}$ it is possible to develop harmonic analysis quite analogous to the one on locally compact Abelian groups.

We first need to define dual objects of a commutative hypergroup $\mathcal{K}$. In contrast to the group case one has to work with three (possibly different) dual objects. The first two of them are defined by

$$\mathcal{X}^b(\mathcal{K}) := \{\alpha \in C^b(\mathcal{K}), \alpha \neq 0, T_x\alpha(y) = \alpha(x)\alpha(y) \text{ for all } x, y \in \mathcal{K}\},$$
$$\widehat{\mathcal{K}} := \{\alpha \in \mathcal{X}^b(\mathcal{K}), \alpha(\tilde{x}) = \overline{\alpha(x)}\}.$$

Elements of $\mathcal{X}^b(\mathcal{K})$ are called characters of $\mathcal{K}$ and the elements of $\widehat{\mathcal{K}}$ are referred to as hermitian characters. The function constant 1 is contained in $\widehat{\mathcal{K}}$ and one easily deduces $\alpha(e) = 1$ and $\|\alpha\|_\infty = 1$ for all $\alpha \in \mathcal{X}^b(\mathcal{K})$.

There are some differences to the group case. In general we have $|\alpha(x)| \not\equiv 1$ and $\alpha\beta$ need not be a character for $\alpha, \beta \in \mathcal{X}^b(\mathcal{K})$. Moreover, it might happen that $\widehat{\mathcal{K}}$ is a proper subset of $\mathcal{X}^b(\mathcal{K})$ and neither $\mathcal{X}^b(\mathcal{K})$ nor $\widehat{\mathcal{K}}$ needs to be a hypergroup again. Thus an analogue of Pontryagin's theorem does not hold for hypergroups, in general. If, however, $\mathcal{K}$ has this property, i.e., $\widehat{\mathcal{K}}$ is a hypergroup and $\widehat{\widehat{\mathcal{K}}}$ is topologically isomorphic to $\mathcal{K}$, then it is called a **Pontryagin hypergroup**.

We may identify the dual objects $\mathcal{X}^b(\mathcal{K})$ and $\widehat{\mathcal{K}}$ with sets of multiplicative functionals on the Banach-$*$-algebra $L^1(\mathcal{K})$. Recall that the structure space (or maximal ideal space) of a Banach-$*$-algebra X is defined by

$$\Delta(X) := \{\phi \in X', \phi \neq 0, \phi(xy) = \phi(x)\phi(y)\},$$

and with $\Delta_s(X)$ we denote its subset of hermitian multiplicative functionals, i.e.,

$$\Delta_s(X) := \{\phi \in \Delta(X), \phi(x^*) = \overline{\phi(x)}\}.$$

Now for $\alpha \in \mathcal{X}^b(\mathcal{K})$ and $\rho \in M(\mathcal{K})$ define

$$\phi_\alpha(\rho) := \int_\mathcal{K} \overline{\alpha(x)} d\rho(x).$$

A simple computation shows $\phi_\alpha(\rho * \tau) = \phi_\alpha(\rho)\phi_\alpha(\tau)$ and, hence, $\phi_\alpha \in \Delta(M(\mathcal{K}))$. Moreover, for $\alpha \in \widehat{\mathcal{K}}$ we have $\phi_\alpha \in \Delta_s(M(\mathcal{K}))$. Restricting ϕ_α to the subalgebra $L^1(\mathcal{K})$ by $\phi_\alpha(f) := \phi_\alpha(fm)$ for $f \in L^1(\mathcal{K})$ we obtain elements of $\Delta(L^1(\mathcal{K}))$ and $\Delta_s(L^1(\mathcal{K}))$ respectively. Hereby, m denotes the Haar measure of $\mathcal{K}$, which exists by commutativity of $\mathcal{K}$.

Theorem 1.4.6. *[Fil00, Satz 1.3.1] The mapping $\alpha \mapsto \phi_\alpha$ establishes an isomorphism between $\mathcal{X}^b(\mathcal{K})$ and $\Delta(L^1(\mathcal{K}))$ and between $\widehat{\mathcal{K}}$ and $\Delta_s(L^1(\mathcal{K}))$.*

Using this theorem we can define a topology on $\mathcal{X}^b(\mathcal{K})$. Indeed $\Delta(L^1(\mathcal{K}))$ carries the Gelfand topology (the weak-$*$ topology induced from $(L^1(\mathcal{K}))'$). Using the isomorphism in Theorem 1.4.6 we transport this topology onto $\mathcal{X}^b(\mathcal{K})$ and take the induced topology on $\widehat{\mathcal{K}}$.

Analogously to the group case we may now define the Fourier transform.

Definition 1.4.3. Let $\rho \in M(\mathcal{K})$. Its **Fourier-Stieltjes transform** is given by

$$\hat{\rho}(\alpha) := \int_{\mathcal{K}} \overline{\alpha(x)} d\rho(x), \quad \alpha \in \mathcal{X}^b(\mathcal{K}).$$

Further, let $f \in L^1(\mathcal{K})$. Its **Fourier transform** is defined by

$$\hat{f}(\alpha) := \widehat{(fm)}(\alpha) = \int_{\mathcal{K}} \overline{\alpha(x)} f(x) dm(x), \quad \alpha \in \mathcal{X}^b(\mathcal{K}). \tag{1.56}$$

The Fourier transform obviously is a linear mapping. It has the following further elementary properties which are rather easy to derive.

Theorem 1.4.7. Let $\rho, \tau \in M(\mathcal{K})$ and $\alpha \in \mathcal{X}^b(\mathcal{K}), x \in \mathcal{K}$. Then the following holds.

(a) (Convolution theorem) $\widehat{(\rho * \tau)} = \hat{\rho}\hat{\tau}$.

(b) $\hat{\epsilon}_x(\alpha) = \overline{\alpha(x)}$.

(c) $\widehat{(T_x f)}(\alpha) = \overline{\alpha(\tilde{x})}\hat{f}(\alpha)$. Hence, if $\alpha \in \widehat{\mathcal{K}}$ then $\widehat{(T_x f)}(\alpha) = \alpha(x)\hat{f}(\alpha)$.

There is also a Riemann-Lebesgue Lemma for commutative hypergroups, see [BH94, Theorem 2.2.4]. For analogues of further classical theorems of harmonic analysis, like the uniqueness theorem and the Plancherel theorem we need to work with a third dual object of $\mathcal{K}$.

Suppose $f \in L^1(\mathcal{K})$. Define the operator

$$T_f h := f * h, \quad h \in L^2(\mathcal{K}).$$

By the Young theorem 1.4.5 T_f is bounded on $L^2(\mathcal{K})$ and according to [Fil00, Satz 1.3.6] the mapping $L^1(\mathcal{K}) \to \mathcal{B}(L^2(\mathcal{K})), f \mapsto T_f$ is injective, where $\mathcal{B}(L^2(\mathcal{K}))$ denotes the bounded operators on $L^2(\mathcal{K})$. It further holds $T_{f*g} = T_f \circ T_g$ and $T_{f^*} = T_f^*$ (recall $f^*(x) = \overline{f(\tilde{x})}$). Thus $B := \{T_f, f \in L^1(\mathcal{K})\}$ is a commutative $*$-subalgebra of the C^*-algebra $\mathcal{B}(L^2(\mathcal{K}))$ and the closure $\mathcal{B}$ of B in the operator norm is a commutative C^*-algebra. It turns out that the structure space $\Delta(\mathcal{B})$ can be identified with the following subset of $\widehat{\mathcal{K}}$, defined by

$$\mathsf{S} := \{\alpha \in \widehat{\mathcal{K}}, |\hat{f}(\alpha)| \leq \|T_f\| \text{ for all } f \in L^1(\mathcal{K})\}.$$

We equip S with the topology induced from $\widehat{\mathcal{K}}$.

Let us shortly make some remarks about the relation of the three dual objects. We have the inclusions

$$\mathsf{S} \subset \widehat{\mathcal{K}} \subset \mathcal{X}^b(\mathcal{K}).$$

In case that $\mathcal{K} = \mathcal{G}$ is a group all these sets coincide. In general, however, the inclusions may be proper. We remark further, that $L^1(\mathcal{K})$ can be identified with $B = \{T_f, f \in L^1(\mathcal{K})\}$ interpreted as subalgebra of $\mathcal{B}(L^1(\mathcal{K}))$. In this sense, $\mathcal{X}^b(\mathcal{K})$ coincides with the structure space of B, $\widehat{K}$ with its subset of hermitian elements and S is the structure space of the completion of B in $\mathcal{B}(L^2(\mathcal{K}))$.

Using the space $\mathcal{K}$ one may formulate a uniqueness theorem, see [BH94, Fil00], and a Plancherel theorem for the Fourier transform on hypergroups.

Theorem 1.4.8. *([BH94, Theorem 2.2.13], [Jew75, Theorem 7.3I]) There exists a unique positive measure π on $\mathcal{K}$ such that*

$$\int_{\mathcal{K}} |f(x)|^2 dm(x) = \int_{\widehat{\mathcal{K}}} |\hat{f}(\alpha)|^2 d\pi(\alpha)$$

for all $f \in L^1(\mathcal{K}) \cap L^2(\mathcal{K})$ with $\operatorname{supp}\pi = \mathsf{S}$. Consequently, the Fourier transform extends to a unitary mapping from $L^2(\mathcal{K})$ onto $L^2(\widehat{\mathcal{K}}) = L^2(\widehat{\mathcal{K}}, \pi)$.

The measure π in the theorem is referred to as the **Plancherel measure**. In case that $\widehat{\mathcal{K}}$ is again a hypergroup, then π is the Haar measure of $\widehat{\mathcal{K}}$. It is a remarkable fact, that also in case $\widehat{\mathcal{K}}$ is not a hypergroup, there is a canonical measure on $\widehat{\mathcal{K}}$.

Using the Plancherel measure π we define the **inverse Fourier transform** for $g \in L^1(\widehat{\mathcal{K}}, \pi)$ by

$$\check{g}(x) := \int_{\widehat{\mathcal{K}}} g(\alpha)\alpha(x) d\pi(\alpha) \qquad (1.57)$$

and analogously for measures $\rho \in M(\widehat{\mathcal{K}})$. Since $|\alpha(x)| \leq 1$ for all $\alpha \in \widehat{\mathcal{K}}, x \in \mathcal{K}$ the integral (1.57) is well-defined for $g \in L^1(\widehat{\mathcal{K}})$. In particular, it holds $\check{g} \in C_0(\mathcal{K})$, see [Fil00, p.48].

Theorem 1.4.9. *([BH94, Theorem 2.2.36],[Fil00, Satz 2.2.3]) Suppose $f \in L^1(\mathcal{K})$ with $\hat{f} \in L^1(\widehat{\mathcal{K}})$. Then it holds $f = (\hat{f})^{\check{}}$ in $L^1(\mathcal{K})$. If in addition f is continuous then*

$$f(x) = \int_{\widehat{\mathcal{K}}} \hat{f}(\alpha)\alpha(x) d\pi(\alpha)$$

holds pointwise.

Let us finally give an example. Denote by $\mathcal{K} = \mathcal{K}_\nu$ the Bessel-Kingman hypergroup of index $\nu > -1/2$. We denote

$$\mathcal{B}_{(\nu)}^\lambda(t) := j_\nu(2\pi\lambda t), \qquad (1.58)$$

where j_ν was defined in Lemma 1.3.7. (The definition applies for arbitrary real-valued $\nu > -1/2$.) For $\nu = \frac{d-2}{2}$ these functions coincide with the spherical Bessel functions, i.e., $\mathcal{B}_{(\nu)} = \mathcal{B}_d$. The product formula (1.43) extends to arbitrary indices $\nu > -1/2$. Indeed, it holds

$$j_\nu(r)j_\nu(s) = \frac{\Gamma(\nu+1)}{\sqrt{\pi}\Gamma(\nu)} \int_0^\pi j_\nu(\sqrt{r^2 + 2rs\cos\theta + s^2})\sin^{2\nu}\theta d\theta. \qquad (1.59)$$

Of course, the proof for general ν requires completely different techniques than in the geometrical setting of Section 1.3. Indeed, one may deduce (1.59) from a series expansion [AAR99, formula (4.10.5)] of the Bessel function J_ν in terms of Gegenbauer polynomials. By (1.59) we have

$$\mathcal{B}_{(\nu)}^\lambda(r)\mathcal{B}_{(\nu)}^\lambda(s) = \tau_r\mathcal{B}_{(\nu)}^\lambda(s) \quad \text{for all } r, s \in \mathbb{R}_+.$$

Therefore, $\mathcal{B}_{(\nu)}^\lambda$ is a character of $\mathcal{K}_\nu$. Since $\mathcal{B}_{(\nu)}^\lambda$ is real-valued and since $\tilde{r} = r$ it is an element of $\widehat{\mathcal{K}_\nu}$. It follows from Theorem 3.5.58 in [BH94] that all elements of $\mathcal{X}^b(\mathcal{K}_\nu)$ are of this form. Moreover, we have $\mathcal{X}^b(\mathcal{K}_\nu) = \widehat{\mathcal{K}_\nu} = \mathsf{S} \cong \mathbb{R}_+$, see [BH94, p.235]. The corresponding Fourier transform is the Hankel transform of index ν, i.e.,

$$H_{(\nu)}(f)(\lambda) := \hat{f}(\lambda) = \int_0^\infty f(r)\mathcal{B}_{(\nu)}(\lambda r)d\mu_{(\nu)}(r).$$

Since $\mathcal{B}_{(\nu)}^\lambda(t) = \mathcal{B}_{(\nu)}^t(\lambda)$ for all $\lambda, t \in \mathbb{R}_+$ the dual $\widehat{\mathcal{K}_\nu}$ is again a hypergroup, which is isomorphic to $\mathcal{K}_\nu$ itself. This implies that $\mathcal{K}_\nu$ is a Pontryagin hypergroup and the Plancherel measure is given by $\mu_{(\nu)}$.

1.4.3   Representations of Hypergroups

As for groups one has to work with representations instead of characters in the non-commutative case. Let us briefly introduce the necessary information.

We assume throughout this section that $\mathcal{K}$ is a general (possibly non-commutative) hypergroup. Given a Hilbert space $\mathcal{H}$ we denote by $\mathcal{B}(\mathcal{H})$ the C^*-algebra of bounded operators on $\mathcal{H}$.

Definition 1.4.4. A mapping $U : M(\mathcal{K}) \to \mathcal{B}(\mathcal{H})$ is called a representation of the hypergroup $\mathcal{K}$ if the following conditions are satisfied.

(i) The mapping $\rho \to U(\rho)$ is a $*$-homomorphism from $M(\mathcal{K})$ into $\mathcal{B}(\mathcal{H})$.

(ii) $U(\epsilon_e) = \text{Id}$.

(iii) The mapping $M(\mathcal{K}) \to \mathbb{C}$, $\rho \mapsto \langle f, U(\rho)g \rangle$ is continuous with respect to the $\sigma(M(\mathcal{K}), C^b(\mathcal{K}))$-topology for all $f, g \in \mathcal{H}$.

As a consequence of (i) it holds (see [Dix82, Proposition 1.3.7], [Ric60, Theorem 4.1.20])

$$\|U(\rho)\| \leq \|\rho\|. \tag{1.60}$$

Moreover, (ii) implies that U is a non-degenerate $*$-representation of $M(\mathcal{K})$ on $\mathcal{B}(\mathcal{H})$, i.e., $U(\rho)f = 0$ for all $\rho \in M(K)$ implies $f = 0$. For convenience, we write $U(x)$ instead of $U(\epsilon_x)$ for $x \in \mathcal{K}$. Property (iii) implies in particular that the function

$$V_g f(x) = \langle f, U(x)g \rangle, \quad x \in \mathcal{K}, f, g \in \mathcal{H},$$

is continuous on $\mathcal{K}$ and together with (1.60) we deduce $V_g f \in C^b(\mathcal{K})$. Moreover, it holds

$$\langle U(\rho)f, g \rangle = \int_{\mathcal{K}} \langle U(x)f, g \rangle d\rho(x) \quad \text{for all } \rho \in M(\mathcal{K}). \tag{1.61}$$

Conversely, if the mapping $x \mapsto \langle U(x)f, g \rangle$ is continuous and $U(\rho)$ is defined for an arbitrary element of $M(\mathcal{K})$ by (1.61) then Property (iii) is satisfied.

We further remark that it is necessary to work with the $\sigma(M(\mathcal{K}), C^b(\mathcal{K}))$ topology on $M(\mathcal{K})$ instead of the weak-$*$ topology because $V_g f$ is in general an element of $C^b(\mathcal{K})$ and not of $C_0(\mathcal{K})$ and we would not be able to conclude (1.61) from Property (iii).

A subspace W of $\mathcal{H}$ is called invariant if $U(\rho)W \subset W$ for all $\rho \in M(\mathcal{K})$. The representation $(U, \mathcal{H})$ of $\mathcal{K}$ is called **irreducible** if the only closed invariant subspaces W are trivial, i.e., $W = \{0\}$ or $W = \mathcal{H}$. The following criterion for irreducibility will be useful.

Lemma 1.4.10. *Suppose $(U, \mathcal{H})$ is a representation of $\mathcal{K}$. Then U is irreducible if and only if the linear span of $\{U(x)f, x \in \mathcal{K}\}$ is dense in $\mathcal{H}$ for all non-zero $f \in \mathcal{H}$.*

Proof: For $f \in \mathcal{H}, f \neq 0$ we denote $W_f := \text{span}\{U(\rho)f, \rho \in M(\mathcal{K})\}$. If $g = \sum_{k=1}^{n} c_k U(\rho_k)f \in W_f$, $\rho_k \in M(\mathcal{K})$ then $U(\rho)g = \sum_{k=1}^{n} c_k U(\rho * \rho_k)f$ is an element of W_f as well. Hence W_f is an invariant subspace of $\mathcal{H}$ and its closure $\overline{W_f}$ is invariant as well. So if $(U, \mathcal{H})$ is irreducible then necessarily $\overline{W_f} = \mathcal{H}$. Conversely, if $\overline{W_f} \neq \mathcal{H}$ for some non-zero f then there is a non-trivial invariant subspace of $\mathcal{H}$ and U cannot be irreducible.

It remains to prove that the closure of $\mathcal{H}_f := \{U(x)f, x \in \mathcal{K}\}$ coincides with $\overline{W_f}$. The relation $\overline{\mathcal{H}_f} \subset \overline{W_f}$ is clear. For the opposite inclusion suppose $g \in \mathcal{H}_f^{\perp}$. We obtain

$$\langle U(\rho)f, g \rangle = \int_{\mathcal{K}} \langle U(x)f, g \rangle d\rho(x) = 0 \quad \text{for all } \rho \in M(\mathcal{K})$$

implying $g \in W_f^{\perp}$ and thus $\mathcal{H}_f^{\perp} \subset W_f^{\perp}$ which yields $\overline{W_f} \subset \overline{H_f}$. $\qquad\square$

Two representations $(U_i, \mathcal{H}_i), i = 1, 2$ are said to be (unitarily) equivalent if there exists a unitary operator $T : \mathcal{H}_1 \to \mathcal{H}_2$ such that $TU_1(\rho) = U_2(\rho)T$ for all $\rho \in$

$M(\mathcal{K})$. The dual $\widehat{\mathcal{K}}$ of a hypergroup is defined to be the set of all equivalence classes of irreducible representations of $\mathcal{K}$. Since a generalization of Schur's lemma is valid for arbitrary $*$-representations of Banach-$*$-algebras (see e.g. [Ric60, Theorem 4.4.12]) we have in particular a version of Schur's lemma for hypergroups. Thus, an irreducible representation of a commutative hypergroup is one-dimensional and we see that the definition of $\widehat{\mathcal{K}}$ coincides with the one in the previous section.

For further details on representations of hypergroups we refer to [Jew75] and [BH94]. We remark however, that only few investigations on representation theory have been done up to now. This might be due to the fact that not many relevant examples of non-commutative hypergroups are known. In the monograph [BH94, p.11] one even finds the following sentence. "Aside from some small hypergroups the nonabelian locally compact groups provide the major source of known examples of non-commutative hypergroups." We will contribute to closing this gap in Chapter 2 by providing examples of non-commutative hypergroups that are important for time-frequency and wavelet analysis.

1.5   Wavelet and Time-Frequency Analysis

This section deals with basics of time-frequency analysis and wavelet analysis and explains the relationship to representation theory of locally compact groups.

1.5.1   The Short Time Fourier Transform

The short time Fourier transform (STFT) represents a basic object in time-frequency analysis, which has important applications, for instance in audio signal processing. Its idea is to measure the frequency content of some function (or distribution) in the neighborhood of some point in space or time, respectively (depending on the physical interpretation of the underlying space $\mathbb{R}^d$).

For $\omega \in \mathbb{R}^d$ we define the modulation operator

$$M_\omega f(y) := e^{2\pi i y \cdot \omega} f(y), \quad y \in \mathbb{R}^d. \tag{1.62}$$

Together with the translation T_x defined in (1.25) it obeys the commutation relation

$$T_x M_\omega = e^{-2\pi i x \cdot \omega} M_\omega T_x. \tag{1.63}$$

On $L^2(\mathbb{R}^d)$ both T_x and M_ω are unitary operators. Now with a non-zero function $g \in L^2(\mathbb{R}^d)$ one defines the **short time Fourier transform** (sometimes also called Gabor transform) of $f \in L^2(\mathbb{R}^d)$ by

$$\mathrm{STFT}_g\, f(x, \omega) := \langle f, M_\omega T_x g \rangle = \int_{\mathbb{R}^d} f(y)\overline{g(y - x)} e^{-2\pi i \omega \cdot t} dy = \mathcal{F}(f\overline{T_x g})(\omega). \tag{1.64}$$

Thinking of g as a window function localized at $x = 0$ we may interpret $\text{STFT}_g f(x, \omega)$ as cutting out the parts of f around x by multiplying with the translated window $\overline{T_x g}$ followed by an application of the Fourier transform. Thus one may think of $\text{STFT}_g f(x, \omega)$ as the amplitude of the frequency ω at space position x. However, although this interpretation seems to be very intuitive one has to be careful not to take it too literally. In particular, several uncertainty principles state that this interpretation does not hold in a pointwise sense, see e.g. Chapter 3.3 in Gröchenig's book [Grö01].

By Lemma 3.1.1 in [Grö01] $\text{STFT}_g f$ is a bounded uniformly continuous function on $\mathbb{R}^d \times \mathbb{R}^d$ for $f, g \in L^2(\mathbb{R}^d)$. Moreover, the STFT satisfies the orthogonality relation [Grö01, Theorem 3.2.1]

$$\int_{\mathbb{R}^d} \int_{\mathbb{R}^d} \text{STFT}_{g_1} f_1(x, \omega) \overline{\text{STFT}_{g_2} f_2(x, \omega)} dx d\omega = \langle f_1, f_2 \rangle \langle g_2, g_1 \rangle \quad (1.65)$$

for $f_1, f_2, g_1, g_2 \in L^2(\mathbb{R}^d)$. This is also known as Moyal's formula. In particular, STFT_g is a bounded operator from $L^2(\mathbb{R}^d)$ into $L^2(\mathbb{R}^d \times \mathbb{R}^d)$, which is injective for any non-zero $g \in L^2(\mathbb{R}^d)$. As a consequence of the orthogonality relation we may invert the STFT as follows [Grö01, Corollary 3.2.3]. Suppose that g, γ are non-zero elements of $L^2(\mathbb{R}^d)$ with $\langle g, \gamma \rangle \neq 0$. Then

$$f = \frac{1}{\langle g, \gamma \rangle} \int_{\mathbb{R}^d} \int_{\mathbb{R}^d} \text{STFT}_g f(x, \omega) M_\omega T_x \gamma \, dx d\omega \quad (1.66)$$

for all $f \in L^2(\mathbb{R}^d)$ in a weak sense. In Chapter 2 we will prove a more general version of this formula.

Although quite simple to derive with Plancherel's theorem, the formula

$$\text{STFT}_g f(x, \omega) = e^{-2\pi i x \cdot \omega} \text{STFT}_{\hat{g}} \hat{f}(\omega, -x) \qquad f, g \in L^2(\mathbb{R}^d) \quad (1.67)$$

is known as the *fundamental identity of time-frequency analysis* [Grö01, p. 40]. Expressed in words, (1.67) states that forming the STFT of f with respect to the window g is essentially the same as taking the STFT of the Fourier transform of f with respect to the Fourier transformed window $\hat{g}$. In particular, (1.67) gives an easy explanation why it is important to take windows g that have good localization properties in both space and Fourier domain. This requirement conflicts, of course, with the uncertainty principle in its various forms (see e.g. [Grö01, FS97]). It states that a function cannot be arbitrarily well localized in both space and Fourier domain. Since Gaussians are the minimizing functions in Heisenberg's uncertainty principle [FS97, Grö01, Rau01], they are often suggested as window functions.

If g is a Schwartz function then also $M_\omega T_x g \in \mathcal{S}(\mathbb{R}^d)$ for all x, ω and we may extend the STFT to $\mathcal{S}'(\mathbb{R}^d)$ by

$$\text{STFT}_g f(x, \omega) = \langle f, M_\omega T_x g \rangle \qquad g \in \mathcal{S}(\mathbb{R}^d), f \in \mathcal{S}'(\mathbb{R}^d)$$

where the bracket $\langle \cdot, \cdot \rangle$ now realizes the duality of $\mathcal{S}(\mathbb{R}^d)$ and $\mathcal{S}'(\mathbb{R}^d)$. The STFT of a tempered distribution is a continuous function that grows at most polynomially [Grö01, Theorem 11.2.5]. Furthermore, the fundamental identity of time-frequency analysis (1.67) extends to $g \in \mathcal{S}(\mathbb{R}^d), f \in \mathcal{S}'(\mathbb{R}^d)$.

The next property, which is sometimes called the *covariance property* of the STFT, states that a time-frequency shift of f amounts essentially to a translation of its STFT, i.e., whenever $\mathrm{STFT}_g\, f$ is defined it holds

$$\mathrm{STFT}_g(T_u M_\eta f)(x, \xi) = e^{-2\pi i u \cdot \eta}\, \mathrm{STFT}_g\, f(x - u, \xi - \eta).$$

For more details on the STFT we refer to [Grö01].

1.5.2 The Continuous Wavelet Transform

A basic object in wavelet theory is the continuous wavelet transform (CWT). We define the following operators acting on functions on $\mathbb{R}^d$ by

$$\begin{aligned} D_a f(y) &= a^{-d/2} f(a^{-1}y) \quad \text{for } a \in (0, \infty), y \in \mathbb{R}^d, \\ U_R f(y) &= f(R^{-1}y), \quad \text{for } R \in SO(d), y \in \mathbb{R}^d. \end{aligned}$$

Clearly, D_a is a dilation operator normalized such that it is unitary on $L^2(\mathbb{R}^d)$ and U_R is a (unitary) rotation by $R \in SO(d)$. These two operations commute, i.e., $D_a U_R = U_R D_a$ for all $a \in (0, \infty), R \in SO(d)$.

Since the one-dimensional case $d = 1$ needs slight modifications in some of the following statements, and since we are mainly interested in the multidimensional case, we assume $d \geq 2$ throughout this section.

Now for a non-zero function $g \in L^2(\mathbb{R}^d)$ the CWT of $f \in L^2(\mathbb{R}^d)$ is defined by

$$\mathrm{CWT}_g\, f(x, a, R) = \langle f, T_x D_a U_R g \rangle = a^{-d/2} \int_{\mathbb{R}^d} f(y) \overline{g\left(R^{-1}\left(\frac{y - x}{a}\right)\right)} dy \quad (1.68)$$

for $x \in \mathbb{R}^d, a \in (0, \infty), R \in SO(d)$. Again, T_x denotes the translation operator defined in (1.25). Clearly, if g is a radial function then the CWT does not depend on $R \in SO(d)$. Sometimes the terminology isotropic wavelet transform is used for this case. If g is not radial then the CWT is called directional wavelet transform. Apparently, the CWT analyzes a given function at different positions in space, various scales and in different orientations (unless g is radial).

In contrast to the STFT an orthogonality relation similar to (1.65) does not hold for arbitrary $g \in L^2(\mathbb{R}^d)$. We call g an admissible wavelet if it satisfies

$$c_g := \int_{\mathbb{R}^d} \frac{|\hat{g}(\xi)|^2}{|\xi|^d} d\xi < \infty, \quad (1.69)$$

where $\hat{g}$ denotes the Fourier transform of g. For such functions we have the following orthogonality relation, see also [AAG00, Theorem 14.2.1].

Theorem 1.5.1. *Suppose g_1, g_2 are admissible wavelets. Then it holds*

$$\int_{\mathbb{R}^d} \int_0^\infty \int_{SO(d)} \mathrm{CWT}_{g_1} f_1(x, a, R) \overline{\mathrm{CWT}_{g_2} f_2(x, a, R)} dR \frac{da}{a^{d+1}} dx \qquad (1.70)$$

$$= |S^{d-1}|^{-1} \langle f_1, f_2 \rangle_{L^2(\mathbb{R}^d)} \int_{\mathbb{R}^d} \overline{\hat{g}_1(\xi)} \hat{g}_2(\xi) \frac{d\xi}{|\xi|^d} \qquad (1.71)$$

where dR denotes the normalized Haar measure of $SO(d)$.

Proof: For completeness we give a proof. It holds

$$(T_x D_a U_R g)\hat{\,}(\xi) = M_{-x} D_{a^{-1}} U_R \hat{g}(\xi) = e^{-2\pi i x \cdot \xi} a^{d/2} \hat{g}(R^{-1} a \xi).$$

Using Plancherel's theorem (1.31) we obtain

$$\mathrm{CWT}_g f(x, a, R) = \langle f, T_x D_a U_R g \rangle = \langle \hat{f}, M_{-x} D_{a^{-1}} U_R \hat{g} \rangle$$
$$= \mathcal{F}^{-1}(\hat{f} \overline{D_{a^{-1}} U_R \hat{g}})(x). \qquad (1.72)$$

Applying Plancherel's theorem once more together with Fubini's theorem we realize that the term in (1.70) coincides with

$$\int_0^\infty \int_{SO(d)} \int_{\mathbb{R}^d} \hat{f}_1(\xi) \overline{\hat{f}_2(\xi)} D_{a^{-1}} U_R \hat{g}_2(\xi) \overline{D_{a^{-1}} U_R \hat{g}_1(\xi)} d\xi dR \frac{da}{a^{d+1}}$$
$$= \int_{\mathbb{R}^d} \hat{f}_1(\xi) \overline{\hat{f}_2(\xi)} \int_0^\infty \int_{SO(d)} \hat{g}_2(aR^{-1}\xi) \overline{\hat{g}_1(aR^{-1}\xi)} dR \frac{da}{a} d\xi. \qquad (1.73)$$

Making the substitution $a \mapsto |\xi|^{-1} a$ and using formula (A.3) for integration on $SO(d)$ we obtain for the inner integral in (1.73)

$$\int_0^\infty \int_{SO(d)} \hat{g}_2(aR^{-1}\xi) \overline{\hat{g}_1(aR^{-1}\xi)} dR \frac{da}{a} = |S^{d-1}|^{-1} \int_0^\infty \int_{S^{d-1}} \hat{g}_2(a\eta) \overline{\hat{g}_1(a\eta)} dS(\eta) \frac{da}{a}$$
$$= |S^{d-1}|^{-1} \int_{\mathbb{R}^d} \hat{g}_2(\xi) \overline{\hat{g}_1(\xi)} \frac{d\xi}{|\xi|^d}.$$

In the last line we used the polar decomposition formula. Again with Plancherel's theorem one concludes that the term in (1.73) coincides with (1.71). By the Cauchy-Schwarz inequality the integral in (1.71) is finite by admissibility of g_1, g_2. □

As a consequence of this theorem one obtains an inversion formula for the CWT. Since it is a special case of Theorem 2.2.7(d) derived later on we omit its explicit form here.

If g is a Schwartz function then also $T_x D_a U_R g \in \mathcal{S}(\mathbb{R}^d)$ for all $x \in \mathbb{R}^d, a \in (0, \infty), R \in SO(d)$. This allows to extend CWT_g to $\mathcal{S}'(\mathbb{R}^d)$ by replacing the scalar product in $L^2(\mathbb{R}^d)$ by the dual pairing of $\mathcal{S}$ and $\mathcal{S}'$, i.e.,

$$\mathrm{CWT}_g f(x, a, R) = \langle f, T_x D_a U_R g \rangle \quad \text{for } g \in \mathcal{S}(\mathbb{R}^d), f \in \mathcal{S}'(\mathbb{R}^d).$$

For more details concerning the continuous wavelet transform and wavelet theory in general we refer to the various books available, e.g. [AAG00, Dau92, Hol95, LMR98, Woj97].

1.5.3 Square-integrable Group Representations

It is rather apparent that the STFT and the CWT share common properties. Indeed, Grossmann, Morlet and Paul realized in [GMP85, GMP86] that both transforms can be explained in the abstract context of square-integrable group representations.

Let us recall some basic facts from the theory of group representations. A unitary representation π of $\mathcal{G}$ is a strongly continuous homomorphism of $\mathcal{G}$ into the group $\mathcal{U}(\mathcal{H})$ of unitary operators on a Hilbert space $\mathcal{H}$. This means that π satisfies

$$\pi(e) = \mathrm{Id}, \quad \pi(xy) = \pi(x)\pi(y), \quad \pi(x^{-1}) = \pi(x)^*$$

and the mapping $x \mapsto \pi(x)f$ from $\mathcal{G}$ into $\mathcal{H}$ is continuous for all $f \in \mathcal{H}$. We remark that strong continuity is already implied by the less restrictive condition of weak continuity, i.e., for all $f, g \in \mathcal{H}$ the mapping $x \mapsto \langle f, \pi(x)g \rangle$ is continuous, see e.g. [Fol95, p.68]. To express the dependence on $\mathcal{H}$ sometimes the notation $(\pi, \mathcal{H})$ is used.

A closed subspace W of $\mathcal{H}$ is called invariant if $\pi(x)W \subset W$ for all $x \in \mathcal{G}$ and π is called irreducible if it has only trivial invariant subspaces.

One may extend a unitary representation $(\pi, \mathcal{H})$ from $\mathcal{G}$ to a non-degenerate $*$-representation of the Banach-$*$-algebra $M(\mathcal{G})$ on $\mathcal{H}$. Indeed for $\tau \in M(\mathcal{G})$ we define the operator

$$\pi(\tau) = \int_{\mathcal{G}} \pi(x) d\tau(x) \tag{1.74}$$

to be interpreted in a weak sense, i.e.,

$$\langle \pi(\tau)f, g \rangle = \int_{\mathcal{G}} \langle \pi(x)f, g \rangle d\tau(x) \quad \text{for all } f, g \in \mathcal{H}. \tag{1.75}$$

By unitarity of π the function $x \mapsto \langle f, \pi(x)g \rangle$ is a bounded continuous function on $\mathcal{G}$ and, hence, (1.75) is well-defined by boundedness of τ. Note that $\pi(\epsilon_x) = \pi(x)$. One has the following theorem.

Theorem 1.5.2. *([Fol95, Theorem 3.12]) The extension of π to $M(\mathcal{G})$ defined by (1.74) is a non-degenerate $*$-representation of the Banach-$*$-algebra $M(\mathcal{G})$. In particular, it holds*

$$\pi(\tau * \nu) = \pi(\tau)\pi(\nu), \quad and \quad \pi(\tau^*) = \pi(\tau)^*, \quad \tau, \nu \in M(\mathcal{G}).$$

It further follows from Theorem 3.12c in [Fol95] that π is irreducible as representation of $\mathcal{G}$ if and only if it is irreducible as representation of $M(\mathcal{G})$. Moreover, Theorem 3.11 in [Fol95] implies that the unitary representations of $\mathcal{G}$ are in one-to-one correspondence with the non-degenerate $*$-representations of $M(\mathcal{G})$ satisfying property (iii) in Definition 1.4.4. Thus interpreting $\mathcal{G}$ as hypergroup we see that the concept of representations of hypergroups coincides with the concept of unitary group representations. Therefore, Lemma 1.4.10 gives also a criterion for irreducibility of a group representation.

Associated to π is the voice transform or wavelet transform defined by

$$V_g f(x) := \langle f, \pi(x)g \rangle, \qquad f, g \in \mathcal{H}, x \in \mathcal{G}. \tag{1.76}$$

The inner products on the right hand side of (1.76) are also called matrix coefficients of π. As already remarked $V_g f$ belongs to $C^b(\mathcal{G})$. Immediately from the definition (1.76) one derives the **covariance property**

$$V_g(\pi(y)f)(x) = \langle \pi(y)f, \pi(x)g \rangle = \langle f, \pi(y^{-1}x)g \rangle = L_y V_g f(x). \tag{1.77}$$

Square-integrability is a key concept for the study of voice transforms.

Definition 1.5.1. A representation π is called **square-integrable** if π is irreducible, and if there exists a non-zero vector $g \in \mathcal{H}$ such that $V_g g \in L^2(\mathcal{G})$, i.e.,

$$\int_{\mathcal{G}} |V_g g(x)|^2 d\mu(x) < \infty. \tag{1.78}$$

Such a vector g is called admissible.

A seminal result in the theory of square-integrable group representations is the following theorem due to Duflo and Moore [DM76, Theorem 3], see also [BT96].

Theorem 1.5.3. *Let $(\pi, \mathcal{H})$ be a square-integrable representation of a locally compact group $\mathcal{G}$. Then there exists a unique self-adjoint, positive and densely defined operator K on $\mathcal{H}$ satisfying the following properties.*

(a) $\mathcal{D}(K) = \{g \in \mathcal{H}, g \text{ is admissible }\}$, where $\mathcal{D}(K)$ denotes the domain of K.

(b) If g is admissible then $V_g f \in L^2(\mathcal{G})$ for all $f \in \mathcal{H}$.

(c) For admissible g_1, g_2 and $f_1, f_2 \in \mathcal{H}$ we have the orthogonality relation

$$\int_{\mathcal{G}} V_{g_1} f_1(x) \overline{V_{g_2} f_2(x)} dx = \langle f_1, f_2 \rangle \langle K g_2, K g_1 \rangle. \tag{1.79}$$

(d) It holds $K = S^{-1/2}$ for some self-adjoint, positive, densely defined operator S that satisfies $\pi(x) S \pi(x)^{-1} = \Delta(x)^{-1} S$.

Furthermore, if $\mathcal{G}$ is unimodular then $K = c\,\mathrm{Id}$ for some constant $c > 0$ and, hence, any non-zero vector in $\mathcal{H}$ is admissible.

The proof of Theorem 1.5.3 relies on the following extension of Schur's lemma which we cite for later reference.

Lemma 1.5.4. *[DM76, Theorem 1] Let $(\pi, \mathcal{H})$ be an irreducible representation of $\mathcal{G}$ and assume χ to be character of $\mathcal{G}$, i.e., $\chi \in C(\mathcal{G})$ with $\chi(xy) = \chi(x)\chi(y)$. Let T be a densely defined closed operator in $\mathcal{H}$ that satisfies*

$$\pi(x) T \pi(x)^{-1} = \chi(x) T \quad \text{for all } x \in \mathcal{G}. \tag{1.80}$$

(An operator T satisfying (1.80) is called semi-invariant with weight χ.) If T' is another operator satisfying (1.80) then $T' = cT$ for some constant $c > 0$.

As an easy consequence of Theorem 1.5.3 we derive an inversion formula and reproducing formula for the voice transform, see also [GMP85]. The latter will become very important in Chapter 4.

Corollary 1.5.5. *Let $(\pi, \mathcal{H})$ be a square-integrable representation of $\mathcal{G}$.*

(a) (Inversion formula) Let $g, \gamma \in \mathcal{D}(K)$ with $\langle Kg, K\gamma \rangle = 1$. Then it holds

$$f = \int_{\mathcal{G}} V_g f(x) \pi(x) \gamma \, dx \quad \text{for all } f \in \mathcal{H}, \tag{1.81}$$

to be read in a weak sense.

(b) For $g_1, f_1 \in \mathcal{D}(K), g_2, f_1 \in \mathcal{H}$ it holds

$$V_{g_1} f_1 * V_{g_2} f_2 = \langle K g_1, K f_2 \rangle V_{g_2} f_1.$$

(c) (Reproducing formula) Let $g \in \mathcal{D}(K)$ with $\|Kg\| = 1$. Then it holds

$$V_g f = V_g f * V_g g \quad \text{for all } f \in \mathcal{H}. \tag{1.82}$$

(d) Suppose $g \in \mathcal{D}(K)$ with $\|Kg\| = 1$. Then the mapping $P : L^2(\mathcal{G}) \to L^2(\mathcal{G})$, $F \mapsto F * V_g g$ is an orthogonal projection from $L^2(\mathcal{G})$ onto the image of V_g.

Proof: (a) Let $f, h \in \mathcal{H}$. By the orthogonality relation (Theorem 1.5.3(c)) it holds

$$\int_{\mathcal{G}} V_g f(x) \langle \pi(x)\gamma, h \rangle dx = \int_{\mathcal{G}} V_g f(x) \overline{V_\gamma h(x)} dx = \langle f, h \rangle \langle K\gamma Kg \rangle = \langle f, h \rangle.$$

Since $h \in \mathcal{H}$ was arbitrary this equation is exactly the weak definition of (1.81). For (b) observe that again by Theorem 1.5.3(c) and (1.77) we have

$$V_{g_1} f_1 * V_{g_2} f_2(x) = \int_{\mathcal{G}} V_{g_1} f_1(y) V_{g_2} f_2(y^{-1}x) dy = \int_{\mathcal{G}} V_{g_1} f_1(y) \overline{V_{f_2}(\pi(x)g_2)(y)} dy$$
$$= \langle Kg_1, Kf_2 \rangle \langle f_1, \pi(x)g_2 \rangle = \langle Kg_1, Kf_2 \rangle V_{g_2} f_1(x).$$

The assertion in (c) follows as immediate consequence.

(d) Let $F \in L^2(\mathcal{G})$. Define $f = \int_{\mathcal{G}} F(y)\pi(y)g\,dy \in \mathcal{H}$ to be read in a weak sense. We obtain

$$V_g f(x) = \int_{\mathcal{G}} F(y) \langle \pi(y)g, \pi(x)g \rangle dy = F * V_g g(x).$$

Together with (a) we conclude that P is surjective onto the image of $\mathcal{H}$ under V_g. Further, if $F = V_g f$ for some $f \in \mathcal{H}$ then by (c) it holds $F = F * V_g f$ and thus P is a projection. Finally, if $F, G \in L^2(\mathcal{G})$ then it follows from $V_g g(x^{-1}) = \overline{V_g g(x)}$ that

$$\langle F * V_g g, G \rangle_{L^2(\mathcal{G})} = \int_{\mathcal{G}} \int_G F(y) V_g g(y^{-1}x) dy \overline{G(x)} dx$$
$$= \int_{\mathcal{G}} F(y) \overline{\int_{\mathcal{G}} V_g g(x^{-1}y) G(x) dx} dy = \langle F, G * V_g g \rangle_{L^2(\mathcal{G})}.$$

Hence, P is orthogonal. $\qquad\square$

Let us now consider examples that illustrate how the STFT and the CWT fit into the setting of square-integrable group representations.

The (reduced) Heisenberg group is topologically the set $\mathbb{H}_d := \mathbb{R}^d \times \mathbb{R}^d \times \mathbb{T}$ where $\mathbb{T} = \{z \in \mathbb{C}, |z| = 1\}$ denotes the torus. The group law on $\mathbb{H}_d$ is given by

$$(x, \omega, \tau)(x', \omega', \tau') = (x + x', \omega + \omega', \tau\tau' e^{\pi i(x' \cdot \omega - x \cdot \omega')}).$$

The Heisenberg group is unimodular and has Haar measure

$$\int_{\mathbb{H}_d} f(h)dh = \int_{\mathbb{R}^d} \int_{\mathbb{R}^d} \int_0^1 f(x, \omega, e^{2\pi it}) dt\, d\omega\, dx.$$

The Schrödinger representation ρ is a unitary representation of the Heisenberg group acting on $L^2(\mathbb{R}^d)$ by

$$\rho(x,\omega,\tau) := \tau e^{\pi i x \cdot \omega} T_x M_\omega = \tau e^{-\pi i x \cdot \omega} M_\omega T_x,$$

where T_x and M_ω are the translation and modulation operators defined in (1.25) and (1.62). To check that this is in fact a representation one uses the commutation relation (1.63). The corresponding voice transform is essentially the short time Fourier transform:

$$V_g f(x,\omega,\tau) = \langle f, \rho(x,\omega,\tau)g \rangle_{L^2(\mathbb{R}^d)} = \overline{\tau} \int_{\mathbb{R}^d} f(t) \overline{e^{-\pi i x \cdot \omega} M_\omega T_x g(t)} dt$$

$$= \overline{\tau} e^{\pi i x \cdot \omega} \int_{\mathbb{R}^d} f(t) \overline{g(t-x)} e^{-2\pi i t \cdot \omega} dt = \overline{\tau} e^{\pi i x \cdot \omega} \operatorname{STFT}_g f(x,\omega). \qquad (1.83)$$

Theorem 1.5.6. *([Grö01, p.182]) The Schrödinger representation is a square-integrable representation of the Heisenberg group.*

The next example involves the similitude group $\mathcal{G} = \mathbb{R}^d \rtimes (\mathbb{R}_+^* \times SO(d))$, where $\mathbb{R}_+^* = (0, \infty)$ denotes the multiplicative group of positive real numbers. We assume $d \geq 2$. The similitude group has left Haar measure

$$\int_{\mathcal{G}} f(x) d\mu(x) = \int_{SO(d)} \int_{\mathbb{R}^d} \int_{\mathbb{R}_+^*} f(x,b,A) \frac{db}{b^{d+1}} dx dA$$

and modular function $\Delta(x,b,A) = b^{-d}$. A unitary representation of $\mathcal{G}$ on $L^2(\mathbb{R}^d)$ is given by

$$\pi(x,b,R) f(t) = b^{-d/2} f\left(b^{-1} R^{-1}(t-x)\right) = T_x D_b U_R f(t), \qquad (1.84)$$

where the notation of Section 1.5.2 is used. The corresponding voice transform is the continuous wavelet transform (1.68).

Theorem 1.5.7. *(Theorem 14.2.1 in [AAG00]) The representation π defined in (1.84) is a square-integrable representation of the similitude group $\mathcal{G}$.*

We remark that the operator K from Theorem 1.5.3 may be given explicitly. Denote $\eta_d(\xi) := |\xi|^{-d/2}$. Then Theorem 1.5.1 implies that

$$(Kg)(x) = \mathcal{F}^{-1}(\hat{g}\eta_d)(x) \qquad (1.85)$$

and $\mathcal{D}(K) = \{g \in L^2(\mathbb{R}^d), \hat{g}\eta_d \in L^2(\mathbb{R}^d)\}$ is the set of admissible vectors.

Chapter 2

Wavelet Transforms of Invariant Functions

Now we turn to our main topic and investigate wavelet and time-frequency analysis of functions, which are invariant under some symmetry group. It sounds reasonable that one can take advantage of the information that the function under consideration possesses symmetry properties. For instance, we are able to reduce the complexity in evaluating or inverting the CWT or the STFT when the wavelet (resp. window function) and the function to analyze are both radial. We will see that the integral over $\mathbb{R}^d$ reduces to an integral over the positive half-line $\mathbb{R}_+$ in these cases.

The key observation is that the building blocks in time-frequency and wavelet analysis – the functions $T_x M_\omega g$ or $T_x D_a U_R g$ – are not radial in general, even when g is radial. In particular, $T_x g$ is only radial if $x = 0$. So in order to exploit the symmetry one would expect that the usual building blocks have to be replaced by functions which are all radial themselves. Moreover, all these functions should be derived from one single function g by applying certain operators to it. So one main issue in radial time-frequency and wavelet analysis is to find such operators, which map radial functions onto radial functions and can still be interpreted as time-frequency shifts or dilation-translation operators. One can imagine that the generalized translation introduced in Section 1.3 plays an important role. In fact, for wavelet analysis one combines the dilation and the generalized translation as in the usual case. For time-frequency analysis, however, the resulting operator is more complicated.

In this chapter we investigate the continuous transforms. As was have seen, both the CWT and the STFT can be treated simultaneously within the representation theory of locally compact groups. Therefore we will work in this abstract setting. The term "wavelet transform" then refers to the abstract wavelet transform V_g associated to an irreducible representation π of some locally compact group $\mathcal{G}$ on a Hilbert space $\mathcal{H}$ defined in (1.76). The symmetry group establishes itself as compact automorphism group $\mathcal{A}$ of $\mathcal{G}$ which acts on $\mathcal{H}$ by means of another representation σ. Additionally, we have to require that π and σ are compatible in some suitable sense. An element of $\mathcal{H}$ is called invariant under $\mathcal{A}$ if $\sigma(A)f = f$ for all $A \in \mathcal{A}$. It turns out that the wavelet transform of invariant elements is itself a function on $\mathcal{G}$ that is invariant under $\mathcal{A}$. Starting with the operator $\pi(x)$ we construct an operator $\widetilde{\pi}(x)$ in a natural way, which maps invariant elements to invariant ones. In the concrete cases of time-frequency analysis and wavelet analysis, $\widetilde{\pi}(x)$ will be the natural substitute for a time-frequency shift or a dilation/translation operator.

Before we start with the abstract theory we briefly look at the CWT of radial functions as a motivating example. This is simple enough to get a feeling for our problem. Then we turn to the abstract setting. After some simple observations we show that the pair of representations (π, σ) generates an irreducible representation of the orbit hypergroup $\mathcal{G}^{\mathcal{A}}$, which is of interest for itself. As main results we provide a different way to compute the wavelet transform of invariant elements and a new inversion formula.

Parts of this chapter will appear in [Rau03b].

2.1 A Motivating Example

We start with the basic example of the continuous wavelet transform of radial functions, which is fairly simple but nevertheless quite illustrative. We use the same notation as in Sections 1.3 and 1.5.2. Let $g \in L^2(\mathbb{R}^d)$ be a radial wavelet for the CWT, i.e., a non-zero radial function g that satisfies (1.69). If $f \in L^2(\mathbb{R}^d)$ is also radial then a simple calculation shows

$$\mathrm{CWT}_g\, f(S^{-1}x, a, R) = \langle f, T_{S^{-1}x} D_a U_R g \rangle = \langle f, U_{S^{-1}} T_x U_S D_a U_R g \rangle$$
$$= \langle U_S f, T_x D_a U_{RS} g \rangle = \langle f, T_x D_a g \rangle = \mathrm{CWT}_g\, f(x, a, \mathrm{Id})$$

for all $S, R \in SO(d), x \in \mathbb{R}^d, a \in \mathbb{R}_+^*$. We denote the restriction of CWT_g to $L^2_{rad}(\mathbb{R}^d)$ by $\widetilde{\mathrm{CWT}}_g$. The calculation above shows that $\widetilde{\mathrm{CWT}}_g$ depends only on $a \in \mathbb{R}_+^*$ and $|x| \in \mathbb{R}_+$ and may therefore be interpreted as a function on $\mathbb{R}_+ \times \mathbb{R}_+^*$. Using Fubini's theorem we derive the following formula:

$$\widetilde{V}_g f(x, a) = \int_{SO(d)} V_g f(S^{-1}x, a, \mathrm{Id}) dS \qquad (2.1)$$
$$= a^{-d/2} \int_{\mathbb{R}^d} f(y) \overline{\int_{SO(d)} g(a^{-1}(y - S^{-1}x)) dS} dy.$$

Let $\mathbf{g} \in L^2(\mathbb{R}_+, \mu_d)$ denote the function associated to g by $g(x) = \mathbf{g}(|x|)$. With formula (A.3) for the integration on $SO(d)$ we observe that

$$\int_{SO(d)} g(y - S^{-1}x) dS = \int_{S^{d-1}} g(y - |x|\xi) dS(\xi) = \tau_r \mathbf{g}(s), \quad r = |x|, s = |y|, \qquad (2.2)$$

where τ denotes the generalized translation defined in (1.22). Since $L_{rad}^2(\mathbb{R}^d)$ is isometrically isomorphic to $L^2(\mathbb{R}_+, \mu_d)$ we conclude

$$\widetilde{\mathrm{CWT}}_g f(s,a) = \langle \mathsf{f}, \tau_s D_a \mathsf{g} \rangle_{L^2(\mathbb{R}_+, \mu_d)} = \int_0^\infty \mathsf{f}(r) \overline{(\tau_s D_a \mathsf{g})(r)} d\mu_d(r). \qquad (2.3)$$

Thus in presence of radial symmetry we may reduce the multidimensional wavelet transform to an integral over $\mathbb{R}_+$.

Starting from the inversion formula for the wavelet transform and using a similar trick as in (2.1) we derive a second inversion formula for the wavelet transform of radial functions

$$\mathsf{f}(r) = \int_{\mathbb{R}_+} \int_{\mathbb{R}_+^*} \widetilde{\mathrm{CWT}}_g f(s,a) \tau_s D_a \mathsf{g}(r) \frac{da}{a^{d+1}} d\mu_d(r) \quad \text{a.e.} \qquad (2.4)$$

where g is admissible and normalized such that $c_g = 1$, see (1.69). Since this inversion formula will also follow from a more general theorem derived in the next section we skip the details of its proof at this place.

We note that formula (2.4) states in particular that we may represent a radial function as a continuous superposition of the radial(!) functions $\tau_s D_a g, s \in \mathbb{R}_+, b \in \mathbb{R}_+^*$.

2.2   General Results

We now turn to the general abstract setting and make the same assumptions as in Section 1.2. In particular, we assume $\mathcal{G}$ to be a locally compact group and $\mathcal{A}$ a *compact* automorphism group (symmetry group) of $\mathcal{G}$. Further, we suppose that we are given an irreducible unitary (strongly continuous) representation π of $\mathcal{G}$ on a Hilbert space $\mathcal{H}$ and a unitary representation σ (not necessarily irreducible) of $\mathcal{A}$ on the same Hilbert space $\mathcal{H}$ such that

$$\pi(A(x))\sigma(A) = \sigma(A)\pi(x). \qquad (2.5)$$

In other words, we require that the representations $\pi_A := \pi \circ A$ are all unitarily equivalent to π and that the intertwining operators $\sigma(A)$ form a representation of $\mathcal{A}$. The condition (2.5) will be essential in the sequel. If for instance $\mathcal{A}$ is a compact subgroup of $\mathcal{G}$ acting by inner automorphisms and $\sigma = \pi|_\mathcal{A}$ then (2.5) holds trivially. Moreover, there is a relation of pairs (π, σ) satisfying (2.5) to representations of the semi-direct product $\mathcal{G} \rtimes \mathcal{A}$ as stated in the following lemma.

Lemma 2.2.1. *(a) Suppose that π, σ are unitary representations of $\mathcal{G}$ and $\mathcal{A}$, respectively, on a Hilbert space $\mathcal{H}$ satisfying (2.5). Then the operators*

$$U(x, A) = \pi(x)\sigma(A), \qquad (x, A) \in \mathcal{G} \rtimes \mathcal{A},$$

form a unitary representation of $\mathcal{G} \rtimes \mathcal{A}$ on $\mathcal{H}$. Moreover, if π is irreducible then also U is irreducible.

(b) Conversely, suppose U is a unitary representation of $\mathcal{G} \rtimes \mathcal{A}$ on $\mathcal{H}$. Define

$$\pi(x) = U(x, e_\mathcal{A}), \qquad \sigma(A) = U(e_\mathcal{G}, A), \qquad x \in \mathcal{G}, A \in \mathcal{A},$$

where $e_\mathcal{G}$ and $e_\mathcal{A}$ denote the unit elements of $\mathcal{G}$ and $\mathcal{A}$, respectively. Then π and σ are unitary representation of $\mathcal{G}$ and $\mathcal{A}$, respectively, which satisfy (2.5).

Proof: (a) Suppose $(x, A), (y, B) \in \mathcal{G} \rtimes \mathcal{A}$. Recall that the group law in $\mathcal{G} \rtimes \mathcal{A}$ is $(x, A) \cdot (y, B) = (xA(y), AB)$. Using (2.5) we obtain

$$\begin{aligned}
U(x, A)U(y, B) &= \pi(x)\sigma(A)\pi(y)\sigma(B) = \pi(x)\pi(A(y))\sigma(A)\sigma(B) \\
&= \pi(xA(y))\sigma(AB) = U(xA(y), AB) = U((x, A)(y, B)).
\end{aligned}$$

Hence, U is a representation of $\mathcal{G} \rtimes \mathcal{A}$. Unitarity of U follows from unitarity of π and σ. Now suppose that π is irreducible. If W is a closed subspace of $\mathcal{H}$ satisfying $U(x, A)W \subset W$ for all $(x, A) \in \mathcal{G} \rtimes \mathcal{A}$, then we have in particular $U(x, e_\mathcal{A})W = \pi(x)W \subset W$ for all $x \in \mathcal{G}$. Irreducibility of π means that W is a trivial subspace of $\mathcal{H}$, which in turn implies that U is irreducible.

(b) Since $\mathcal{G}$ and $\mathcal{A}$ can be interpreted as subgroups of $\mathcal{G} \rtimes \mathcal{A}$ it is clear that π and σ are unitary representations of $\mathcal{G}$ and $\mathcal{A}$, respectively. Moreover, we have

$$\sigma(A)\pi(x) = U(e_\mathcal{G}, A)U(x, e_\mathcal{A}) = U(A(x), A) = \pi(A(x))\sigma(A).$$

This shows (2.5). $\hfill\square$

We remark that we cannot deduce irreducibility of π from irreducibility of U. For $f \in \mathcal{H}$ we let $f_A = \sigma(A)f$ and

$$\mathcal{H}_\mathcal{A} := \{f \in \mathcal{H}, \sigma(A)f = f \text{ for all } A \in \mathcal{A}\},$$

the closed subspace of invariant elements. Note that closedness follows from $\mathcal{H}_\mathcal{A} = \cap_{A \in \mathcal{A}} \ker(\mathrm{Id} - \sigma(A))$.

Lemma 2.2.2. *The weakly defined operator*

$$Q_\mathcal{A} : \mathcal{H} \to \mathcal{H}, \qquad Q_\mathcal{A}f = \int_\mathcal{A} \sigma(A)f \, dA \qquad (2.6)$$

is an orthogonal projection from $\mathcal{H}$ onto $\mathcal{H}_\mathcal{A}$.

Proof: Let $f, g \in \mathcal{H}$. We have

$$|\langle Q_\mathcal{A}f, g \rangle| \leq \int_\mathcal{A} |\langle \sigma(A)f, g \rangle| \, dA \leq \|f\| \|g\|.$$

Taking the supremum over all f, g with $\|f\| = \|g\| = 1$ shows that $Q_{\mathcal{A}}$ is bounded with operator norm $\|Q_{\mathcal{A}}\| \leq 1$. By invariance of the Haar measure of $\mathcal{A}$ we further obtain for all $B \in \mathcal{A}$

$$\sigma(B)Q_{\mathcal{A}}f = \int_{\mathcal{A}} \sigma(B)\sigma(A)f\,dA = \int_{\mathcal{A}} \sigma(BA)f\,dA = Q_{\mathcal{A}}f$$

to be read in a weak sense. We conclude $Q_{\mathcal{A}}f \in \mathcal{H}_{\mathcal{A}}$. Moreover, if $f \in \mathcal{H}_{\mathcal{A}}$ then clearly $Q_{\mathcal{A}}f = f$, which together with the first observation yields $Q_{\mathcal{A}}^2 = Q_{\mathcal{A}}$. Moreover, by compactness of $\mathcal{A}$ it holds $dA^{-1} = dA$ and, hence, for all $f, g \in \mathcal{H}$ we obtain

$$\langle Q_{\mathcal{A}}f, g \rangle = \int_{\mathcal{A}} \langle \sigma(A)f, g \rangle = \int_{\mathcal{A}} \langle f, \sigma(A^{-1})g \rangle dA = \int_{\mathcal{A}} \langle f, \sigma(A)g \rangle dA = \langle f, Q_{\mathcal{A}}g \rangle$$

implying $Q_{\mathcal{A}}^* = Q_{\mathcal{A}}$. Thus, $Q_{\mathcal{A}}$ is an orthogonal projection from $\mathcal{H}$ onto $H_{\mathcal{A}}$. $\qquad \square$

It might happen that $\mathcal{H}_{\mathcal{A}} = \{0\}$. So in order to avoid speaking about trivialities we always assume in the sequel that $\mathcal{H}_{\mathcal{A}} \neq \{0\}$.

As in Section 1.5.3 the wavelet transform or voice transform is defined by

$$V_g f(x) := \langle f, \pi(x)g \rangle.$$

It maps $\mathcal{H}$ into $C^b(\mathcal{G})$. With an element $g \in \mathcal{H}_{\mathcal{A}}$ we denote by $\widetilde{V}_g$ the restriction of V_g to $\mathcal{H}_{\mathcal{A}}$.

Lemma 2.2.3. *Suppose that (2.5) holds.*

(a) *For $f, g \in \mathcal{H}$ we have $(V_g f)_A(x) = V_{g_A} f_A(x)$.*

(b) *Consequently, for $f, g \in \mathcal{H}_{\mathcal{A}}$ the voice transform $\widetilde{V}_g f$ is invariant under $\mathcal{A}$, i.e., $\widetilde{V}_g f(A^{-1}x) = \widetilde{V}_g f(x)$ for all $A \in \mathcal{A}, x \in \mathcal{G}$.*

(c) *For $x \in \mathcal{G}$ define the operator*

$$\widetilde{\pi}(x) := \int_{\mathcal{A}} \pi(Ax)dA$$

where the integral is understood weakly, i.e., $\langle f, \widetilde{\pi}(x)g \rangle = \int_{\mathcal{A}} \langle f, \pi(Ax)g \rangle dA$ for all $f, g \in \mathcal{H}$. Then (with $g \in \mathcal{H}_{\mathcal{A}}$) it holds

$$\widetilde{V}_g f(x) = \langle f, \widetilde{\pi}(x)g \rangle_{\mathcal{H}_{\mathcal{A}}} \qquad \text{for all } f, g \in \mathcal{H}_{\mathcal{A}}, \, x \in \mathcal{G}.$$

(d) *The operators $\widetilde{\pi}(x)$, $x \in \mathcal{G}$ depend only on the orbit $\mathcal{A}x$, i.e., $\widetilde{\pi}(Ax) = \widetilde{\pi}(x)$ for all $A \in \mathcal{A}$. Moreover, it holds $\sigma(A)\widetilde{\pi}(x) = \widetilde{\pi}(x)\sigma(A)$ for all $A \in \mathcal{A}$, $x \in \mathcal{G}$. Consequently, $\widetilde{\pi}(x)$ maps $\mathcal{H}_{\mathcal{A}}$ into $\mathcal{H}_{\mathcal{A}}$.*

(e) For all $f \in \mathcal{H}_A$ and $x \in \mathcal{G}$ it holds

$$\widetilde{\pi}(x)f = Q_A \pi(x)f.$$

Proof: (a) Using (2.5) we obtain

$$V_g f(A^{-1}x) = \langle f, \pi(A^{-1}x)g \rangle = \langle f, \sigma(A^{-1})\pi(x)\sigma(A)g \rangle$$
$$= \langle \sigma(A)f, \pi(x)\sigma(A)g \rangle = V_{g_A} f_A(x).$$

(b) If $g_A = g$ and $f_A = f$ then as a consequence of (a) we get $V_g f(A^{-1}x) = V_g f(x)$.
(c) Using (b) we have for $f, g \in \mathcal{H}_A$

$$V_g f(x) = \int_{\mathcal{A}} V_g f(Ax) dA = \int_{\mathcal{A}} \langle f, \pi(Ax)g \rangle dA.$$

This is nothing else than (c).
(d) Let $A \in \mathcal{A}$. Using the translation invariance of the Haar measure of $\mathcal{A}$ we immediately get $\widetilde{\pi}(Ax) = \widetilde{\pi}(x)$. Furthermore for $f \in \mathcal{H}$ we obtain using (2.5)

$$\sigma(A)\widetilde{\pi}(x)f = \sigma(A) \int_{\mathcal{A}} \pi(Bx)f dB = \int_{\mathcal{A}} \sigma(AB)\pi(x)\sigma(B)^{-1} f dA$$
$$= \int_{\mathcal{A}} \sigma(B)\pi(x)\sigma((A^{-1}B)^{-1})f dA = \int_{\mathcal{A}} \pi(Bx)\sigma(A)f dA = \widetilde{\pi}(x)\sigma(A)f,$$

where all expressions are understood in a weak sense. If, in addition, f is invariant then it follows $\widetilde{\pi}(x)f = \widetilde{\pi}(x)\sigma(A)f = \sigma(A)\widetilde{\pi}(x)f$, which means that also $\widetilde{\pi}(x)f$ is invariant.
(e) Using (2.5) we obtain

$$Q_A \pi(x)f = \int_{\mathcal{A}} \sigma(A)\pi(x)f dA = \int_{\mathcal{A}} \pi(A(x))\sigma(A)f dA = \int_{\mathcal{A}} \pi(A(x))f dA$$
$$= \widetilde{\pi}(x)f. \qquad \qquad \qquad \qquad \qquad \qquad \square$$

According to (b) the wavelet transform $\widetilde{V_g}f$ can be interpreted as a function on the space of all orbits $\mathcal{K} = \mathcal{G}^{\mathcal{A}}$, which has the structure of a hypergroup by Theorem 1.4.3. By (c) the operators $\widetilde{\pi}(x)$ only depend on the elements $\mathcal{A}x$ of $\mathcal{K}$. One may ask whether they arise from a representation of the hypergroup $\mathcal{K}$. To treat this question we recall that π extends to a non-degenerate $*$-representation of $M(\mathcal{G})$ by (1.74). Furthermore, $M(\mathcal{K}) \cong M_A(\mathcal{G})$ can be interpreted as closed subalgebra of $M(\mathcal{G})$. Observe that $\widetilde{\pi}(x) = \widetilde{\pi}(\mathcal{A}x) = \pi(\epsilon_{\mathcal{A}x})$. Thus, we may denote by $\widetilde{\pi}$ also the restriction of π to $M_A(\mathcal{G}) \cong M(\mathcal{K})$ and for measures $\rho \in M(\mathcal{K})$ we have $\widetilde{\pi}(\rho)f = \int_{\mathcal{K}} \widetilde{\pi}(x)f d\rho(x)$. Recall also that the involution on $\mathcal{K}$ is given by $(\mathcal{A}x)^{\sim} = \mathcal{A}x^{-1}$.

Theorem 2.2.4. *The mapping* $\tilde{\pi} : M(\mathcal{K}) \to \mathcal{B}(\mathcal{H}_{\mathcal{A}})$, *where* $\mathcal{B}(\mathcal{H}_{\mathcal{A}})$ *denotes the space of all bounded operators on* $\mathcal{H}_{\mathcal{A}}$, *is an irreducible representation of the hypergroup* $(\mathcal{K}, *, \tilde{\ })$.

Proof: It is proven in Lemma 2.2.3 that $\tilde{\pi}(x)$ is a mapping from $\mathcal{H}_{\mathcal{A}}$ into $\mathcal{H}_{\mathcal{A}}$. Since by construction of $\tilde{\pi}$ we have $\tilde{\pi}(\rho)f = \int_{\mathcal{K}} \tilde{\pi}(x)f d\rho(x)$ for a general measure $\rho \in M(\mathcal{K})$, also $\tilde{\pi}(\rho)$ maps $\mathcal{H}_{\mathcal{A}}$ into $\mathcal{H}_{\mathcal{A}}$ by closedness of $\mathcal{H}_{\mathcal{A}}$.

Interpreting $\mathcal{G}$ as a hypergroup, the representation π regarded as a representation of $M(\mathcal{G})$ satisfies all axioms of Definition 1.4.4 of a hypergroup representation. Since $\tilde{\pi}$ is derived by first restricting π to the closed subalgebra $M_{\mathcal{A}}(\mathcal{G})$ containing $\epsilon_e = \epsilon_{\mathcal{A}e}$ followed by restricting each operator $\tilde{\pi}(\rho), \rho \in M_{\mathcal{A}}(\mathcal{G})$, to the closed subspace $\mathcal{H}_{\mathcal{A}}$ of $\mathcal{H}$, it follows that also $\tilde{\pi}$ satisfies all axioms of a representation of a hypergroup. Hereby, it is crucial that the topology and also the involution on $\mathcal{K}$ be inherited from $\mathcal{G}$.

It remains to prove the irreducibility. Suppose $f \in \mathcal{H}_{\mathcal{A}}$. Since π is assumed to be irreducible we conclude with Lemma 1.4.10 that the linear span W_f of $\{\pi(x)f, x \in \mathcal{G}\}$ is dense in $\mathcal{H}$. This implies that $Q_{\mathcal{A}}W_f$ is dense in $Q_{\mathcal{A}}(\mathcal{H}) = \mathcal{H}_{\mathcal{A}}$. By Lemma 2.2.3(e) we have $Q_{\mathcal{A}}W_f = \text{span}\{\tilde{\pi}(\mathcal{A}x)f, \mathcal{A}x \in \mathcal{K}\}$ and we conclude once again with Lemma 1.4.10 that the representation $\tilde{\pi}$ of $\mathcal{K}$ on $\mathcal{H}_{\mathcal{A}}$ is irreducible. $\qquad\square$

Now we are ready to state a **covariance principle** for $\tilde{V}_g$ similar to (1.77). Recall that $\mathcal{K}$ possesses a generalized left translation $\mathcal{L}$ defined in (1.13).

Theorem 2.2.5. *Let* $f, g \in \mathcal{H}_{\mathcal{A}}, y \in \mathcal{G}$. *Then*

$$\tilde{V}_g(\tilde{\pi}(y)f) = \mathcal{L}_y \tilde{V}_g f. \tag{2.7}$$

Proof: Using that $\tilde{\pi}$ is a $*$-representation of $M(\mathcal{K})$ we obtain

$$\tilde{V}_g(\tilde{\pi}(y)f)(x) = \langle \pi(\epsilon_{\mathcal{A}y})f, \pi(\epsilon_{\mathcal{A}x})g \rangle = \langle f, \pi(\epsilon_{(\mathcal{A}y)^-})\pi(\epsilon_{\mathcal{A}x})g \rangle = \langle f, \pi(\epsilon_{(\mathcal{A}y)^-} * \epsilon_{\mathcal{A}x})g \rangle$$
$$= \epsilon_{\mathcal{A}y^{-1}} * \epsilon_{\mathcal{A}x}(\tilde{V}_g f) = \mathcal{L}_y \tilde{V}_g f(x). \qquad\square$$

For the following we make the additional assumption that π is square-integrable. Let K be the operator of the Duflo-Moore theorem 1.5.3 and define $\mathcal{D}_{\mathcal{A}}(K) := \mathcal{D}(K) \cap \mathcal{H}_{\mathcal{A}}$, the space of admissible invariant vectors.

Lemma 2.2.6. *(a) The operator K commutes with the action of $\mathcal{A}$, i.e., $\sigma(A)K = K\sigma(A)$ for all $A \in \mathcal{A}$. Hence, K maps $\mathcal{D}_{\mathcal{A}}(K)$ into $\mathcal{H}_{\mathcal{A}}$.*

(b) The space $\mathcal{D}_{\mathcal{A}}(K)$ is dense in $\mathcal{H}_{\mathcal{A}}$. Hence, if $\mathcal{H}_{\mathcal{A}}$ is non-trivial then also $\mathcal{D}_{\mathcal{A}}(K)$ is non-trivial.

Proof: (a) By Theorem 1.5.3(d) we have $K = S^{-1/2}$ for some self-adjoint, positive, densely defined operator S that satisfies $\pi(x)S\pi(x)^{-1} = \Delta(x)^{-1}S$. Replacing x by $A(x)$ for $A \in \mathcal{A}$ and using (2.5) yields

$$\sigma(A)\pi(x)\sigma(A)^{-1}S\sigma(A)\pi(x)^{-1}\sigma(A)^{-1} = \Delta(A(x))^{-1}S.$$

Using the invariance of the modular function Δ under $\mathcal{A}$ (Lemma 1.2.1) we obtain

$$\pi(x)\left(\sigma(A)^{-1}S\sigma(A)\right)\pi(x)^{-1} = \Delta(x)^{-1}\sigma(A)^{-1}S\sigma(A).$$

From the positiveness of S if follows that also $\sigma(A)^{-1}S\sigma(A)$ is a positive operator. Since Δ is a character of $\mathcal{G}$ we conclude with Lemma 1.5.4 that there exists a number $\lambda(A) > 0$ such that $\sigma(A)^{-1}S\sigma(A) = \lambda(A)S$. This implies $K\sigma(A) = \lambda^{-1/2}(A)\sigma(A)K$ for all $A \in \mathcal{A}$. Using the Duflo-Moore theorem 1.5.3 and the invariance of the Haar measure of $\mathcal{G}$ under $\mathcal{A}$, we obtain for all $f \in \mathcal{H}$ and $g \in \mathcal{D}(K)$:

$$\|Kg\|^2\,\|f\|^2 = \int_{\mathcal{G}} |\langle f, \pi(x)g\rangle|^2 dx = \int_{\mathcal{G}} |\langle \sigma(A)f, \pi(A(x))\sigma(A)g\rangle|^2 dx$$

$$= \int_{\mathcal{G}} |\langle \sigma(A)f, \pi(x)\sigma(A)g\rangle|^2 dx = \|K\sigma(A)g\|^2\,\|\sigma(A)f\|^2$$

$$= \|\lambda^{-1/2}(A)\sigma(A)Kg\|^2\,\|f\|^2 = |\lambda(A)|^{-1}\|Kg\|^2\,\|f\|^2.$$

Together with the positiveness of $\lambda(A)$ we conclude $\lambda(A) = 1$ and (a) is shown.

(b) By (a) the domain $\mathcal{D}(K)$ is invariant under $\sigma(A)$ for all $A \in \mathcal{A}$ and, hence, it holds $Q_{\mathcal{A}}(\mathcal{D}(K)) \subset \mathcal{D}(K)$. This implies in particular that $\mathcal{D}_{\mathcal{A}}(K) = Q_{\mathcal{A}}(\mathcal{D}(K))$. Since $\mathcal{D}(K)$ is dense in $\mathcal{H}$ its image $\mathcal{D}_{\mathcal{A}}(K)$ under the projection $Q_{\mathcal{A}}$ is dense in $\mathcal{H}_{\mathcal{A}} = Q_{\mathcal{A}}(\mathcal{H})$. $\square$

Let us collect some further properties of the restriction $\widetilde{V}_g$ of V_g to $\mathcal{H}_{\mathcal{A}}$ in a theorem as follows. Recall that $L^2(\mathcal{K}) = L^2(\mathcal{K}, m) \cong L^2_{\mathcal{A}}(\mathcal{G})$ with the measure m given in (1.6).

Theorem 2.2.7. *Suppose $g \in \mathcal{D}_{\mathcal{A}}(K)$ with $\|Kg\| = 1$.*

(a) For $\gamma \in \mathcal{D}_{\mathcal{A}}(K)$ and $f, h \in \mathcal{H}_{\mathcal{A}}$ it holds

$$\langle \widetilde{V}_g f, \widetilde{V}_\gamma h\rangle_{L^2(\mathcal{K})} = \langle K\gamma, Kg\rangle_{\mathcal{H}_{\mathcal{A}}}\langle f, h\rangle_{\mathcal{H}_{\mathcal{A}}}. \tag{2.8}$$

In particular, $\widetilde{V}_g$ is an isometry from $\mathcal{H}_{\mathcal{A}}$ onto $L^2(\mathcal{K})$.

(b) For $f \in \mathcal{H}_{\mathcal{A}}$ we have the reproducing formula

$$\widetilde{V}_g f = \widetilde{V}_g f * \widetilde{V}_g g. \tag{2.9}$$

(c) The adjoint operator of $\widetilde{V}_g : \mathcal{H}_{\mathcal{A}} \to L^2(\mathcal{K})$ is given by

$$\widetilde{V}_g^* : L^2(\mathcal{K}) \to \mathcal{H}_{\mathcal{A}}, \quad \widetilde{V}_g^* F = \int_{\mathcal{K}} F(\mathcal{A}x)\widetilde{\pi}(\mathcal{A}x)g\,dm(\mathcal{A}x) \tag{2.10}$$

where the integral is understood in a weak sense.

(d) Suppose $\gamma \in \mathcal{D}_{\mathcal{A}}(K)$ with $\langle K\gamma, Kg\rangle_{\mathcal{H}_{\mathcal{A}}} = 1$ and $f \in \mathcal{H}_{\mathcal{A}}$. Then the following inversion formula holds weakly:

$$f = \widetilde{V}_{\gamma}^{*}\widetilde{V}_{g}f = \int_{\mathcal{K}} \widetilde{V}_{g}f(\mathcal{A}x)\widetilde{\pi}(\mathcal{A}x)\gamma \, dm(\mathcal{A}x). \tag{2.11}$$

Proof: (a) is an immediate consequence of the orthogonality relation (1.79) and Lemma 2.2.6 and the reproducing formula in (b) follows from (1.82). For (c) let QF denote the right hand side of (2.10). Then for $h \in \mathcal{H}_{\mathcal{A}}$

$$\langle QF, h\rangle_{\mathcal{H}_{\mathcal{A}}} = \int_{\mathcal{K}} F(\mathcal{A}x)\langle\widetilde{\pi}(\mathcal{A}x)g, h\rangle dm(\mathcal{A}x) = \langle F, \widetilde{V}_{g}h\rangle_{L^{2}(\mathcal{K})}.$$

Hence $Q = \widetilde{V}_{g}^{*}$. For (e) let $h \in \mathcal{H}_{\mathcal{A}}$. Using (2.8) we obtain

$$\langle \widetilde{V}_{\gamma}^{*}\widetilde{V}_{g}f, h\rangle_{\mathcal{H}_{\mathcal{A}}} = \int_{\mathcal{K}} \widetilde{V}_{g}f(\mathcal{A}x)\langle\widetilde{\pi}(\mathcal{A}x)\gamma, h\rangle dm(\mathcal{A}x) = \int_{\mathcal{K}} \widetilde{V}_{g}f(\mathcal{A}x)\overline{\widetilde{V}_{\gamma}h(\mathcal{A}x)}dm(\mathcal{A}x)$$
$$= \langle K\gamma, Kg\rangle_{\mathcal{H}_{\mathcal{A}}}\langle f, h\rangle_{\mathcal{H}_{\mathcal{A}}} = \langle f, h\rangle_{\mathcal{H}_{\mathcal{A}}}.$$

Since h is arbitrary the theorem is proved. $\qquad\square$

Looking at the reproducing formula (2.9) one should think of formula (1.16) for the convolution. Clearly, the choice $\gamma = g$ is possible in (d). We remark that the inversion formula (2.11) states on one hand that we can reconstruct an element $f \in \mathcal{H}_{\mathcal{A}}$ from its transform $\widetilde{V}_{g}f$. Of course, to this end the values of $\widetilde{V}_{g}f$ do not have to be computed on the whole group $\mathcal{G}$ but only for a single element from each orbit $\mathcal{A}x$. On the other hand formula (2.11) states that we can represent any $f \in \mathcal{H}_{\mathcal{A}}$ by a continuous superposition of elements $\widetilde{\pi}(\mathcal{A}x)g$. We emphasize that each of these elements is contained in $\mathcal{H}_{\mathcal{A}}$.

2.3 Examples

2.3.1 The Similitude Group and Radial Functions

Let us return to our motivating example and discuss it in more detail in the context of the previous section. We consider the similitude group $\mathcal{G} = \mathbb{R}^{d} \rtimes (\mathbb{R}_{+}^{*} \times SO(d))$ as in Section 1.5.3. We assume that $d \geq 2$. For the case $d = 1$ (which is not very interesting in our context) some modifications have to be done at certain places in the sequel. Again we consider the unitary irreducible representation of $\mathcal{G}$ on $L^{2}(\mathbb{R}^{d})$ given by

$$\pi(x, b, R)f(t) = b^{-d/2}f\left(b^{-1}R^{-1}(t-x)\right) = T_{x}D_{b}U_{R}f(t), \qquad (x, b, R) \in \mathcal{G},$$

whose corresponding voice transform is the continuous wavelet transform (1.68). We already know that π is square-integrable and the domain of the operator K is exactly the space of those functions $g \in L^2(\mathbb{R}^d)$ that satisfy the admissibility condition (1.69). A compact subgroup of $\mathcal{G}$ is given by $\mathcal{A} := \{(0, 1, A) \mid A \in SO(d)\} \cong SO(d)$, whose elements act on $\mathcal{G}$ as inner automorphisms, i.e., for $B \in \mathcal{A}$ and $\zeta = (x, b, A) \in \mathcal{G}$ we have

$$B(\zeta) = B\zeta B^{-1} = (0, 1, B)(x, b, A)(0, 1, B^{-1}) = (Bx, b, BAB^{-1}).$$

With $\sigma := \pi|_{\mathcal{A}}$ it holds $\pi(B(\zeta)) = \pi(B\zeta B^{-1}) = \sigma(B)\pi(\zeta)\sigma(B)^{-1}$. Thus, condition (2.5) is satisfied. Clearly, $\sigma(B)f(t) = f(B^{-1}t)$ and, once again, $\mathcal{H}_{\mathcal{A}}$ coincides with the space $L^2_{rad}(\mathbb{R}^d)$ of all radial square-integrable functions on $\mathbb{R}^d$.

The space $\mathcal{K} = \mathcal{G}^{\mathcal{A}}$ is the collection of all orbits $\mathcal{A}\zeta = \{(Bx, b, BAB^{-1})|B \in SO(d)\}$. The operator $\tilde{\pi}(\zeta), \zeta = (x, b, A)$ on $\mathcal{H}_{\mathcal{A}} = L^2_{rad}(\mathbb{R}^d)$ turns out to be

$$\tilde{\pi}(\zeta)f(t) = \int_{SO(d)} \pi(Bx, b, BAB^{-1})f(t)dB = b^{-d/2}\int_{SO(d)} f(b^{-1}(t - Bx))dB$$
$$= \tau_x D_b f(t), \tag{2.12}$$

where τ_x denotes the generalized translation (1.21) on $\mathbb{R}^d$. Hence, $\tilde{\pi}(\mathcal{A}(x, b, A))$ depends only on $|x|$ and on b and, therefore, we may always choose $A = \text{Id}$, the identity matrix. In fact, the set $\{\epsilon_{\mathcal{A}(x,b,\text{Id})} \mid x \in \mathbb{R}^d, b \in \mathbb{R}^*_+\}$ generates a subhypergroup $\mathcal{K}'$ of $\mathcal{K}$, i.e., the generated measure algebra $M(\mathcal{K}')$ is a closed subalgebra of $M(\mathcal{K})$. In other words if $F \in C^b_{\mathcal{A}}(\mathcal{G})$ then we have

$$\epsilon_{\mathcal{A}(x,b,\text{Id})} * \epsilon_{\mathcal{A}(y,c,\text{Id})}(F) = \int_{SO(d)} F(Ax + y, bc, \text{Id})dA = \int_{SO(d)} \epsilon_{\mathcal{A}(Ax+y,bc,\text{Id})}(F)dA.$$

The representation $\tilde{\pi}$ of $M(\mathcal{K})$ restricted to $M(\mathcal{K}')$ generates the same algebra of operators on $L^2_{rad}(\mathbb{R}^d)$. Clearly, an orbit $\mathcal{A}(x, b, \text{Id}) = \{(Bx, b, \text{Id}), B \in SO(d)\}$ depends only on $|x|$ and $b \in \mathbb{R}^*_+$. Hence, it holds

$$\mathcal{K}' \cong \mathbb{R}_+ \times \mathbb{R}^*_+$$

and we may write $\tilde{\pi}(r, b) = \tilde{\pi}(\mathcal{A}(x, b, \text{Id}))$ with $r = |x|$. The hypergroup $\mathcal{K}'$ can also be seen as a semi-direct product of a Bessel-Kingman-Hypergroup and the group $\mathbb{R}^*_+$ as explained in [Rös97].

Let us remark that the dilation D_a and generalized translation obey the commutation rule

$$D_a \tau_r = \tau_{ar} D_a, \qquad r \in \mathbb{R}_+, a \in \mathbb{R}^*_+. \tag{2.13}$$

Indeed, $D_a T_x = \pi(0, a, \text{Id})\pi(x, 1, \text{Id}) = \pi(ax, a, \text{Id})$ and hence

$$D_a \tau_{|x|} = \int_{SO(d)} \pi(0, a, \text{Id})\pi(Ax, 1, \text{Id})dA = \int_{SO(d)} \pi(Aax, a, \text{Id})dA = \tau_{a|x|} D_a.$$

For $g \in L^2_{rad}(\mathbb{R}^d)$ the restriction $\widetilde{V}_g$ to $L^2_{rad}(\mathbb{R}^d)$ can be computed by formula (2.3). The projection m' of the Haar measure of $\mathcal{G}$ onto $\mathcal{K}'$ is given by

$$\int_{\mathcal{K}'} F(y)dm'(y) \;=\; \int_{\mathbb{R}_+} \int_{\mathbb{R}_+^*} F(r,b)\frac{db}{b^{d+1}}d\mu_d(r),$$

where $d\mu_d(r) = |S^{d-1}|r^{d-1}dr$ as in (1.17).

Denoting $r = |x|, s = |y|$ and F_0 the function on $\mathbb{R}_+ \times \mathbb{R}_+^* \cong \mathcal{K}'$ corresponding to F by $F(x,b) = F_0(|x|,b)$, we obtain with a similar calculation as in the proof of Lemma 1.3.3(f) for the generalized translation on $\mathcal{K}'$

$$\mathcal{T}_{(y,c)}F(x,b) \;=\; \int_{SO(d)} F(Ay + cx, cb)dA \;=\; \frac{1}{|S^{d-1}|} \int_{S^{d-1}} F(s\xi + cx, cb)dS(\xi)$$

$$= \frac{|S^{d-2}|}{|S^{d-1}|} \int_0^{\pi} F_0(\sqrt{s^2 - 2rsc\cos\phi + c^2r^2}, cb) \sin^{d-2}(\phi)d\phi \;=\; \mathcal{T}_{(s,c)}F_0(r,b).$$

From $\mathcal{A}(x,b,\mathrm{Id})^{-1} = \mathcal{A}(-b^{-1}x, b^{-1}, \mathrm{Id})$ it follows that the left translation is given by

$$\mathcal{L}_{(r,b)} \;=\; \mathcal{T}_{(b^{-1}r, b^{-1})}, \quad r \in \mathbb{R}_+, b \in \mathbb{R}_+^*.$$

Theorems 2.2.5 and 2.2.7 immediately yield the following properties of $\widetilde{V}_g$. Some of them were already noted in Section 2.1. Recall that the Duflo-Moore operator K is given by (1.85).

Theorem 2.3.1. *Suppose $g \in L^2_{rad}(\mathbb{R}^d) \cap \mathcal{D}(K)$ with $\|Kg\|_2 = 1$.*

(a) *For $\gamma \in L^2_{rad}(\mathbb{R}^d) \cap \mathcal{D}(K)$ and $f,h \in L^2_{rad}(\mathbb{R}^d)$ it holds*

$$\langle \widetilde{V}_g f, \widetilde{V}_\gamma h \rangle_{L^2(\mathcal{K}', m')} \;=\; \langle K\gamma, Kg \rangle_{L^2_{rad}(\mathbb{R}^d)} \langle f, h \rangle_{L^2_{rad}(\mathbb{R}^d)}.$$

(b) *The adjoint operator of $\widetilde{V}_g$ is given by*

$$\widetilde{V}_g^* : L^2(\mathcal{K}', m') \to L^2_{rad}(\mathbb{R}^d)$$

$$\widetilde{V}_g^* F(t) \;=\; \int_{\mathcal{K}'} F(x)\overline{\widetilde{\pi}(x)}g(t)dm'(x) \;=\; \int_{\mathbb{R}_+} \int_{\mathbb{R}_+^*} F(r,b)\tau_r D_b g(t)\frac{db}{b^{d+1}}d\mu_d(r).$$

(c) *(Inversion) Suppose $\gamma \in L^2_{rad}(\mathbb{R}^d)$ with $\langle K\gamma, Kg \rangle = 1$ and $f \in L^2_{rad}(\mathbb{R}^d)$. Then we have*

$$f(t) \;=\; \widetilde{V}_\gamma^* \widetilde{V}_g f(t) \;=\; \int_{\mathbb{R}_+} \int_{\mathbb{R}_+^*} \widetilde{V}_g f(r,b)\tau_r D_b \gamma(t)\frac{db}{b^{d+1}}d\mu_d(r) \quad a.e. \;. \qquad (2.14)$$

(d) *(Covariance property) If $f \in L^2_{rad}(\mathbb{R}^d)$ and $r \in \mathbb{R}_+, b \in \mathbb{R}_+^*$ then it holds*

$$\widetilde{V}_g(\widetilde{\pi}(r,b)f) \;=\; \mathcal{L}_{(r,b)}(\widetilde{V}_g f).$$

Observe that setting $\gamma = g$ in (2.14) yields (2.4).

2.3.2   The Heisenberg Group and Radial Functions

Our second example is connected to the STFT. Let $\mathbb{H}_d$ denote the (reduced) Heisenberg group as in Section 1.5.3. We consider again the Schrödinger representation ρ of $\mathbb{H}_d$ acting on $L^2(\mathbb{R}^d)$ by

$$\rho(x,\omega,\tau) \; := \; \tau e^{\pi i x \cdot \omega} T_x M_\omega \; = \; \tau e^{-\pi i x \cdot \omega} M_\omega T_x.$$

As seen in (1.83) the corresponding voice transform is essentially the short time Fourier transform and we already know that the Schrödinger representation is square-integrable (Theorem 1.5.6). Due to the orthogonality relation (1.65), the corresponding operator K in the Duflo-Moore theorem 1.5.3 is the identity, thus $\mathcal{D}(K) = \mathcal{H} = L^2(\mathbb{R}^d)$.

The automorphisms of $\mathbb{R}^d \times \mathbb{R}^d$ that extend to automorphisms of the Heisenberg group $\mathbb{H}_d$ are the elements of the symplectic group $Sp(d)$. The latter is defined as the subgroup of $GL(2d, \mathbb{R})$ leaving invariant the symplectic form

$$[(x,\omega),(x',\omega')] \; := \; x' \cdot \omega - x \cdot \omega'.$$

For more details on the Heisenberg group and its relation to the symplectic group the reader is referred to Chapter 9 of Gröchenig's excellent book [Grö01].

A compact subgroup of $Sp(d)$ is given by

$$\mathcal{A} := \left\{ \begin{pmatrix} A & 0 \\ 0 & A \end{pmatrix} \mid A \in SO(d) \right\} \cong SO(d).$$

An element $A \in SO(d) \cong \mathcal{A}$ acts on $\mathbb{H}_d$ by $A(x,\omega,\tau) = (Ax, A\omega, \tau)$. In the sequel we assume $d \geq 2$. In the case $d = 1$ (which is not a very illustrative example) one has to replace $SO(1) = \{1\}$ by $O(1) = \{\pm 1\} \cong Z_2$ and put adjustments in some of the following statements.

As in the previous example we choose the natural representation σ of $SO(d)$ on $L^2(\mathbb{R}^d)$ given by $\sigma(A)f(t) = f(A^{-1}t)$ for $A \in SO(d), t \in \mathbb{R}^d$. Using the orthogonality of $A \in SO(d)$ we obtain

$$\begin{aligned}
\rho(Ax, A\omega, \tau)\sigma(A)f(t) &= \tau e^{-\pi i(Ax \cdot A\omega)} e^{2\pi i A\omega \cdot t} f(A^{-1}(t - Ax)) \\
&= \tau e^{-\pi i(x \cdot \omega)} e^{2\pi i \omega \cdot A^{-1}t} f(A^{-1}t - x) = \sigma(A)\rho(x,\omega,\tau)f(t).
\end{aligned}$$

Thus, condition (2.5) is satisfied. As in the previous example we have $\mathcal{H}_{\mathcal{A}} = L^2_{rad}(\mathbb{R}^d)$. The action of $\tilde{\rho}(x,\omega,\tau) = \rho(\epsilon_{\mathcal{A}(x,\omega,\tau)})$ on $L^2(\mathbb{R}^d)$ is given by

$$\tilde{\rho}(x,\omega,\tau)f(t) = \int_{SO(d)} \rho(Ax, A\omega, \tau)f(t)dA = \tau e^{\pi i x \cdot \omega} \int_{SO(d)} e^{2\pi i A\omega \cdot t} f(t - Ax)dA.$$

$$(2.15)$$

We already know from the general theory in Section 2.2 that $\tilde{\rho}(x,\omega,\tau)$ maps $L^2_{rad}(\mathbb{R}^d)$ onto $L^2_{rad}(\mathbb{R}^d)$, see Lemma 2.2.3(d).

Lemma 2.3.2. *(a) Let $(x, \omega, \tau), (x', \omega', \tau') \in \mathbb{H}_d$. Both elements are contained in the same orbit under $\mathcal{A}$ if and only if $\tau = \tau', |x| = |x'|, |\omega| = |\omega'|$ and $x \cdot \omega = x' \cdot \omega'$. Hence with*

$$\mathcal{N} := \mathbb{R}_+ \times \mathbb{R}_+ \times [-1, 1] \times \mathbb{T}$$

the orbit space $\mathbb{H}_d^{SO(d)}$ is parametrized by

$$\mathcal{K} := \mathcal{N} \setminus \{(r, s, t, \tau) \in \mathcal{N}, t \neq 1 \text{ and } (r = 0 \text{ or } s = 0)\} \tag{2.16}$$

(b) Consequently, $\tilde{\rho}(x, \omega, \tau)$ depends only on $|x|, |\omega|, x \cdot \omega$ and τ. If $f \in L^2_{rad}(\mathbb{R}^d)$ with corresponding function f on $\mathbb{R}_+$ then the following formula applies

$$\tilde{\rho}(x, \omega, \tau) f(t) = \tau e^{\pi i x \cdot \omega} \frac{|S^{d-2}|}{|S^{d-1}|} \times$$

$$\times \int_0^\pi \mathsf{f}(\sqrt{\theta^2 - 2r\theta \cos\phi + r^2}) e^{2\pi i \theta s \cos\alpha \cos\phi} \mathcal{B}_{d-1}(\theta s \sin\alpha \sin\phi) \sin^{d-2}\phi \, d\phi$$

$$=: \tilde{\rho}(r, s, \cos\alpha, \tau) \mathsf{f}(\theta),$$

where $r = |x|, s = |\omega|, x \cdot \omega = rs \cos\alpha, \theta = |t|$ and $\mathcal{B}_{d-1}$ is the spherical Bessel function defined in (1.34).

Proof: (a) Both elements are contained in the same orbit under $\mathcal{A} \cong SO(d)$ if and only if there exists a matrix $A \in SO(d)$ such that $(Ax, A\omega, \tau) = (x', \omega', \tau')$. Assume first that both elements belong to the same orbit. Then it holds necessarily $\tau = \tau', |x| = |x'|$ and $|\omega| = |\omega'|$ implying that there exist elements $B, C \in SO(d)$ (not unique) such that $x' = Bx$ and $\omega' = C\omega$. Hence, we end up with the equations $A^{-1}Bx = x$ and $A^{-1}C\omega = \omega$. This means $D := A^{-1}B \in I(x)$ and $E := A^{-1}C \in I(\omega)$ where $I(x) := \{R \in SO(d), Rx = x\}$ denotes the isotropy subgroup of $x \in \mathbb{R}^d$ (which is isomorphic to $SO(d-1)$). Using the orthogonality of A we finally obtain

$$x' \cdot \omega' = Bx \cdot C\omega = ADx \cdot AE\omega = Dx \cdot E\omega = x \cdot \omega.$$

Suppose conversely that $\tau = \tau', |x| = |x'| \neq 0$ and $|\omega| = |\omega'| \neq 0$ and $x' \cdot \omega' = x \cdot \omega$. Without loss of generality we may assume that $x = x' \in S^{d-1}$ and $\omega, \omega' \in S^{d-1}$. Since the sphere S^{d-1} is a two-point homogeneous space whose metric is given by $\cos d(x, y) = x \cdot y$ there exists a matrix $A \in I(x)$ such that $A\omega = \omega'$. This implies $(Ax, A\omega, \tau) = (x', \omega', \tau')$. The cases $x = 0$ or $\omega = 0$ are trivial.

In order to have an explicit correspondence between $\mathbb{H}_d^\mathcal{A}$ and the set in (2.16) we fix unit vectors $\eta, \xi \in S^{d-1}$ with $\eta \cdot \xi = 0$. For an element $(r, s, t, \tau) \in \mathcal{K}$ we set

$$(x, \omega, \tau) = (r\eta, s(\cos(\alpha)\eta + \sin(\alpha)\xi), \tau) \in \mathbb{H}_d \tag{2.17}$$

where $\cos\alpha = t$. Conversely, given $(x,\omega,\tau)\in\mathbb{H}_d$, the orbit $\mathcal{A}(x,\omega,\tau)$ can be identified with

$$(r,s,\cos\alpha,\tau) = \left(|x|,|\omega|,\frac{x\cdot\omega}{|x|\,|\omega|},\tau\right) \in \mathcal{K}.$$

The first assertion of (b) is an immediate consequence of (a). It can also be easily verified directly. For the proof of the second assertion we choose $\eta,\omega'\in S^{d-1}$ with $\eta\cdot\omega' = 0$ such that $x = r\eta$ and $\omega = s(\cos\alpha\,\eta + \sin\alpha\,\omega')$. This is possible according to (a). Using the integration formula (A.3) we obtain

$$
\begin{aligned}
\int_{SO(d)} & e^{2\pi i A\omega\cdot t} f(t - Ax)dA = \int_{SO(d)} e^{2\pi i \omega\cdot A^{-1}t} f(A^{-1}t - x)dA \\
&= \frac{1}{|S^{d-1}|}\int_{S^{d-1}} e^{2\pi i \theta\omega\cdot\xi} f(\theta\xi - x)dS(\xi) \\
&= \frac{1}{|S^{d-1}|}\int_0^\pi \int_{S^{d-2}} e^{2\pi i \theta s(\cos\alpha\,\eta+\sin\alpha\,\omega')\cdot(\cos\phi\,\eta+\sin\phi\,\xi')} \times \\
&\qquad\qquad \times\; f(\theta(\cos\phi\,\eta + \sin\phi\,\xi') - r\eta)dS^{d-2}(\xi')\sin^{d-2}\phi\,d\phi \\
&= \frac{1}{|S^{d-1}|}\int_0^\pi \mathsf{f}(\sqrt{\theta^2 - 2r\theta\cos\phi + r^2})e^{2\pi i \theta s \cos\alpha\cos\phi} \times \\
&\qquad\qquad \times \int_{S^{d-2}} e^{2\pi i \theta s \sin\alpha \sin\phi\,\omega'\cdot\xi'}dS^{d-2}(\xi')\sin^{d-2}\phi\,d\phi \\
&= \frac{|S^{d-2}|}{|S^{d-1}|}\int_0^\pi \mathsf{f}(\sqrt{\theta^2 - 2r\theta\cos\phi + r^2})e^{2\pi i \theta s \cos\alpha\cos\phi}\mathcal{B}_{d-1}(\theta s \sin\alpha\sin\phi)\sin^{d-2}\phi\,d\phi \\
&=: \Omega(r,s,\cos\alpha)\mathsf{f}(\theta). \qquad\qquad\qquad\qquad\qquad\qquad\qquad\qquad\qquad\qquad (2.18)
\end{aligned}
$$

This finishes the proof. $\qquad\qquad\qquad\qquad\qquad\qquad\qquad\qquad\qquad\qquad\qquad\qquad$ $\square$

We obtain an easy corollary which nevertheless does not seem to be present in the literature.

Corollary 2.3.3. Let $f,g \in L^2_{rad}(\mathbb{R}^d)$. Then the short time Fourier transform $\mathrm{STFT}_g f(x,\omega)$ depends only on $|x|,|\omega|$ and $x\cdot\omega$.

The operator $\Omega(r,s,\cos\alpha)$ may be viewed as a generalized combined translation and modulation. For special values it simplifies a little,

$$\Omega(0,s,1)\mathsf{f}(\theta) = \mathsf{f}(\theta)\mathcal{B}_d(s\theta),$$

$$\Omega(r,0,1)\mathsf{f}(\theta) = \frac{|S^{d-2}|}{|S^{d-1}|}\int_0^\pi \mathsf{f}(\sqrt{\theta^2 - 2r\theta\cos\phi + r^2})\sin^{d-2}\phi\,d\phi = \tau_r\mathsf{f}(\theta).$$

Here, τ_r denotes the generalized translation (1.22). We note that the first relation follows directly from the definition (2.15) of $\widetilde{\pi}(x,\omega,\tau)$ together with the definition (1.34) of the spherical Bessel functions.

With $g \in L^2_{rad}(\mathbb{R}^d)$ let us consider now the restriction $\widetilde{V}_g$ of V_g to $L^2_{rad}(\mathbb{R}^d)$. Interpreting it as a function on $\mathcal{K}$ it holds

$$V_g(x,\omega,\tau) = \widetilde{V}_g f(r,s,\cos\alpha,\tau) = \int_{\mathbb{R}^d} f(t)\overline{\tilde{\rho}(r,s,\cos\alpha,\tau)g(t)}dt$$

$$= \overline{\tau}e^{-\pi irs\cos\alpha}\int_0^\infty \mathsf{f}(\theta)\overline{\Omega(r,s,\cos\alpha)\mathsf{g}(\theta)}d\mu_d(\theta),$$

where $(x,\omega,\tau) = (r\eta, s(\cos(\alpha)\eta + \sin(\alpha)\xi), \tau)$ with fixed $\eta, \xi \in S^{d-1}$ such that $\eta \cdot \xi = 0$. The integral is the **STFT of a radial function**,

$$\mathrm{STFT}_g\, f(x,\omega) = \int_0^\infty \mathsf{f}(\theta)\overline{\Omega(r,s,\cos\alpha)\mathsf{g}(\theta)}d\mu_d(\theta) =: \widetilde{\mathrm{STFT}}_\mathsf{g}\mathsf{f}(r,s,\cos\alpha) \qquad (2.19)$$

where $r = |x|, s = |\omega|, rs\cos\alpha = x \cdot \omega$. The transform $\widetilde{\mathrm{STFT}}_\mathsf{g}\mathsf{f}$ can be interpreted as the natural analogue of the STFT on the Bessel-Kingman hypergroup of index $\nu = (d-2)/2$. Since it is inherited from the usual STFT on $\mathbb{R}^d$ it has many desired properties. In particular, there is an analogue to the fundamental identity of time-frequency analysis (1.67).

Corollary 2.3.4. *Suppose* $\mathsf{g}, \mathsf{f} \in L^2(\mathbb{R}_+, \mu_d)$. *Then it holds*

$$\widetilde{\mathrm{STFT}}_\mathsf{g}\mathsf{f}(r,s,\cos\alpha) = e^{-2\pi irs\cos\alpha}\widetilde{\mathrm{STFT}}_{\widehat{\mathsf{g}}}\widehat{\mathsf{f}}(s,r,-\cos\alpha)$$

for all $r, s \in \mathbb{R}_+, \alpha \in [0,\pi]$. *Hereby* $\widehat{\mathsf{f}}$ *denotes the Hankel transform of* f.

Proof: The assertion follows immediately from the fundamental identity of time-frequency analysis (1.67) and from the relationship (2.19) of $\widetilde{\mathrm{STFT}}_g$ with STFT_g together with the fact that the Hankel transform of f corresponds to the Fourier transform of the associated radial function f by Lemma 1.3.5. $\qquad\square$

The generalized translation $\mathcal{T}$ on the hypergroup $\mathbb{H}_d^{SO(d)}$, which appears in the covariance principle for $\widetilde{V}_g$, is given by

$$\mathcal{T}_{(x,\omega,\tau)}F(x',\omega',\tau') = \int_{SO(d)} F(Ax + x', A\omega + \omega', \tau\tau'e^{\pi i(x' \cdot A\omega - Ax \cdot \omega')})dA,$$

where dA denotes the Haar measure on $SO(d)$ and $(x,\omega,\tau), (x',\omega',\tau') \in \mathbb{H}_d$. We associate to a function F on $\mathbb{H}_d$ invariant under $\mathcal{A} \cong SO(d)$ a function F_0 on $\mathcal{K}$ by

$$F_0(r,s,\cos\alpha,\tau) = F(r\eta, s(\cos(\alpha)\eta + \sin(\alpha)\xi), \tau), \quad (r, s, \cos(\alpha), \tau) \in \mathcal{K},$$

where $\eta, \xi \in S^{d-1}$ are as above. As before we use the same symbol for the generalized translation for invariant functions on $\mathbb{H}_d$ and for functions on $\mathcal{K}$. Let us compute $\mathcal{T}$ explicitly for the cases $d = 2$ and $d = 3$.

Lemma 2.3.5. *Let $F \in C^b_\mathcal{A}(\mathbb{H}_d)$ with associated function F_0 on $\mathcal{K}$ and $(r, s, \cos\alpha, \tau)$, $(r', s', \cos\alpha', \tau') \in \mathcal{K}$.*

(a) For $d = 2$ it holds

$$\mathcal{T}_{(r,s,\cos\alpha,\tau)} F_0(r', s', \cos\alpha', \tau')$$
$$= \frac{1}{2\pi} \int_0^{2\pi} F_0\Big(\sqrt{r^2 + 2rr'\cos\theta + (r')^2}, \sqrt{s^2 + 2ss'\cos(\theta + \alpha' - \alpha) + (s')^2},$$
$$\frac{rs\cos\alpha + r's'\cos\alpha' + r's\cos(\alpha - \theta) + rs'\cos(\alpha' + \theta)}{\sqrt{r^2 + 2rr'\cos\theta + (r')^2}\sqrt{s^2 + 2ss'\cos(\theta + \alpha' - \alpha) + (s')^2}},$$
$$\tau\tau' e^{\pi i(r's\sin(\theta + \alpha) - rs'\sin(\alpha' + \theta))}\Big) d\theta. \tag{2.20}$$

(b) Let $d = 3$ and define with $\theta \in [0, \pi), \phi, \psi \in [0, 2\pi)$

$$p_1 := \sqrt{r^2 + 2rr'\cos\theta + (r')^2}$$
$$p_2 := \big(s^2 + (s')^2 + 2ss'(\cos\alpha\cos\alpha'\cos\theta + \sin(\alpha - \alpha')\sin\theta\cos\psi$$
$$\qquad + \sin\alpha\sin\alpha'(\cos\theta\cos\phi\cos\psi - \sin\phi\sin\psi)))^{1/2}$$
$$p_3 := rs\cos\alpha + r's'\cos\alpha' + rs'(\cos\alpha'\cos\theta - \sin(\alpha')\sin\theta\cos\phi)$$
$$\qquad + r's(\cos\alpha\cos\theta + \sin\alpha\sin\theta\cos\psi)$$
$$p_4 := r's(\cos\alpha\cos\theta + \sin\alpha\sin\theta\cos\psi) - rs'(\cos\alpha'\cos\theta - \sin\alpha'\sin\theta\cos\phi).$$

Then it holds

$$\mathcal{T}_{(r,s,\cos\alpha,\tau)} F_0(r', s', \cos\alpha', \tau')$$
$$= \frac{1}{8\pi^2} \int_0^{2\pi} \int_0^{2\pi} \int_0^\pi F_0\Big(p_1, p_2, \frac{p_3}{p_1 p_2}, e^{\pi i p_4}\Big) \sin\theta\, d\theta\, d\phi\, d\psi.$$

Proof: (a) We choose $\eta = e_1 := (1, 0)^T$ and $\xi = e_2 := (0, 1)^T$. The group $SO(2)$ is parametrized by $[0, 2\pi)$ by means of the matrices $A(\theta)$ defined in (A.4) and has Haar measure $\int_{SO(2)} G(A)dA = (2\pi)^{-1} \int_0^{2\pi} G(A(\theta))d\theta$. After some calculations involving trigonometric identities we obtain

$$
\int_{SO(2)} F(Ax + x', A\omega + \omega', \tau\tau' e^{\pi i(x' \cdot A\omega - Ax \cdot \omega')}) dA
$$

$$
= \frac{1}{2\pi} \int_0^{2\pi} F\left(\begin{pmatrix} r\cos\theta + r' \\ -r\sin\theta \end{pmatrix}, \begin{pmatrix} s\cos(\alpha - \theta) + s'\cos\alpha' \\ s\sin(\alpha - \theta) + s'\sin\alpha' \end{pmatrix}, \right.
$$

$$
\left. \tau\tau' e^{\pi i(r' s \sin(\alpha + \theta) - rs' \sin(\alpha' + \theta))} \right) d\theta, \tag{2.21}
$$

where as before $(x, \omega, \tau) = (re_1, s(\cos\alpha e_1 + \sin\alpha e_2), \tau)$ and

$$
(x', \omega', \tau') = (r'e_1, s'(\cos\alpha' e_1 + \sin\alpha' e_2), \tau').
$$

Note that $F(x, \omega, \tau) = F_0(|x|, |\omega|, \frac{x \cdot \omega}{|x||\omega|}, \tau)$. Computing the Euclidean norms of the first and second entry of F in (2.21) and their scalar product gives (2.20).

For (b) we proceed similarly using formula (A.8) for integration on $SO(3)$. With the matrices $A(\theta), B(\phi)$ defined in (A.5),(A.7) we obtain

$$
v_1 := B(\phi)A(\theta)B(\psi)x + x' = \begin{pmatrix} r\cos\theta + r' \\ -r\cos\theta\sin\theta \\ r\sin\phi\sin\theta \end{pmatrix},
$$

$$
v_2 := B(\phi)A(\theta)B(\psi)\omega + \omega'
$$

$$
= \begin{pmatrix} s(\cos\alpha\cos\theta + \sin\alpha\sin\theta\cos\psi) + s'\cos\alpha' \\ s(-\cos\alpha\cos\phi\sin\theta + \sin\alpha(\cos\phi\cos\theta\cos\psi - \sin\phi\sin\psi)) + s'\sin\alpha' \\ s(\cos\alpha\sin\phi\sin\theta - \sin\alpha(\sin\phi\cos\theta\cos\psi + \cos\phi\sin\psi)) \end{pmatrix}.
$$

Hereby, we used again the identification (2.17) with $\eta = (1, 0, 0)^T$ and $\xi = (0, 1, 0)^T$. Elementary but tedious calculations show that $p_1 = |v_1|$, $p_2 = |v_2|$ and $p_3 = v_1 \cdot v_2$. The expression p_4 is computed similarly. Together with the integration formula (A.8) this gives the desired result. □

Since $(x, \omega, \tau)^{-1} = (-x, -\omega, \overline{\tau})$ the involution of the hypergroup $\mathcal{K}$ is given by $(r, s, \cos\alpha, \tau)^{\sim} = (r, s, \cos\alpha, \overline{\tau})$ and hence

$$
\mathcal{L}_{(r,s,\cos\alpha,\tau)} = \mathcal{T}_{(r,s,\cos\alpha,\overline{\tau})}.
$$

Let us compute the projection m of the Haar measure of $\mathbb{H}_d$, i.e., the Haar measure of the hypergroup $\mathcal{K} \cong \mathbb{H}_d^{SO(d)}$.

Lemma 2.3.6. *The Haar measure* m *of* $\mathcal{K}$ *is given by*

$$
\int_{\mathcal{K}} F(x) dm(x)
$$

$$
= \frac{|S^{d-2}|}{|S^{d-1}|} \int_0^1 \int_0^\infty \int_0^\infty \int_0^\pi F_0(r, s, \cos\alpha, e^{2\pi i t}) \sin^{d-2}(\alpha) d\alpha d\mu_d(r) d\mu_d(s) dt.
$$

Proof: Suppose $F \in C_c(\mathbb{H}_d)$ is invariant under $\mathcal{A} \cong SO(d)$ with function F_0 on $\mathcal{K}$ given by $F(x, \omega, \tau) = F_0(|x|, |\omega|, \frac{x \cdot \omega}{|x||\omega|}, \tau)$. By the polar decomposition formula we obtain

$$
\int_{\mathbb{H}_d} F(h)dh = \int_0^1 \int_{\mathbb{R}^d} \int_{\mathbb{R}^d} F(x, \omega, e^{2\pi it})dxd\omega dt
$$

$$
= \frac{1}{|S^{d-1}|^2} \int_0^1 \int_0^\infty \int_0^\infty \int_{S^{d-1}} \int_{S^{d-1}} F(r\xi, sy, e^{2\pi it})dS(\xi)dS(y)d\mu_d(r)d\mu_d(s)dt
$$

$$
= \frac{1}{|S^{d-1}|^2} \int_0^1 \int_0^\infty \int_0^\infty \int_{S^{d-1}} \int_0^\pi \int_{S_\eta} F_0(r, s, (\cos(\alpha)\eta + \sin(\alpha)\zeta) \cdot \eta, e^{2\pi it})
$$

$$
dS^{d-2}(\zeta) \sin^{d-2}(\alpha)d\alpha dS(\xi)d\mu_d(r)d\mu_d(s)dt
$$

$$
= \frac{|S^{d-2}|}{|S^{d-1}|} \int_0^1 \int_0^\infty \int_0^\infty \int_0^\pi F_0(r, s, \cos(\alpha), e^{2\pi it}) \sin^{d-2}(\alpha)d\alpha d\mu_d(r)d\mu_d(s)dt.
$$

Hereby the rule (A.1) for integration on the sphere was used (with a change of variable $t \mapsto \cos \alpha$). $\qquad \square$

Now we are ready to apply Theorems 2.2.5 and 2.2.7 to our situation. Recall also that here, the Duflo-Moore operator K is the identity.

Theorem 2.3.7. Let $g \in L^2_{rad}(\mathbb{R}^d)$ such that $\|g\|_2 = 1$.

(a) For $\gamma, f, h \in L^2_{rad}(\mathbb{R}^d)$ then

$$
\langle \widetilde{V}_g f, \widetilde{V}_\gamma h \rangle_{L^2(\mathcal{K})} = \langle \gamma, g \rangle_{L^2_{rad}(\mathbb{R}^d)} \langle f, g \rangle_{L^2_{rad}(\mathbb{R}^d)}
$$

In particular, $\widetilde{V}_g$ is an isometry from $L^2_{rad}(\mathbb{R}^d)$ onto $L^2(\mathcal{K})$.

(b) The adjoint operator of $\widetilde{V}_g$ is given by

$$
\widetilde{V}_g^* : L^2(\mathcal{K}) \to L^2_{rad}(\mathbb{R}^d),
$$

$$
\widetilde{V}_g^* F(t) = \int_{\mathcal{K}} F(r, s, \cos \alpha, \tau) \tilde{\rho}(r, s, \cos \alpha, \tau)g(t)dm(r, s, \alpha, \tau).
$$

(c) Suppose $\gamma \in L^2_{rad}(\mathbb{R}^d)$ with $\langle \gamma, g \rangle = 1$. Then the following inversion formula holds:

$$
f(t) = \widetilde{V}_\gamma^* \widetilde{V}_g f(t)
$$

$$
= \int_{\mathcal{K}} \widetilde{V}_g f(r, s, \cos \alpha, \tau) \tilde{\rho}(r, s, \cos \alpha, \tau)\gamma(t)dm(r, s, \alpha, \tau) \quad a.e. \ .
$$

(d) (Covariance property) Let $f \in L^2_{rad}(\mathbb{R}^d)$ and $(r, s, \cos\alpha, \tau) \in \mathcal{K}$. Then it holds

$$\widetilde{V}_g(\tilde{\rho}(r, s, \cos\alpha, \tau)f) = \mathcal{L}_{(r,s,\cos\alpha,\tau)}\widetilde{V}_g f.$$

Of course, these results immediately imply corresponding results for the STFT. We only state one property explicitly.

Corollary 2.3.8. *Suppose $g, \gamma \in L^2_{rad}(\mathbb{R}^d)$ with $\langle \gamma, g \rangle = 1$. Then $\widetilde{\mathrm{STFT}}_g$ is inverted on $L^2_{rad}(\mathbb{R}^d)$ by the formula*

$$f(t) =$$
$$\frac{|S^{d-2}|}{|S^{d-1}|} \int_0^\infty \int_0^\infty \int_0^\pi \widetilde{\mathrm{STFT}}_g f(r, s, \cos\alpha)\Omega(r, s, \cos\alpha)\gamma(t)\sin^{d-2}(\alpha)\,d\alpha\,d\mu_d(s)\,d\mu_d(r)$$

for almost all $t \in \mathbb{R}^d$.

Proof: It holds

$$\widetilde{V}_g f(r, s, \cos\alpha, \tau)\tilde{\rho}(r, s, \cos\alpha, \tau)\gamma = \widetilde{\mathrm{STFT}}_g f(r, s, \cos\alpha)\Omega(r, s, \cos\alpha)\gamma,$$

the latter being independent of τ. Thus the assertion follows from the previous theorem. □

With this example we have settled the foundations for radial time-frequency analysis. There are many open questions left such as investigating radial Gabor frames. Having the formula (2.18) for the generalized combined translation and modulation Ω in mind it would probably be very involved to get results in this direction by direct methods. Nevertheless we will attack this problem with abstract methods in Chapter 4.

2.3.3 Reflection Symmetries

Let us consider the two previous examples in which the symmetry group $\mathcal{A} = SO(d)$ is replaced by a finite reflection group W. This is a finite subgroup of $O(d)$. Sometimes W is called Coxeter group or (in the context of Lie algebras) Weyl group [GB85]. Of course, W acts in the same way as $SO(d)$ on the similitude group and on the Heisenberg group. The representation σ is again the natural action on $L^2(\mathbb{R}^d)$. The Haar measure of W is given by a sum, $\int_W f(w)d\mu(w) = |W|^{-1}\sum_{w \in W} f(w)$.

A Coxeter group can be described using a root system. For $0 \neq u \in \mathbb{R}^d$ denote by S_u the reflection at the hyperplane perpendicular to u, i.e.,

$$S_u x := x - 2\frac{x \cdot u}{|u|^2}u, \quad x \in \mathbb{R}^d.$$

A root system is a finite set $R \subset \mathbb{R}^d \setminus \{0\}$ such that $S_u v \in R$ for all $u, v \in R$. The Coxeter group $W = W(R)$ associated to the root system R is the (finite!) subgroup of $O(d)$ generated by the reflections $\{S_u, u \in R\}$. For a classification of all Coxeter groups see [GB85]. The complement of the union of the hyperplanes $\bigcup_{u \in R} u^{\perp}$ splits into several open connected components, called Weyl chambers. For an arbitrary *closed* Weyl chamber C it holds $\bigcup_{w \in W} w(C) = \mathbb{R}^d$. Hence, a function on $\mathbb{R}^d$, which is invariant under $W(R)$ is determined by its values on a closed Weyl chamber C and, hence, can be regarded as a function on C.

Applying the results of Section 2.2 to this setting yields wavelet analysis and time-frequency analysis on Weyl chambers. In order to come up with explicit formulae one certainly has to treat each class of reflection groups for its own. However, we will not further pursue this direction.

2.4   Notes

The radial wavelet transform in (2.3) was already introduced by Trimèche [Tri95, Tri97] and, independently, by Rösler [Rös95a] in the more general context of Bessel-Kingman hypergroups, see Example 1.4.1(a). As substitute for the usual translation, the Bessel-Kingman translation is used, of course. If $\nu = \frac{d-2}{2}$ then the radial wavelet transform (2.3) in $\mathbb{R}^d$ coincides (up to normalization) with the wavelet transform on the Bessel-Kingman hypergroup of index ν in [Tri95, Tri97, Rös95a]. Although the authors probably knew this fact, we remark that it was not noted in their papers. The construction of Rösler uses representation theory like in our approach. A semi-direct product of a Bessel-Kingman hypergroup $\mathcal{K}_\nu$ with $\mathbb{R}_+^*$ is constructed. The continuous wavelet transform on $\mathcal{K}_\nu$ arises as the voice transform associated to a certain representation of this semi-direct product hypergroup $\mathcal{K}_\nu \rtimes \mathbb{R}_+^*$. If $\nu = \frac{d-2}{2}$ then $\mathcal{K}_\nu \rtimes \mathbb{R}_+^*$ coincides with the hypergroup $\mathcal{K}'$ considered in Section 2.3.1. We remark that Trimèche studied continuous wavelet transforms on a more general class of hypergroups than just the Bessel-Kingman hypergroups [Tri97]. Also Hinz contributed to this kind of investigations [Hin00].

The radial STFT in (2.19) has not yet been considered. Of course, it can be interpreted as the STFT on the Bessel-Kingman hypergroup of index $\nu = \frac{d-2}{2}$. We conjecture that one can generalize this transform to Bessel-Kingman hypergroups of arbitrary real-valued index $\nu > -1/2$. To do this one probably just has to replace the number $\nu = \frac{d-2}{2}$ in definition (2.18) of the generalized combined translation and modulation Ω by an arbitrary real-valued index $\nu > -1/2$. So the corresponding operators would be defined as

$$(\Omega_\nu(r, s\cos\alpha)\mathsf{f})(\theta) :=$$

$$\frac{\Gamma(\nu+1)}{\sqrt{\pi}\Gamma(\nu)} \int_0^\pi \mathsf{f}(\sqrt{\theta^2 - 2r\theta\cos\phi + r^2})e^{2\pi i\theta s\cos\alpha\cos\phi}\mathcal{B}_{(\nu-1/2)}(\theta s\sin\alpha\sin\phi)\sin^{2\nu}\phi\,d\phi.$$

The functions $\mathcal{B}_{(\nu)}$ were defined in (1.58). We leave the study of this operator and the corresponding transform to future investigations.

Recently, there were done other attempts to define a STFT on certain classes of commutative hypergroups, in particular on Bessel-Kingman hypergroups [CG03b, Dac01, Dac03, Tri97]. We discuss these approaches in detail in Appendix B. It turns out that all of them lack some important properties. So our radial STFT (2.19) can be considered the most natural approach to the STFT on the Bessel-Kingman hypergroup of index $\frac{d-2}{2}$. However, it is not clear how the radial STFT can be generalized to other hypergroups.

The orbit hypergroup $\mathbb{H}_d^{SO(d)}$ appearing in connection with the radial STFT has not yet been considered in the literature, at least up to our knowledge. Like the hypergroup $\mathcal{K}'$ from Section 2.3.1 it is a non-commutative hypergroup. Actually, the investigations done in this chapter lead in a natural way to certain non-commutative hypergroups. It seems that, apart from [Rös97], this is the first time that useful examples of non-commutative hypergroups (that are not groups) appear. Indeed, nearly all known examples in hypergroup theory are commutative. The few non-commutative proper hypergroups provided in the literature are usually very artificial finite hypergroups. We remark that even in the article by Vrem [Vre79] on harmonic analysis on compact hypergroups the only example of a non-commutative hypergroup consists of three elements. Maybe with their relevance in time-frequency and wavelet analysis in mind, research on non-commutative hypergroups will be forced in the future.

One topic in this direction would be the following. With $\tilde{\pi}$ we constructed an irreducible representation of the orbit hypergroup $\mathcal{K} = \mathcal{G}^{\mathcal{A}}$ starting with an irreducible representation π of $\mathcal{G}$. The question arises whether all irreducible representations of $\mathcal{K}$ are obtained in this way. Very related to this question are investigations by Hermann [Her95] concerning representations of double coset hypergroups. Starting with an irreducible representation of some group H, Hermann constructed an irreducible representation of a double coset hypergroup $H//K$ [Her95, Lemma 1]. Moreover, Hermann proved that whenever the algebra $L^1(H)$ is symmetric then one obtains all irreducible representations of $H//K$ in this way [Her95]. We remark that the method of induced representations of hypergroups also developed by Hermann [Her92] was used for the proof. Now the orbit hypergroup $\mathcal{G}^{\mathcal{A}}$ is isomorphic to the double coset hypergroup $(\mathcal{G} \rtimes \mathcal{A})//\mathcal{A}$, see Example 1.4.1(b). Specializing to $H = \mathcal{G} \rtimes \mathcal{A}$ and bearing Lemma 2.2.1 in mind it turns out that Hermann's construction is precisely the same as ours. In particular, in case $L^1(\mathcal{G} \rtimes \mathcal{A})$ is symmetric one can describe all irreducible representations of $\mathcal{G}^{\mathcal{A}}$ by means of the irreducible representations of $\mathcal{G} \rtimes \mathcal{A}$. We remark that also investigations done by Mosak in the context of $[FIA]_B^-$-groups are related to questions of this kind [Mos72].

Let us make another remark related to Section 2.3.2. The $SO(d)$ (resp. $O(d)$) is not the maximal compact subgroup of $Sp(d)$, i.e., the group of automorphisms of $\mathbb{R}^d \times \mathbb{R}^d$, which extend to automorphism of $\mathbb{H}_d$. Indeed, the maximal subgroup $Sp(d) \cap O(2d)$ is strictly larger than $O(d)$. Identifying the Heisenberg group with $\mathbb{C}^d \times \mathbb{T}$ via $(z, \tau) = (x + iy, \tau) \in \mathbb{C}^d \times \mathbb{T}$ for some element $(x, y, \tau) \in \mathbb{R}^d \times \mathbb{R}^d \times \mathbb{T}$, one may actually identify $Sp(d) \cap O(2d)$ with $U(d)$, the group of unitary matrices. The space of orbits can be identified with $\mathbb{R}_+ \times \mathbb{T}$, and the corresponding orbit hypergroup becomes commutative, see [BJR98, BR96]. It essentially coincides with the Laguerre hypergroup of index $d - 1$, see [NT97]. The commutativity, however, makes this setting rather irrelevant for our context, since the operator $\widetilde{\pi}$ would be an irreducible representation of $\mathbb{H}_d^{U(d)}$ on the space $\mathcal{H}_{U(d)}$ of invariant elements by Theorem 2.2.4. By Schur's lemma $\mathcal{H}_{U(d)}$ would be one-dimensional.

Chapter 3

Radial Multiresolution in 3D

In the previous chapter we investigated the continuous wavelet transform and the (continuous) STFT of functions, which are invariant under symmetry groups. Now the question arises whether it is possible to obtain discretizations. This is, of course, important for applications, since any algorithm can handle only discrete sets (actually only finite sets). Moreover, discretizations are of great interest for theoretical reasons, like for the study of function spaces. So the problem consists in finding a set of the form $B_g := \{\tilde{\pi}(x_i)g, i \in I\}$, $x_i \in \mathcal{G}$, with discrete index set I and $g \in \mathcal{H}_{\mathcal{A}}$, which forms an orthonormal or Riesz basis of $\mathcal{H}_{\mathcal{A}}$. To be less restrictive, one could also require that B_g be only a frame for $\mathcal{H}_{\mathcal{A}}$. The rest of this thesis is dedicated to giving solutions to this problem.

While we postpone the general abstract case to the next chapter, we only consider (discrete) radial wavelets in this chapter. There have already been some investigations by Epperson and Frazier [EF95], who constructed discrete radial wavelets in $\mathbb{R}^d$. Although they probably did not know about Bessel-Kingman hypergroups and the corresponding translation, the same wavelet concept as in the previous chapter is underlying their approach. They use dyadic dilations whereas the translations are taken from a sampling lattice in $\mathbb{R}_+$ that is determined by the positive zeros of the related spherical Bessel functions $\mathcal{B}_d$. The spatial lattice is equidistant only in the special cases $d = 1$ and $d = 3$.

Up to the author's knowledge, radial multiresolution analyses have not yet been considered. Actually, there seems to be no general rigorous approach available for the construction of orthogonal radial wavelet bases in arbitrary dimension. This problem is related to the open question of finding a Poisson summation formula compatible with the generalized translation. We mention that the construction of [EF95] requires some restrictions on the involved radial wavelets. Beside other conditions, the support of their Hankel transforms have to be contained in $(0, 1]$.

In this chapter we construct radial multiresolution analyses and orthogonal radial wavelet bases in $\mathbb{R}^3$. In dimension $d = 3$, the special form of the corresponding spherical Bessel function $\mathcal{B}_3(t) = \frac{\sin(2\pi t)}{2\pi t}$ allows to carry out the constructions along the same lines as in the well-known Euclidean setting. Hereby, it is crucial that the zeros of $\mathcal{B}_3$ are equidistant.

Instead of working on $L^2_{rad}(\mathbb{R}^3)$ we may, of course, use the isomorphic $L^2(\mathbb{R}_+, \mu_3)$. The latter is abbreviated by $L^2(\mathcal{K})$ throughout this chapter, where $\mathcal{K}$ stands for the Bessel-Kingman hypergroup of index $\nu = 1/2$ corresponding to the dimension $d = 3$. Indeed, our concept of a radial multiresolution analysis is based on $L^2(\mathcal{K})$. The scale spaces $(V_j)_{j \in \mathbb{Z}} \subset L^2(\mathcal{K})$ are obtained by dyadic dilations from V_0, which in turn is spanned by equidistant generalized translates of a fixed "radial" scaling function $\phi \in L^2(\mathcal{K})$. The latter is characterized by a two-scale relation, but in contrast to classical MRAs, ϕ itself is not contained in V_0. Moreover, the scale spaces are not shift-invariant with respect to the generalized translation. We put some emphasis on the construction of orthogonal MRAs. From a given orthogonal MRA we derive an orthogonal wavelet basis for $L^2(\mathcal{K})$. By construction, this "radial" basis has a direct interpretation as an orthogonal wavelet basis for $L^2_{rad}(\mathbb{R}^3)$. We note that periodicity arguments similar to the classical case are used at several places. However, such arguments are not available in arbitrary dimensions.

We give a simple construction of radial scaling functions and the corresponding wavelets by means of even classical scaling functions. This immediately provides a huge variety of possible radial wavelets. Moreover, essentially any radial scaling function can be constructed in this way. It in particular implies that in contrast to the classical case, there do not exist any real-valued orthonormal radial scaling functions with compact support. Finally, we formulate the corresponding algorithm and compare it with the classical discrete wavelet transform.

The basic idea to develop a radial MRA in $\mathbb{R}^3$ goes back to Margit Rösler and the content of this chapter actually stems from a joint work with her that will appear in [RR03].

3.1 Radial Multiresolution Analysis in $\mathbb{R}^3$

Radial analysis in $\mathbb{R}^3$ corresponds to the Bessel-Kingman hypergroup $\mathcal{K}_\nu$ with $\nu = 1/2$. For convenience we put

$$d\omega(r) := d\mu_3(r) = 4\pi r^2 dr, \quad L^2(\mathcal{K}) := L^2(\mathbb{R}_+, \omega).$$

Norms and scalars products are usually taken with respect to $L^2(\mathcal{K})$ in this chapter if nothing else is stated. Notice that the spherical Bessel function $\mathcal{B}_3(t) = \frac{\sin(2\pi t)}{2\pi t}$ (Lemma 1.3.7(b)) is even on $\mathbb{R}$. Hence, it is natural to assume the Hankel transform

$$\widehat{f}(\lambda) = \int_0^\infty \mathcal{B}_3(\lambda r) f(r) d\omega(r)$$

of $f \in L^2(\mathcal{K})$ to be continued to an even function on $\mathbb{R}$ as well. We will always do so throughout this chapter. Furthermore, we mention that by a change of variables, the generalized translation (1.21) on $\mathcal{K}$ can be written in the simple form

$$\tau_r f(s) = \frac{1}{2rs} \int_{|r-s|}^{r+s} f(t)t \, dt, \quad r, s > 0. \tag{3.1}$$

We recall also the different formulae for the Hankel transform and the generalized translation derived at the end of Section 1.3.

The non-negative zeros of the spherical Bessel function $\mathcal{B}_3$ are located at the numbers

$$t_k := \frac{k}{2}, \quad k \in \mathbb{N},$$

and the normalized Fourier-Bessel functions

$$\rho_k(r) := M_k \mathcal{B}_3(t_k r) = \frac{\sin(k\pi r)}{\sqrt{2\pi} r} \quad \text{with } M_k := \sqrt{\frac{\pi}{2}} \, k \tag{3.2}$$

form an orthonormal basis of the Hilbert space $L^2([0,1], \omega|_{[0,1]})$. This is equivalent to the obvious fact that the functions

$$s_k(r) := 2\sqrt{\pi} r \rho_k(r) = \sqrt{2} \sin(k\pi r), \quad k \in \mathbb{N}$$

form an orthonormal basis for $L^2[0,1] := L^2([0,1], dr)$. It will be of importance in the following that s_k is 2-periodic for all $k \in \mathbb{N}$.

Let us turn to the definition of a radial multiresolution analysis (MRA) for $\mathbb{R}^3$. It is close to the well-known definition of Mallat [Mal89, Dau92] for $\mathbb{R}$. For convenience, we introduce the notation

$$T^{(k)} := \tau_{t_k} = \tau_{k/2}, \quad (k \in \mathbb{N}).$$

If $f \in L^2(\mathcal{K})$ then according to Corollary 1.3.9

$$(M_k T^{(k)} f)^{\wedge}(\lambda) = \rho_k(\lambda) \widehat{f}(\lambda) = \frac{s_k(\lambda)}{2\sqrt{\pi}\lambda} \widehat{f}(\lambda). \tag{3.3}$$

The norm in $l^2(\mathbb{N})$ is given by $\|\alpha\|_2^2 := \sum_{k=1}^{\infty} |\alpha_k|^2$ for some sequence $\alpha = (\alpha_k)_{k \in \mathbb{N}}$ as usual. When there is no danger of confusion we denote both the norm in $L^2(\mathcal{K})$ and the one in $l^2(\mathbb{N})$ by $\|\cdot\|_2$.

Definition 3.1.1. (Radial Multiresolution Analysis) A radial MRA for $\mathbb{R}^3$ is a sequence $\{V_j\}_{j \in \mathbb{Z}}$ of closed linear subspaces of $L^2(\mathcal{K})$ such that the following properties are satisfied.

(1) $V_j \subseteq V_{j+1}$ for all $j \in \mathbb{Z}$.

(2) $\bigcap_{j=-\infty}^{\infty} V_j = \{0\}$.

(3) $\bigcup_{j=-\infty}^{\infty} V_j$ is dense in $L^2(\mathcal{K})$.

(4) $f \in V_j$ if and only if $f(2\,\cdot) \in V_{j+1}$.

(5) There exists a function $\phi \in L^2(\mathcal{K})$ such that

$$B_\phi := \{ M_k T^{(k)} \phi, \ k \in \mathbb{N} \}$$

is a Riesz basis of V_0, that is, span B_ϕ is dense in V_0 and there exist constants $A, B > 0$ such that

$$A\|\alpha\|_2^2 \le \|\sum_{k=1}^{\infty} \alpha_k M_k T^{(k)} \phi \|_2^2 \le B\|\alpha\|_2^2$$

for all $\alpha = (\alpha_k)_{k \in \mathbb{N}} \in l^2(\mathbb{N})$.

The function ϕ in (5) is called a scaling function for the MRA $\{V_j\}$. We remark explicitly that in contrast to the classical case, ϕ itself is not contained in V_0, and V_0 is not shift invariant. Indeed, if $f \in V_0$ then $T^{(k)} f$ is not contained in V_0 for all k. This will be shown in Corollary 3.1.4 below.

Our first aim is to determine an orthonormal basis for V_0 from its Riesz basis, i.e., a function $\phi^\heartsuit \in L^2(\mathcal{K})$ such that $B_{\phi^\heartsuit}$ constitutes an orthonormal basis for V_0. For $\phi \in L^2(\mathcal{K})$ we define

$$P_\phi(\lambda) := \sum_{n=-\infty}^{\infty} |\widehat{\phi}(\lambda + 2n)|^2,$$

which is even and 2-periodic on $\mathbb{R}$.

Proposition 3.1.1. *Let $\phi \in L^2(\mathcal{K})$ and $A, B > 0$. Then*

$$A\|\alpha\|_2^2 \le \|\sum_{k=1}^{\infty} \alpha_k M_k T^{(k)} \phi \|_2^2 \le B\|\alpha\|_2^2 \quad \text{for all } \alpha \in l^2(\mathbb{N}) \tag{3.4}$$

if and only if

$$A \le P_\phi(\lambda) \le B \quad \text{for almost all } \lambda \in \mathbb{R}. \tag{3.5}$$

Proof: Let $\alpha \in l^2(\mathbb{N})$ be an arbitrary finite sequence. Define

$$\widetilde{\alpha} := \sum_{k=1}^{\infty} \alpha_k s_k \in L^2[0,1]. \tag{3.6}$$

We may regard $\widetilde{\alpha}$ as an odd, 2-periodic function on $\mathbb{R}$. By the Plancherel theorem for the Hankel transform and (3.3) we obtain

$$\| \sum_{k=1}^{\infty} \alpha_k M_k T^{(k)} \phi \|_2^2 = \| \sum_{k=1}^{\infty} \alpha_k \rho_k \, \widehat{\phi} \|_2^2 = \int_0^{\infty} | \sum_{k=1}^{\infty} \alpha_k s_k(\lambda)|^2 |\widehat{\phi}(\lambda)|^2 d\lambda$$

$$= \frac{1}{2} \int_{-\infty}^{\infty} |\widetilde{\alpha}(\lambda)|^2 |\widehat{\phi}(\lambda)|^2 d\lambda = \frac{1}{2} \int_{-1}^{1} |\widetilde{\alpha}(\lambda)|^2 P_\phi(\lambda) d\lambda = \int_0^1 |\widetilde{\alpha}(\lambda)|^2 P_\phi(\lambda) d\lambda.$$

Since the functions s_k, $k \in \mathbb{N}$, form an orthonormal basis of $L^2[0,1]$, we have $\|\alpha|l^2(\mathbb{N})\| = \|\widetilde{\alpha}|L^2[0,1]\|$. As the finite sequences form a dense subspace of $l^2(\mathbb{N})$, this implies the assertion. $\qquad \square$

Setting $A = B = 1$ immediately yields the following result.

Corollary 3.1.2. *For* $\phi \in L^2(\mathcal{K})$ *the following statements are equivalent.*

(1) *The set* $B_\phi = \{M_k T^{(k)} \phi : k \in \mathbb{N}\}$ *is orthonormal in* $L^2(\mathcal{K})$.
(2) *It holds* $P_\phi \equiv 1$ *a.e.*.

For $\phi \in L^2(\mathcal{K})$, we put

$$V_\phi := \overline{\operatorname{span} B_\phi},$$

where the closure is taken in $L^2(\mathcal{K})$. The set B_ϕ is a Riesz basis of V_ϕ if and only if there exist constants $A, B > 0$ such that the equivalent conditions of Proposition 3.1.1 are satisfied. This will be a standard requirement in the sequel, and we therefore introduce a separate notation.

Definition 3.1.2. A function $\phi \in L^2(\mathcal{K})$ satisfies condition (RB) if B_ϕ is a Riesz basis of V_ϕ.

As before, we will often consider functions of $L^2[0,1]$ as odd, 2-periodic functions on $\mathbb{R}$. We therefore define

$$Z := \{\alpha : \mathbb{R} \to \mathbb{C} | \ \alpha|_{[0,1]} \in L^2[0,1], \ \alpha(-x) = -\alpha(x), \ \alpha(x+2) = \alpha(x) \text{ a.e.}\}.$$

Together with the inner product and norm of $L^2[0,1]$ this is a Hilbert space.

Lemma 3.1.3. *Let* $\phi \in L^2(\mathcal{K})$ *satisfy (RB). Then a function* f *is contained in* V_ϕ *if and only if*

$$\widehat{f}(\lambda) = \frac{\beta(\lambda)}{\lambda} \widehat{\phi}(\lambda) \qquad \text{with } \beta \in Z.$$

The function $f \in V_\phi$ *corresponding to* $\beta = \sum_{k=1}^{\infty} \alpha_k s_k \in Z$ *with* $(\alpha_k)_{k \in \mathbb{N}} \in l^2(\mathbb{N})$ *is given by* $f = 2\sqrt{\pi} \sum_{k=1}^{\infty} \alpha_k M_k T^{(k)} \phi$.

Proof: By (3.3) we have

$$\frac{s_k(\lambda)}{\lambda}\widehat{\phi}(\lambda) \;=\; 2\sqrt{\pi}(M_k T^{(k)}\phi)^{\wedge}(\lambda).$$

The translates $M_k T^{(k)}\phi$ form a Riesz basis of V_ϕ and, hence, $\{(M_k T^{(k)}\phi)^{\wedge},\, k \in \mathbb{N}\}$ is a Riesz basis of $\widehat{V_\phi}$. As $(s_k)_{k\in\mathbb{N}}$ is an orthonormal basis of Z, this implies the assertion.
□

Lemma 3.1.3 is of particular interest when ϕ is the scaling function of a MRA $\{V_j\}$. Then it holds $V_0 = V_\phi$, and we easily deduce the previously mentioned lack of shift invariance.

Corollary 3.1.4. *Let $\{V_j\}_{j\in\mathbb{Z}}$ be a radial MRA. Then $f \in V_0$ implies that $T^{(k)}f = \tau_{k/2}f \notin V_0$ for all $k \in \mathbb{N}$. Similarly, $f \in V_j$ implies that $\tau_{2^{-(j+1)}k}f \notin V_j$ for all $k \in \mathbb{N}$.*

Proof: After rescaling it is enough to consider V_0. Recall that $M_k(T^{(k)}f)^{\wedge} = \rho_k\widehat{f}$. But if $\beta \in Z$, then $\rho_k\beta \notin Z$ for all k, because periodicity is lost. The characterization of $V_\phi = V_0$ according to the previous lemma thus shows that $f \in V_0$ implies $T^{(k)}f \notin V_0$.
□

Now we are in the position to determine an orthonormal basis of V_ϕ.

Theorem 3.1.5. *(Orthogonalization) Suppose $\phi \in L^2(\mathcal{K})$ satisfies condition (RB). Define $\phi^{\heartsuit} \in L^2(\mathcal{K})$ by its Hankel transform*

$$\widehat{\phi^{\heartsuit}} := \frac{\widehat{\phi}}{\sqrt{P_\phi}}. \tag{3.7}$$

Then $B_{\phi^{\heartsuit}} = \{M_k T^{(k)}\phi^{\heartsuit},\, k \in \mathbb{N}\}$ forms an orthonormal basis of $V_\phi = V_{\phi^{\heartsuit}}$.

If ϕ is a scaling function of a MRA $\{V_j\}$, then $V_\phi = V_{\phi^{\heartsuit}} = V_0$, and we call $\phi^{\heartsuit}$ an orthonormal scaling function for $\{V_j\}_{j\in\mathbb{Z}}$.

Proof: By definition of $\phi^{\heartsuit}$ we have $P_{\phi^{\heartsuit}} \equiv 1$ a.e. and, hence, $B_{\phi^{\heartsuit}}$ is orthonormal according to Corollary 3.1.2. It remains to prove that $V_{\phi^{\heartsuit}}$ coincides with V_ϕ. To this end we have to verify that $M_k T^{(k)}\phi^{\heartsuit} \in V_\phi$ and $M_k T^{(k)}\phi \in V_{\phi^{\heartsuit}}$ for all $k \in \mathbb{N}$. By Lemma 3.1.3 and relation (3.3) the first inclusion is equivalent to the existence of a function $\beta \in Z$ such that

$$\beta(\lambda)\hat{\phi}(\lambda) \;=\; \lambda\rho_k(\lambda)\hat{\phi}^{\heartsuit}(\lambda) \;=\; \frac{s_k(\lambda)}{2\sqrt{\pi}}\frac{\hat{\phi}(\lambda)}{\sqrt{P_\phi(\lambda)}},$$

which in turn is implied by the condition $s_k(P_\phi)^{-1/2} \in Z$. Similarly, $M_k T^{(k)} \phi \in V_{\phi^\heartsuit}$ is implied by $s_k \sqrt{P_\phi} \in Z$ and both conditions are satisfied by our assumption on P_ϕ.
                                                       □

Let us return to our definition of a radial MRA for $\mathbb{R}^3$. We denote by

$$D_a f(r) = a^{-3/2} f(a^{-1} r)$$

the unitary dilation operator on $L^2(\mathcal{K})$. It is related to the Hankel transform by

$$(D_a f)^\wedge = D_{1/a} \hat{f}.$$

Now suppose we start with a function $\phi \in L^2(\mathcal{K})$ satisfying condition (RB) with Riesz constants $A, B > 0$. We define corresponding scale spaces $\{V_j\}_{j \in \mathbb{Z}}$ by

$$V_0 := V_\phi, \quad V_j := D_{2^{-j}} V_0.$$

Then in particular, the spaces V_j, $j \in \mathbb{Z}$, satisfy axiom (4) of Definition 3.1.1. Furthermore, we set

$$\phi_{j,k}(r) := D_{2^{-j}} \big(M_k T^{(k)} \phi \big)(r) = 8^{j/2} M_k(T^{(k)} \phi)(2^j r) \quad j \in \mathbb{Z}, k \in \mathbb{N}. \tag{3.8}$$

Then $\langle \phi_{j,k}, \phi_{j,l} \rangle = \langle \phi_{0,k}, \phi_{0,l} \rangle$ for all j, k, l. Thus, $\{\phi_{j,k}, k \in \mathbb{N}\}$ forms a Riesz basis of V_j with the same Riesz constants A, B as for $j = 0$. In particular, it holds

$$V_j = \overline{\operatorname{span}\{\phi_{j,k}, k \in \mathbb{N}\}}.$$

Moreover, if $B_\phi = \{\phi_{0,k} : k \in \mathbb{N}\}$ is an orthonormal basis for V_0, then $\{\phi_{j,k}, k \in \mathbb{N}\}$ is an orthonormal basis of V_j.

Recall now axiom (1) of Definition 3.1.1, which requires that the V_j, $j \in \mathbb{Z}$, are nested. As in the classical case, this condition can be reformulated in terms of a two-scale relation for ϕ.

Proposition 3.1.6. *For ϕ satisfying (RB) and $\{V_j\}_{j \in \mathbb{Z}}$ as above, the following statements are equivalent.*

(1) $V_j \subseteq V_{j+1}$ *for all* $j \in \mathbb{Z}$.

(2) $V_{-1} \subseteq V_0$.

(3) *There exists a function* $\gamma \in Z$ *such that*

$$\sin(2\pi\lambda)\widehat{\phi}(2\lambda) = \gamma(\lambda)\widehat{\phi}(\lambda). \tag{3.9}$$

In this case, the coefficients $(h_k)_{k \in \mathbb{N}} \in l^2(\mathbb{N})$ in the two-scale relation

$$\phi_{-1,1} = \sum_{k=1}^{\infty} h_k \phi_{0,k} \qquad (3.10)$$

are the coefficients in the Fourier sine series of $\gamma \in Z$:

$$\gamma = \frac{1}{2} \sum_{k=1}^{\infty} h_k s_k.$$

Proof: Rescaling by the factor 2^j shows that (1) and (2) are equivalent. For (2), we need at least $\phi_{-1,1} \in V_0$. According to Lemma 3.1.3 this is equivalent to the existence of a function $\beta \in Z$ such that for almost all λ,

$$\sqrt{8}\,\rho_1(2\lambda)\widehat{\phi}(2\lambda) = \widehat{\phi_{-1,1}}(\lambda) = \frac{\beta(\lambda)}{\lambda}\widehat{\phi}(\lambda). \qquad (3.11)$$

Setting $\gamma(\lambda) = \sqrt{\pi}\beta(\lambda)$ we obtain the two-scale relation (3.9). Moreover, if $\phi_{-1,1}$ has the expansion (3.10), then $\beta = (2\sqrt{\pi})^{-1}\sum_{k=1}^{\infty} h_k s_k$ by Lemma 3.1.3. This gives the stated connection between γ and $\phi_{-1,1}$.

It remains to show that $\phi_{-1,1} \in V_0$ (or equivalently, relation (3.9)) already implies that $\phi_{-1,k} \in V_0$ for all $k \in \mathbb{N}$. As above, the latter is equivalent to

$$\sin(2k\pi\lambda)\widehat{\phi}(2\lambda) = k\gamma_k(\lambda)\widehat{\phi}(\lambda) \qquad (3.12)$$

with $\gamma_k \in Z$. The relation between γ_k and $\phi_{-1,k}$ is now given by

$$\gamma_k = \frac{1}{2k}\sum_{l=1}^{\infty} h_l^{(k)} s_k, \qquad \phi_{-1,k} = \sum_{l=1}^{\infty} h_l^{(k)} \phi_{0,l}.$$

Comparison of (3.9) with (3.12) yields

$$\gamma_k(\lambda) = \gamma(\lambda)\frac{\sin(2k\pi\lambda)}{k\sin(2\pi\lambda)} = \gamma(\lambda)U_{k-1}(\cos 2\pi\lambda), \qquad (3.13)$$

where

$$U_k(x) = \frac{\sin(k+1)t}{(k+1)\sin t}, \qquad x = \cos t,$$

denotes the k-th Chebychev polynomial of the second kind, normalized such that $U_k(1) = 1$. Thus, given $\gamma \in Z$ we define

$$\gamma_k(\lambda) := \gamma(\lambda)U_{k-1}(\cos 2\pi\lambda).$$

As U_{k-1} is bounded on $[-1,1]$ and $\lambda \mapsto U_{k-1}(\cos 2\pi\lambda)$ is even the function γ_k is contained in Z as well and, hence, $\phi_{-1,k} \in V_0$. $\qquad \square$

Let us now consider the remaining axioms (2) and (3) of a radial multiresolution analysis.

Theorem 3.1.7. *Let* $\phi \in L^2(\mathcal{K})$ *satisfy condition (RB) and assume that the scale spaces*

$$V_j = \overline{\text{span}\{\phi_{j,k} : k \in \mathbb{N}\}}, \quad j \in \mathbb{Z},$$

satisfy $V_{-1} \subseteq V_0$. *Suppose further that* $|\widehat{\phi}|$ *is continuous at* 0. *Then* $\{V_j\}_{j \in \mathbb{Z}}$ *is a radial MRA if and only if* $\widehat{\phi}(0) \neq 0$. *Moreover,* ϕ *is an orthonormal scaling function if and only if* $|\widehat{\phi}(0)| = 1$.

We remark that continuity of $\widehat{\phi}$ at 0 is, for instance, guaranteed if $\phi \in L^2(\mathcal{K}) \cap L^1(\mathcal{K})$.

Proof: We have to check axioms (2) and (3). This may be done by slight modifications of standard arguments in the affine case. The condition on $\widehat{\phi}$ at 0 will be needed only for (3). We define an orthonormal scaling function $\phi^\heartsuit$ according to Theorem 3.1.5. The orthogonal projection P_j of $L^2(\mathcal{K})$ onto V_j is then given by

$$P_j f = \sum_{k=1}^{\infty} \langle f, \phi_{j,k}^\heartsuit \rangle \phi_{j,k}^\heartsuit$$

where the functions $\phi_{j,k}^\heartsuit$, $j \in \mathbb{Z}, k \in \mathbb{N}$, are defined as in (3.8). For (2) we need to show that $\lim_{j \to -\infty} \|P_j f\|_2 = 0$ for all $f \in L^2(\mathcal{K})$. Since functions with compact support are dense in $L^2(\mathcal{K})$, we may assume that $\text{supp} f$ is contained in a compact interval $[0, R]$. Parseval's equation implies

$$\|P_j f\|_2^2 = \sum_{k=1}^{\infty} |\langle f, \phi_{j,k}^\heartsuit \rangle|^2 \leq \sum_{k=1}^{\infty} \left| \int_0^R f(r) \overline{\phi_{j,k}^\heartsuit(r)} d\omega(r) \right|^2$$

$$\leq \|f\|_2^2 \int_0^R \sum_{k=1}^{\infty} |\phi_{j,k}^\heartsuit(r)|^2 d\omega(r) = \|f\|_2^2 \int_0^{2^j R} \sum_{k=1}^{\infty} |\phi_{0,k}^\heartsuit(r)|^2 d\omega(r).$$

Using the explicit formula (3.1) for the generalized translation we further deduce

$$\sum_{k=1}^{\infty} |\phi_{0,k}^\heartsuit(r)|^2 = \sum_{k=1}^{\infty} |M_k T^{(k)} \phi^\heartsuit(r)|^2 = \sum_{k=1}^{\infty} \left| \frac{\sqrt{2\pi}}{r} \int_{|k/2-r|}^{k/2+r} \phi^\heartsuit(t) t \, dt \right|^2.$$

Now assume that j is sufficiently small so that $2^j R < 1/4$. Then for $r \in [0, 2^j R]$, the integration domains $[k/2 - r, k/2 + r]$ do not overlap and we obtain

$$\sum_{k=1}^{\infty} |\phi_{0,k}^\heartsuit(r)|^2 \leq \frac{C}{r} \sum_{k=1}^{\infty} \int_{k/2-r}^{k/2+r} |\phi^\heartsuit(t)|^2 t^2 dt \leq \frac{C'}{r} \|\phi^\heartsuit\|_2^2$$

with suitable constants $C, C' > 0$ independent of j. Hence, for j sufficiently small we get

$$\|P_j f\|_2^2 \leq C'' \int_0^{2^j R} \frac{1}{r} d\omega(r),$$

which tends to 0 as $j \to -\infty$. This proves (2).

As to (3), suppose first that $\widehat{\phi}(0) \neq 0$ and let $h \in \left(\cup_{j=-\infty}^{\infty} V_j \right)^{\perp}$, i.e., $P_j h = 0$ for all $j \in \mathbb{Z}$. We claim that $h = 0$. Indeed, for $\epsilon > 0$ there exists a function $f \in L^2(\mathcal{K})$ such that the support of its Hankel transform $\widehat{f}$ is compact and $\|f - h\|_2 \leq \epsilon$. This implies

$$\|P_j f\|_2 = \|P_j(f - h)\|_2 \leq \epsilon \quad \text{for all } j \in \mathbb{Z}.$$

By the Riesz basis assumption on ϕ, we further have

$$A \sum_{k=1}^{\infty} |\langle f, \phi_{j,k} \rangle|^2 \leq \|P_j f\|_2^2 \leq B \sum_{k=1}^{\infty} |\langle f, \phi_{j,k} \rangle|^2, \tag{3.14}$$

see e.g. [Woj97, Lemma 2.7] or [Chr03, Theorem 5.4.1]. Further, if $\operatorname{supp} \widehat{f} \in [0, R]$ then it holds

$$\langle f, \phi_{j,k} \rangle = \langle \widehat{f}, \widehat{\phi}_{j,k} \rangle = \int_0^R \widehat{f}(\lambda) \rho_k^{(j)}(\lambda) \overline{\widehat{\phi}(2^{-j}\lambda)} \, d\omega(\lambda)$$

where

$$\rho_k^{(j)} := D_{2^j} \rho_k.$$

Note that the functions $\{\rho_k^{(j)}, k \in \mathbb{N}\}$ form an orthonormal basis of $L^2([0, 2^j], \omega|_{[0,2^j]}) =: Z_j$. Suppose now that j is sufficiently large, i.e., $2^j \geq R$. Then we have

$$\langle f, \phi_{j,k} \rangle = \langle \widehat{f} \, \overline{\widehat{\phi}(2^{-j} \cdot)}, \rho_k^{(j)} \rangle_{Z_j}.$$

Thus, Parseval's equation for Z_j yields

$$\sum_{k=1}^{\infty} |\langle f, \phi_{j,k} \rangle|^2 = \left\| \widehat{f} \, \overline{\widehat{\phi}(2^{-j} \cdot)} | Z_j \right\|^2 = \int_0^R |\widehat{f}(\lambda)|^2 |\widehat{\phi}(2^{-j}\lambda)|^2 d\omega(\lambda).$$

As $|\widehat{\phi}|$ is assumed to be continuous at 0, the functions $\lambda \to |\widehat{\phi}(2^{-j}\lambda)|$ converge to the constant $|\widehat{\phi}(0)| > 0$ uniformly on $[0, R]$ as $j \to \infty$. We conclude

$$\epsilon \geq \limsup_{j \to \infty} \|P_j f\|_2 \geq \sqrt{A} \, |\widehat{\phi}(0)| \|\widehat{f}\|_2 \geq \sqrt{A} \, |\widehat{\phi}(0)| (\|h\|_2 - \epsilon).$$

Since ϵ is arbitrarily small, this shows that $h = 0$ and, hence, axiom (3) is satisfied. Vice versa, axiom (3) implies that

$$\lim_{j \to \infty} P_j f = f \quad \text{for all } f \in L^2(\mathcal{K}).$$

If $\widehat{f}$ is compactly supported, then the same calculation as above shows that

$$\lim_{j \to \infty} \|P_j f\|_2 \leq \sqrt{B} \, |\widehat{\phi}(0)| \, \|\widehat{f}\|_2,$$

which forces $\widehat{\phi}(0) \neq 0$.

If the functions $\phi_{j,k}$, $k \in \mathbb{N}$, are orthonormal then we may choose $A = B = 1$ in (3.14). The converse is also true. Indeed, by assumption the $\phi_{j,k}$, $k \in \mathbb{N}$, form a Riesz basis of V_j, in particular an exact frame, whose frame operator is the identity if $A = B = 1$ (see e.g. [HW89, Theorem 2.1.3]). It then follows from Corollary 2.1.7 in [HW89] that the $\phi_{j,k}, k \in \mathbb{N}$, are orthonormal. (Alternatively, the claim follows from Theorem 5.4.1 in [Chr03], which states that the Riesz constants coincide with the frame constants. Hence, by $A = B = 1$ the $\phi_{j,k}$ must be orthonormal.) Furthermore, in case $A = B = 1$ we obtain (for f as just before)

$$\lim_{j \to \infty} \|P_j f\|_2 = \left|\widehat{\phi}(0)\right| \|\widehat{f}\|_2 = \left|\widehat{\phi}(0)\right| \|f\|_2.$$

Thus, (3) is satisfied exactly if $\left|\widehat{\phi}(0)\right| = 1$. $\square$

Let us now write the two scale relation (3.9) in a slightly different form, namely

$$\widehat{\phi}(2\lambda) = G(\lambda)\,\widehat{\phi}(\lambda) \tag{3.15}$$

with

$$G(\lambda) := \frac{\gamma(\lambda)}{\sin(2\pi\lambda)}.$$

The filter function G is obviously 2-periodic and even. As for a classical multiresolution analysis one proves the following.

Lemma 3.1.8. *Suppose that $\phi \in L^2(\mathcal{K})$ is an orthonormal scaling function of a radial MRA. Then the associated filter function G satisfies*

$$|G(\lambda)|^2 + |G(\lambda+1)|^2 = 1 \quad a.e.. \tag{3.16}$$

Consequently, G is essentially bounded and contained in $L^2[0,1]$, which allows to expand it into a cosine series,

$$G(\lambda) = \sqrt{2} \sum_{n=0}^{\infty} g_n \cos(n\pi\lambda).$$

If, in addition, $\phi \in L^1(\mathcal{K})$ then (3.16) holds pointwise and

$$G(0) = 1, \ G(1) = 0$$

which implies

$$\sqrt{2} \sum_{n=0}^{\infty} g_n = 1, \quad \sqrt{2} \sum_{n=0}^{\infty} (-1)^n g_n = 0.$$

Proof: In view of Corollary 3.1.2 we have

$$
\begin{aligned}
1 &= \sum_{n=-\infty}^{\infty} |\widehat{\phi}(\lambda + 2n)|^2 = \sum_{n=-\infty}^{\infty} |G(\lambda/2 + n)|^2 \, |\widehat{\phi}(\lambda/2 + n)|^2 \\
&= |G(\lambda/2)|^2 \sum_{n=-\infty}^{\infty} |\widehat{\phi}(\lambda/2 + 2n)|^2 + |G(\lambda/2 + 1)|^2 \sum_{n=-\infty}^{\infty} |\widehat{\phi}(\lambda/2 + 2n + 1)|^2 \\
&= |G(\lambda/2)|^2 + |G(\lambda/2 + 1)|^2
\end{aligned}
$$

almost everywhere. If $\phi \in L^1(\mathcal{K})$ then $\widehat{\phi}$ is continuous and $\widehat{\phi}(0) \neq 0$ by Theorem 3.1.7. Hence, it then holds $G(0) = 1$ by (3.15) and $G(1) = 0$ is an immediate consequence of (3.16). $\qquad\square$

3.2   Orthogonal Radial Wavelets

In this section we construct wavelets for a given radial MRA $\{V_j\}_{j \in \mathbb{Z}}$ in $\mathbb{R}^3$ with **orthonormal** scaling function ϕ and filter function G. As usual, the wavelet space W_j is defined as the orthogonal complement of V_j in V_{j+1},

$$ W_j := V_{j+1} \ominus V_j . $$

Thus $L^2(\mathcal{K})$ decomposes as an orthogonal Hilbert sum

$$ L^2(\mathcal{K}) = \bigoplus_{j=-\infty}^{\infty} W_j. $$

Recall the definition of Z in Section 4 and the characterization of $V_0 = V_\phi$ according to Lemma 3.1.3,

$$ f \in V_0 \iff \widehat{f}(\lambda) = \frac{\beta(\lambda)}{\lambda} \, \widehat{\phi}(\lambda) \quad \text{with } \beta \in Z. \tag{3.17} $$

We define

$$ Z_0 := \{ \alpha \in Z, \; \alpha(\lambda + 1) = -\alpha(\lambda) \text{ for almost all } \lambda \}, $$

which is a closed subspace of Z with respect to $\| \cdot |L^2[0,1]| \|$. Then $W_{-1} = V_0 \ominus V_{-1}$ is characterized as follows.

Proposition 3.2.1. *Let ϕ be an orthonormal radial scaling function with corresponding filter function G. Then the mapping $J_\phi : Z_0 \to W_{-1}$ defined by*

$$ \widehat{J_\phi(\alpha)}(\lambda) := (2\sqrt{\pi})^{-1} \frac{\alpha(\lambda)}{\lambda} \, \overline{G(\lambda + 1)} \, \widehat{\phi}(\lambda) \tag{3.18} $$

is an isometric isomorphism.

Proof: Let us first calculate the norm of $J_\phi(\alpha)$, $\alpha \in Z_0$, as follows

$$\|J_\phi(\alpha)\|_2^2 = \|\widehat{J_\phi(\alpha)}\|_2^2 = (4\pi)^{-1} \int_0^\infty \frac{|\alpha(\lambda)|^2}{\lambda^2} |G(\lambda+1)|^2 |\widehat{\phi}(\lambda)|^2 d\omega(\lambda)$$

$$= \int_{-\infty}^\infty |\alpha(\lambda)|^2 |G(\lambda+1)|^2 |\widehat{\phi}(\lambda)|^2 d\lambda = \int_{-1}^1 |\alpha(\lambda)|^2 |G(\lambda+1)|^2 d\lambda,$$

where we used that α and G are 2-periodic and ϕ is an orthonormal scaling function implying $P_\phi \equiv 1$ by Corollary 3.1.2. By assumption on α, we have $\alpha(\lambda-1) = \alpha(\lambda+1) = -\alpha(\lambda)$. Thus we obtain by Lemma 3.1.8

$$\int_{-1}^1 |\alpha(\lambda)|^2 |G(\lambda+1)|^2 d\lambda = \int_0^1 |\alpha(\lambda-1)|^2 |G(\lambda)|^2 d\lambda + \int_0^1 |\alpha(\lambda)|^2 |G(\lambda+1)|^2 d\lambda$$

$$= \int_0^1 |\alpha(\lambda)|^2 d\lambda = \|\alpha\|_2^2.$$

This proves that J_ϕ is isometric from Z_0 into $L^2(\mathcal{K})$.

In order to show that this mapping is actually an isomorphism onto W_{-1} notice first that the Hankel transform is a unitary isomorphism of $L^2(\mathcal{K})$, so $\widehat{W}_{-1} = \widehat{V}_0 \ominus \widehat{V}_{-1}$. Rescaling of (3.17) by the factor 2 and relation (3.15) imply that $h \in L^2(\mathcal{K})$ is contained in $\widehat{V}_{-1}$ if and only if there exists some $\widetilde{\beta} \in Z$ such that

$$h(\lambda) = \frac{\widetilde{\beta}(2\lambda)}{\lambda} \widehat{\phi}(2\lambda) = \frac{\widetilde{\beta}(2\lambda)}{\lambda} G(\lambda) \widehat{\phi}(\lambda).$$

Thus $\beta \in Z$ corresponds to $f \in W_{-1}$ according to (3.17) if and only if

$$\int_0^\infty \frac{\widetilde{\beta}(2\lambda)}{\lambda} G(\lambda) \frac{\overline{\beta(\lambda)}}{\lambda} |\widehat{\phi}(\lambda)|^2 d\omega(\lambda) = 0 \quad \text{for all } \widetilde{\beta} \in Z.$$

Up to a constant factor, the integral on the left equals

$$\int_{-\infty}^\infty \widetilde{\beta}(2\lambda) G(\lambda) \overline{\beta(\lambda)} |\widehat{\phi}(\lambda)|^2 d\lambda = \sum_{n\in\mathbb{Z}} \int_0^1 \widetilde{\beta}(2\lambda) \overline{\beta(\lambda+2n)} G(\lambda+2n) |\widehat{\phi}(\lambda+2n)|^2 d\lambda$$

$$+ \sum_{n\in\mathbb{Z}} \int_0^1 \widetilde{\beta}(2\lambda) \overline{\beta(\lambda+2n+1)} G(\lambda+2n+1) |\widehat{\phi}(\lambda+2n+1)|^2 d\lambda$$

$$= \int_0^1 \widetilde{\beta}(2\lambda) \big(\overline{\beta(\lambda)} G(\lambda) + \overline{\beta(\lambda+1)} G(\lambda+1) \big) d\lambda,$$

where we used the periodicity and symmetry properties of $\beta, \widetilde{\beta}$ and G as well as Corollary 3.1.2. Since $\widetilde{\beta} \in Z$ is arbitrary, we conclude that the vectors $(\beta(\lambda), \beta(\lambda+1))^T$ and

$(G(\lambda), G(\lambda+1))^T$ must be orthogonal in $\mathbb{C}^2$ for almost all λ. This means that

$$\begin{pmatrix} \beta(\lambda) \\ \beta(\lambda+1) \end{pmatrix} = \alpha(\lambda) \begin{pmatrix} \overline{G(\lambda+1)} \\ -\overline{G(\lambda)} \end{pmatrix} \tag{3.19}$$

for some function $\alpha : [0,1] \to \mathbb{C}$. Taking the norm in $\mathbb{C}^2$ on both sides of (3.19) and using (3.15) yields

$$|\beta(\lambda)|^2 + |\beta(\lambda+1)|^2 = |\alpha(\lambda)|^2,$$

which by $\beta \in Z$ implies that α belongs to $L^2[0,1]$. Since β and G are 2-periodic on $\mathbb{R}$, β is odd and G is even, an extension of α to $\mathbb{R}$ must be 2-periodic and odd. Hence, we conclude $\alpha \in Z$ with $\beta(\lambda) = \alpha(\lambda)\overline{G(\lambda+1)}$. Using the 2-periodicity of β and G in the second component of (3.19), we further deduce that $\beta(\lambda) = -\alpha(\lambda+1)\overline{G(\lambda+1)}$ and therefore $\alpha(\lambda+1) = -\alpha(\lambda)$ apart from the zero-set of $G(\lambda+1)$. If $G(\lambda+1) = 0$ we simply define $\alpha(\lambda+1) := -\alpha(\lambda)$, which clearly does not affect (3.18). Moreover, in this case $\alpha(\lambda)$ is well-defined, since $G(\lambda) = 1$ by (3.15) except possibly on a null-set. Thus $\widehat{f}$ is of the claimed form. Conversely, if $\alpha \in Z_0$, then $\beta(\lambda) := \alpha(\lambda)\overline{G(\lambda+1)} \in Z$, and (3.19) is satisfied. Thus, we finally proved that J_ϕ is an isomorphism. $\qquad\square$

It is now easy to obtain an orthonormal basis of W_{-1}. Recall that the functions s_k, $k \in \mathbb{N}$, form an orthonormal basis of Z. Moreover, $\{s_{2k-1}, \ k \in \mathbb{N}\}$ is an orthonormal basis of Z_0. Thus by the previous result, the functions

$$e_k := J_\phi(s_{2k-1}), \quad k \in \mathbb{N},$$

constitute an orthonormal basis of W_{-1}. Define $\psi \in L^2(\mathcal{K})$ by

$$\widehat{\psi}(2\lambda) = \overline{G(\lambda+1)}\,\widehat{\phi}(\lambda). \tag{3.20}$$

In view of (3.18) we have

$$\widehat{e}_k(\lambda) = \rho_{2k-1}(\lambda)\widehat{\psi}(2\lambda)$$

and together with (3.3) we conclude

$$e_k = \frac{M_{2k-1}}{2} T^{(2k-1)} D_2 \psi.$$

To obtain an orthonormal basis of W_0, we just have to rescale. Extending the notation $T^{(r)} := \tau_{r/2}$ to $r \in \mathbb{N}/2$ and using the relation $D_a \tau_x = \tau_{ax} D_a$, see (2.13), we obtain that an orthonormal basis of W_0 is given by the functions

$$\psi_k := D_{1/2} e_k = \frac{M_{2k-1}}{2} T^{((2k-1)/2)} \psi, \quad k \in \mathbb{N}.$$

We call ψ a wavelet for the radial multiresolution $(V_j)_{j \in \mathbb{Z}}$.

Definition 3.2.1. For $j \in \mathbb{Z}$ and $k \in \mathbb{N}$, we define the "radial" wavelets

$$\psi_{j,k}(r) := D_{2^{-j}}\psi_k(r) = 8^{j/2}\frac{M_{2k-1}}{2}T^{((2k-1)/2)}\psi(2^j r).$$

We state in a theorem what we have just proven.

Theorem 3.2.2. (i) *For each $j \in \mathbb{Z}$, the set $\{\psi_{j,k} : k \in \mathbb{N}\}$ constitutes an orthonormal basis of W_j.*

(ii) *The set $\{\psi_{j,k} : j \in \mathbb{Z}, k \in \mathbb{N}\}$ forms an orthonormal wavelet basis of $L^2(\mathcal{K})$.*

Corollary 3.2.3. *The functions*

$$\Psi_{j,k}(x) := \psi_{j,k}(|x|), \quad x \in \mathbb{R}^3, j \in \mathbb{Z}, k \in \mathbb{N}$$

form an orthonormal basis for the space $L^2_{rad}(\mathbb{R}^3)$.

3.3 Construction of Radial Scaling Functions and Wavelets

We still do not have particular examples of radial scaling functions and radial wavelets at hand. The analogy of our constructions to those on $\mathbb{R}$, however, leads to the following close relationship to classical scaling functions.

Theorem 3.3.1. *Suppose $\phi_{\mathbb{R}}$ is a classical scaling function on $\mathbb{R}$ which is even and such that its (classical, one-dimensional) Fourier transform $\mathcal{F}(\phi_{\mathbb{R}})$ is continuous at 0 and satisfies $\mathcal{F}(\phi_{\mathbb{R}}) \in L^2(\mathcal{K})$. Define $\phi \in L^2(\mathcal{K})$ via its Hankel transform,*

$$\widehat{\phi}(\lambda) := \mathcal{F}(\phi_{\mathbb{R}})(\lambda/2). \tag{3.21}$$

Then ϕ is a radial scaling function.

Conversely, if ϕ is a scaling function for a radial MRA such that $\widehat{\phi}$ is continuous at 0, then $\widehat{\phi} \in L^2(\mathbb{R})$ and the function $\phi_{\mathbb{R}}$ defined by (3.21) (where $\widehat{\phi}$ is extended to an even function on $\mathbb{R}$) is a classical scaling function on $\mathbb{R}$.

Moreover, ϕ is an orthonormal radial scaling function if and only if $\phi_{\mathbb{R}}$ is an orthonormal classical scaling function.

Proof: Let us start with the first assertion. As $\phi_{\mathbb{R}}$ is a classical scaling function, we have by equation (5.3.2) in [Dau92]

$$A \leq \sum_{k \in \mathbb{Z}} |\mathcal{F}(\phi_{\mathbb{R}})(\xi + k)|^2 \leq B \quad \text{a.e.}$$

with suitable constants $0 < A \leq B < \infty$. (Note that constants differ from [Dau92] due to our normalization of the Fourier transform.) Moreover, $\phi_{\mathbb{R}}$ is orthonormal if and only if $A = B = 1$. Since $\phi_{\mathbb{R}}$ is assumed to be even, definition (3.21) is compatible with the even extension of $\widehat{\phi}$. By Proposition 3.1.1, the set $\{M_k T^{(k)} \phi, \, k \in \mathbb{N}\}$ forms a Riesz basis for $V_0 = \overline{\mathrm{span}\{B_\phi\}}$, which is an orthonormal basis if and only if $\phi_{\mathbb{R}}$ is orthonormal, see Corollary 3.1.2. Moreover, by equation (5.3.18) in [Dau92], there exists a 1-periodic function $m_0 \in L^2([-1/2, 1/2])$ such that $\mathcal{F}(\phi_{\mathbb{R}})(\xi) = m_0(\xi/2)\mathcal{F}(\phi_{\mathbb{R}})(\xi/2)$, and m_0 is necessarily even in our case. Hence, with $\gamma(\lambda) := m_0(\lambda/2)\sin(2\pi\lambda)$, which clearly is contained in Z, we have $\sin(2\pi\lambda)\widehat{\phi}(2\lambda) = \gamma(\lambda)\widehat{\phi}(\lambda)$. This is exactly the radial two-scale equation (3.9). Since $\widehat{\phi}$ is continuous at 0, the condition $\widehat{\phi}(0) \neq 0$ of Theorem 3.1.7 is automatically satisfied (see e.g. Remark 3 on p.144 in [Dau92]) and, thus, we finally obtain that ϕ is a radial scaling function.

For the converse part notice first that continuity of $\widehat{\phi}$ at 0 already implies that $\widehat{\phi} \in L^2(\mathbb{R})$. We further proceed similar as before, using Proposition 5.3.1 and Proposition 5.3.2 in [Dau92] and the corresponding results presented in this chapter. Hereby, it is important to note that the filter function of $\phi_{\mathbb{R}}$ is given by

$$m_0(\xi) := \frac{\gamma(2\xi)}{\sin(4\pi\xi)} = G(2\xi).$$

By Lemma 3.1.8 m_0 is contained in $L^2([-1/2, 1/2])$.   □

This theorem supplies a variety of radial scaling functions since there are many classical scaling functions on $\mathbb{R}$ which satisfy the assumptions of the theorem. However, concerning orthonormal radial scaling functions with compact support, a famous theorem of Daubechies implies the following negative result.

Corollary 3.3.2. *There do not exist any real-valued orthonormal radial scaling functions with compact support.*

Proof: The proof of Theorem 8.1.4 in [Dau92] shows that an even, real-valued and compactly supported scaling function is necessarily the Haar function $\chi_{[-1/2,1/2]}$, i.e., the characteristic function of the interval $[-1/2, 1/2]$. However, its Fourier transform $\mathcal{F}(\chi_{[-1/2,1/2]})(\xi) = \frac{\sin(\pi\xi)}{\pi\xi}$ is not contained in $L^2(\mathcal{K})$.   □

If ϕ corresponds to an even classical scaling function $\phi_{\mathbb{R}}$ according to Theorem 3.3.1, then by (1.46) the generalized translates $\phi_{0,k} = M_k \tau_{k/2} \phi$ may be expressed according to the formula

$$\phi_{0,k}(r) = \frac{1}{\sqrt{8\pi}r}\left(\phi_{\mathbb{R}}(2r - k) - \phi_{\mathbb{R}}(2r + k)\right). \tag{3.22}$$

As an example, we consider the radial analogue of the Shannon wavelets. We define the scaling function via its Hankel transform,

$$\widehat{\phi}(\lambda) = \chi_{[0,1]}(\lambda), \qquad \phi(r) = \frac{\sin(2\pi r)}{2\pi^2 r^3} - \frac{\cos(2\pi r)}{\pi r^2}.$$

Constructing the associated wavelet according to formula (3.20) yields (after a short calculation)

$$\widehat{\psi}(\lambda) = \chi_{[1,2]}(\lambda), \qquad \psi(r) = \frac{\sin(4\pi r) - \sin(2\pi r) - 4\pi r \cos(4\pi r) + 2\pi r \cos(2\pi r)}{2\pi^2 r^3}.$$

The translates of the scaling function and the wavelet turn out to be

$$\phi_{0,k}(r) = \frac{1}{\sqrt{8\pi r}} \left(\frac{\sin \pi(2r - k)}{\pi(2r - k)} - \frac{\sin \pi(2r + k)}{\pi(2r + k)} \right), \quad k \in \mathbb{N},$$

$$\psi_{0,k}(r) = \frac{1}{\sqrt{8\pi r}} \left(p(2r - k) - p(2r + k) \right), \quad k \in \mathbb{N},$$

with $p(x) = \dfrac{\sin(2\pi x) - \sin(\pi x)}{\pi x}$.

3.4   Algorithms

For the use of our radial multiresolution in applications we need to formulate decomposition and reconstruction algorithms. The first step in such an algorithm consists of projecting the function f into a scale space V_j for some suitable j. We obtain a representation

$$P_j f = \sum_{k=1}^{\infty} c_k^{(j)} \phi_{j,k}.$$

We will discuss below, how to approximate the coefficients $c_k^{(j)}$.

So from now on we assume that we have given a function $f \in V_j$ in terms of its coefficients $c_k^{(j)}$. The decomposition algorithm consists of decomposing f into V_{j-1} and W_{j-1}, i.e., of calculating the coefficients $c_k^{(j-1)}$ and $d_k^{(j-1)}$ in the representation

$$f = \sum_{k=1}^{\infty} c_k^{(j-1)} \phi_{j-1,k} + \sum_{k=1}^{\infty} d_k^{(j-1)} \psi_{j-1,k}. \tag{3.23}$$

Such a representation exists, since by construction $\{\phi_{j-1,k}, \psi_{j-1,k}, k \in \mathbb{N}\}$ is also a basis of V_j. A reconstruction algorithm determines the coefficients $c_k^{(j)}$ when f is given in terms of $c_k^{(j-1)}$ and $d_k^{(j-1)}$, $k \in \mathbb{N}$.

We still assume that ϕ is an orthonormal scaling function (and, hence, ψ is an orthonormal wavelet). Let

$$q_\ell^{(k)} := \langle \phi_{1,k}, \phi_{0,l} \rangle = \langle \phi_{j,k}, \phi_{j-1,l} \rangle, \qquad r_\ell^{(k)} := \langle \phi_{1,k}, \psi_{0,l} \rangle = \langle \phi_{j,k}, \psi_{j-1,l} \rangle, \qquad k, \ell \in \mathbb{N}.$$

By Parseval it holds

$$\phi_{j,k} = \sum_{\ell=1}^{\infty} q_\ell^{(k)} \phi_{j-1,\ell} + \sum_{\ell=1}^{\infty} r_\ell^{(k)} \psi_{j-1,\ell} \qquad (3.24)$$

and

$$\phi_{j-1,\ell} = \sum_{k=1}^{\infty} \overline{q_\ell^{(k)}} \phi_{j,k}, \qquad \psi_{j-1,\ell} = \sum_{k=1}^{\infty} \overline{r_\ell^{(k)}} \phi_{j,k}. \qquad (3.25)$$

With (3.24) we obtain

$$f = \sum_{k=1}^{\infty} c_k^{(j)} \phi_{j,k} = \sum_{\ell=1}^{\infty} \left(\sum_{k=1}^{\infty} c_k^{(j)} q_\ell^{(k)} \right) \phi_{j-1,\ell} + \sum_{\ell=1}^{\infty} \left(\sum_{k=1}^{\infty} c_k^{(j)} r_\ell^{(k)} \right) \psi_{j-1,\ell}.$$

Comparison with (3.23) yields the **decomposition formulae**

$$c_\ell^{(j-1)} = \sum_{k=1}^{\infty} c_k^{(j)} q_\ell^{(k)}, \qquad d_\ell^{(j-1)} = \sum_{k=1}^{\infty} c_k^{(j)} r_\ell^{(k)}.$$

Further, using (3.25) we obtain as well

$$f = \sum_{\ell=1}^{\infty} c_\ell^{(j-1)} \phi_{j-1,\ell} + \sum_{\ell=1}^{\infty} d_\ell^{(j-1)} \psi_{j-1,\ell} = \sum_{k=1}^{\infty} \sum_{\ell=1}^{\infty} \left(c_\ell^{(j-1)} \overline{q_\ell^{(k)}} + d_\ell^{(j-1)} \overline{r_\ell^{(k)}} \right) \phi_{j,k}.$$

Again by comparison of coefficients we get the **reconstruction formula**

$$c_k^{(j)} = \sum_{\ell=1}^{\infty} c_\ell^{(j-1)} \overline{q_\ell^{(k)}} + \sum_{\ell=1}^{\infty} d_\ell^{(j-1)} \overline{r_\ell^{(k)}}.$$

It turns out that the coefficients $q_\ell^{(k)}$ and $r_\ell^{(k)}$ are determined in terms of the numbers g_n in the cosine expansion of G, i.e., the coefficients in

$$G(\lambda) = \sqrt{2} \sum_{n=0}^{\infty} g_n \cos(n\pi\lambda).$$

Theorem 3.4.1. *For $\ell, k \in \mathbb{N}$ it holds*

$$q_\ell^{(k)} = \begin{cases} \overline{g_{k-2\ell}} - g_{2\ell+k} & \text{for } 2\ell < k, \\ 2g_0 - g_{4\ell} & \text{for } 2\ell = k, \\ \overline{g_{2\ell-k}} - g_{2\ell+k} & \text{for } 2\ell > k, \end{cases}$$

$$r_\ell^{(k)} = \begin{cases} (-1)^{k-1}(\overline{g_{k-2\ell+1}} - g_{k+2\ell-1}) & \text{for } 2\ell - 1 < k, \\ 2g_0 - g_{4\ell-2} & \text{for } 2\ell - 1 = k, \\ (-1)^{k-1}(\overline{g_{2\ell-1-k}} - g_{2\ell-1+k}) & \text{for } 2\ell - 1 > k. \end{cases}$$

Proof: Using the Plancherel theorem, relation (3.15) and Corollary (3.1.2) we obtain

$$q_\ell^{(k)} = \langle \phi_{1,k}, \phi_{0,\ell} \rangle = \langle \hat{\phi}_{1,k}, \hat{\phi}_{0,\ell} \rangle$$

$$= 8^{-1/2} \int_0^\infty \rho_k(\lambda/2) \rho_\ell(\lambda) \overline{G(\lambda/2)} |\hat{\phi}(\lambda/2)|^2 d\omega(\lambda)$$

$$= \sqrt{2} \int_0^1 s_k(\lambda) s_\ell(2\lambda) \overline{G(\lambda)} \sum_{n=-\infty}^\infty |\hat{\phi}(\lambda + 2n)|^2 d\lambda$$

$$= 4 \sum_{n=0}^\infty \overline{g_n} \int_0^1 \sin(k\lambda\pi) \sin(2\ell\lambda\pi) \cos(n\pi\lambda) d\lambda. \qquad (3.26)$$

An easy calculation using trigonometric identities shows

$$\int_0^1 \sin(k\lambda\pi) \sin(t\lambda\pi) \cos(n\pi\lambda) d\lambda = \begin{cases} \frac{1}{4}(\delta_{n,|t-k|} - \delta_{n,t+k}) & \text{for } n > 0, \\ \frac{1}{4}(2\delta_{0,t-k} - \delta_{0,t+k}) & \text{for } n = 0. \end{cases} \qquad (3.27)$$

Setting $t = 2\ell$ and inserting into (3.26) yields the assertion for $q_\ell^{(k)}$. We proceed similarly for $r_\ell^{(k)}$:

$$r_\ell^{(j)} = \langle \phi_{1,k}, \psi_{0,\ell} \rangle = \sqrt{2} \int_0^1 s_k(\lambda) s_{2\ell-1}(\lambda) G(\lambda + 1) d\lambda$$

$$= 4 \sum_{n=0}^\infty g_n \int_0^1 \sin(k\pi\lambda) \sin((2\ell - 1)\pi\lambda) \cos(n\pi(\lambda + 1)) d\lambda$$

$$= 4 \sum_{n=0}^\infty g_n (-1)^n \int_0^1 \sin(k\pi\lambda) \sin((2\ell - 1)\pi\lambda) \cos(n\pi\lambda) d\lambda.$$

Setting $t = 2\ell - 1$ in (3.27) and inserting into the last expression gives the result for $r_\ell^{(k)}$. $\qquad \square$

Let us consider the case where only finitely many coefficients g_k are different from zero. Although this is not possible for real-valued orthonormal scaling functions this assumption makes it easier to compare the radial wavelet algorithm with the classical one. Of course, in applications one can only handle finitely many coefficients anyway. So let us assume $\operatorname{supp} g \subset [0, N]$, i.e., $g_k = 0$ for $k \notin \{0, \ldots, N\}$. Elementary considerations show the following. Leaving k fixed yields

$$
\begin{array}{rcll}
q_\ell^{(k)} & = & 0 \quad \text{for } \ell \notin [\frac{k-N}{2}, \frac{k+N}{2}] & \text{if } k > N, \\
q_\ell^{(k)} & = & 0 \quad \text{for } \ell \notin [1, \frac{k+N}{2}] & \text{if } k \leq N, \\
r_\ell^{(k)} & = & 0 \quad \text{for } \ell \notin [\frac{k-N+1}{2}, \frac{N+k+1}{2}] & \text{if } k > N+1, \\
r_\ell^{(k)} & = & 0 \quad \text{for } \ell \notin [1, \frac{N+k+1}{2}] & \text{if } k \leq N+1.
\end{array}
$$

If ℓ is fixed then

$$
\begin{array}{rcll}
q_\ell^{(k)} & = & 0 \quad \text{for } k \notin [2\ell - N, 2\ell + N] & \text{if } 2\ell > N, \\
q_\ell^{(k)} & = & 0 \quad \text{for } k \notin [1, N + 2\ell] & \text{if } 2\ell \leq N, \\
r_\ell^{(k)} & = & 0 \quad \text{for } k \notin [2\ell - 1 - N, 2\ell - 1 + N] & \text{if } 2\ell - 1 > N, \\
r_\ell^{(k)} & = & 0 \quad \text{for } k \notin [1, 2\ell - 1 + N] & \text{if } 2\ell - 1 \leq N.
\end{array}
$$

With

$$
h_k := \begin{cases} g_{|k|} & \text{for } 1 \leq |k| \leq N, \\ 2g_0 & \text{for } k = 0, \\ 0 & \text{otherwise.} \end{cases}
$$

it holds $G(\lambda) = \frac{1}{\sqrt{2}} \sum_{k=-N}^{N} h_k e^{ik\pi\lambda}$. Because of the conditions on it, G is also the filter function for an ordinary multiresolution analysis on $\mathbb{R}$ with coefficients h_k. Now, if $2\ell > N$ resp. $2\ell - 1 > N$ then it is easy to see that

$$
\begin{aligned}
q_\ell^{(2\ell+k)} &= \overline{h_k} \qquad \text{for all } k \geq -2\ell + 1, \\
r_\ell^{(2\ell-1+k)} &= (-1)^k h_k \quad \text{for all } k \geq -2\ell + 2.
\end{aligned}
$$

Similarly, if $k > N$ then

$$
q_\ell^{(k)} = \overline{h_{k-2\ell}}, \qquad r_\ell^{(k)} = (-1)^{k+1} h_{k+1-2\ell} \qquad \text{for all } \ell \in \mathbb{N}.
$$

Hence, for $2\ell > N + 1$ the decomposition formulae become

$$
c_\ell^{(j-1)} = \sum_{k=1}^{\infty} c_k^{(j)} \overline{h_{k-2\ell}}, \qquad d_\ell^{(j-1)} = \sum_{k=1}^{\infty} c_k^{(j)} (-1)^{k+1} h_{k+1-2\ell},
$$

and for $k > N + 1$ the reconstruction formula is

$$
c_k^{(j)} = \sum_{\ell=1}^{\infty} c_\ell^{(j-1)} h_{k-2\ell} + \sum_{\ell=1}^{\infty} d_\ell^{(j-1)} (-1)^{k+1} \overline{h_{k+1-2\ell}}.
$$

These formulae are well-known. Indeed, they are the decomposition and reconstruction formulae of the classical discrete wavelet transform. So our approach leads to the classical algorithm if we are far enough away from the origin. If we are close to the origin we have derived an algorithm to handle the boundary point 0.

Let us finally discuss how to obtain an approximation of the coefficients $c_k^{(j)}$ representing $P_j f$, when the function f is given. We assume for the moment that the classical scaling function $\phi_{\mathbb{R}}$ related to ϕ as in Theorem 3.3.1 is interpolatory, i.e., $\phi_{\mathbb{R}}(k) = D\delta_{0,k}$ for $k \in \mathbb{Z}, D \neq 0$. For example the Shannon wavelet satisfies this property. According to (3.22) we have

$$P_j f(x) = \sum_{k=1}^{\infty} c_k^{(j)} \frac{8^{j/2}}{\sqrt{8\pi} 2^j x} \left(\phi_{\mathbb{R}}(2^{j+1}x - k) - \phi_{\mathbb{R}}(2^{j+1}x + k) \right).$$

Letting $x = 2^{-(j+1)}\ell, \ell \in \mathbb{N}$, and reordering yields

$$\sum_{k=1}^{\infty} c_k^{(j)} \phi_{\mathbb{R}}(\ell - k) = 2^{-1}\sqrt{8\pi} 8^{-j/2}\ell P_j f(2^{-(j+1)}\ell) + \sum_{l=1}^{\infty} c_k^{(j)} \phi_{\mathbb{R}}(\ell + k).$$

Using the interpolatory condition we obtain

$$c_\ell^{(j)} = \sqrt{2\pi} D^{-1} 8^{-j/2}\ell P_j f(2^{-(j+1)}\ell) \approx \sqrt{2\pi} D^{-1} 8^{-j/2}\ell f(2^{-(j+1)}\ell).$$

We note that by axiom (3) it is reasonable that $P_j f(x) \approx f(x)$ if j is large enough. So we suggest to use the formula

$$c_k^{(j)} = C 8^{-j/2} k f(2^{-(j+1)}k)$$

with $C = \sqrt{2\pi} D^{-1}$ as a heuristic approximation method. As in the classical case this should work also reasonable even if $\phi_{\mathbb{R}}$ is not interpolatory.

Chapter 4

Coorbit Space Theory

In this chapter we treat the general problem of discretizing the continuous transform $\tilde{V}_g$ of Chapter 2. One cannot expect that this problem can be solved in full generality with methods similar to those in the previous chapter. Instead we take an abstract approach based on coorbit space theory, which was developed by Feichtinger and Gröchenig in the late 1980's [FG88, FG89a, FG89b, FG92c, Grö91, Grö88]. It is also known under the name of Feichtinger-Gröchenig theory.

The main ingredient in coorbit space theory is the reproducing formula $V_g f = V_g f * V_g g$ (1.82). Under the assumption that π is integrable, i.e., $V_g g$ is contained in a suitable weighted L^1-space, the convolution with $V_g g$ is a bounded operator on a class of function spaces Y on $\mathcal{G}$. This allows to define certain Banach spaces – the so-called coorbit spaces $\mathrm{Co}Y$ – as retract of Y under the transform V_g, i.e., $\mathrm{Co}Y = \{f, V_g f \in Y\}$. In particular, the coorbit spaces associated to the CWT are the homogeneous Besov and Triebel-Lizorkin spaces [Tri83b, Tri92] and the ones associated to the STFT are the modulation spaces [Fei83c, Grö01]. Thus many classical function spaces can be described in this way.

Coorbit space theory allows to treat the discretization problem not only on the Hilbert space level but also on the more general coorbit spaces. The main ingredient in the discretization method is again the reproducing formula (1.82). Indeed, one discretizes the convolution in a suitable way in order to obtain an approximation of the identity on the image of V_g. Under certain conditions this approximation operator is invertible by the von Neumann series and one obtains Banach frames and atomic decompositions of the form $\{\pi(x_i)g\}_{i \in I}$ for the coorbit spaces. In the particular cases, these are (irregular) Banach frames and atomic decompositions of wavelet type for Besov and Triebel-Lizorkin spaces or of Gabor type for modulation spaces.

Of course, Feichtinger and Gröchenig did not consider invariances under automorphism groups in their theory. So in order to apply coorbit space theory to the setting in Chapter 2 we have to generalize it. Fortunately, it turns out that every single ingredient can be adapted with more or less effort. Finally, we will derive Banach frames and atomic decompositions of the form $\{\widetilde{\pi}(x_i)g\}_{i \in I}$ for subspaces $\mathrm{Co}Y_{\mathcal{A}}$ of coorbit spaces consisting of elements invariant under $\mathcal{A}$. Applied to wavelet analysis we are able to derive radial wavelet frames in arbitrary dimension of the same type as in [EF95], but with more flexibility. Concerning time-frequency analysis, the abstract theory allows to construct radial Gabor frames for modulation spaces. Frames of this type have not yet been constructed. It is remarkable that the construction uses the operator Ω defined in (2.18), and in view of its complicated explicit form it seems hard to come up with a similar result by direct methods.

Instead of referring to the original papers of Feichtinger and Gröchenig we develop the whole coorbit space theory from scratch. The chapter is organized as follows. Section 4.1 collects some background about certain translation invariant function spaces on groups. As our next ingredient we prove in Section 4.2 the existence of certain locally finite coverings of arbitrary locally compact groups with sets that are invariant under the automorphism group $\mathcal{A}$. This covering property is a generalization of a classical result in [EG67, MB79, Fei81a]. Associated to such a covering one may construct a certain partition of unity on $\mathcal{G}$, which we call invariant bounded uniform partition of unity (IBUPU) in generalization of BUPUs introduced by Feichtinger in [Fei81a]. In Section 4.3 we introduce certain spaces of sequences associated to an invariant covering of a group and to a function space Y, and prove some of their basic properties. These sequence spaces are a very important tool for the discretization method. We remark that in contrast to the classical theory we need to work with two possibly different types of sequence spaces. If the automorphism group is trivial then both types coincides. As another tool we will need Wiener amalgam spaces introduced by Feichtinger in [Fei83b, Fei81b, Fei90]. We state some classical results concerning these in Section 4.4. In our context we also need to consider their subspaces of elements that are invariant under $\mathcal{A}$. We prove some of their properties that will become important for the discretization method later on.

After all these preparations we introduce the coorbit spaces in Section 4.5 and the corresponding subspaces of invariant elements. Among other basic properties we show that the reproducing formula extends to the coorbit spaces. Most of the proofs are analogous to the ones in the classical theory in [FG88, FG89a, Grö91]. A particular role is played by the space $\mathcal{H}_w^1 = \mathrm{Co}L_w^1$ since its (anti-) dual serves as reservoir for the definition of the general coorbits. Moreover, $\mathcal{H}_w^1$ possesses a certain minimality property, which is well-known for the special case of the Feichtinger algebra S_0 [Fei81c]. Section 4.6 is devoted to the discretization machinery. First, we briefly introduce the notion of Banach frames and atomic decomposition. Then we state our three main theorems analogous to the corresponding theorems in [FG89a, Grö91]. Their proof requires some effort and, therefore, we split it into several lemmas. Finally, Section 4.7 treats examples. We discuss in detail how to obtain radial Gabor frames for radial modulation spaces and radial wavelet frames for radial homogeneous Besov and

Triebel-Lizorkin spaces. In order to treat the problem related to radial Gabor frames we explicitly construct a particular covering of $\mathbb{R}^d \times \mathbb{R}^d$ with sets that are invariant under $SO(d)$. Since $SO(d)$ acts on both components of $\mathbb{R}^d \times \mathbb{R}^d$ simultaneously, that is, $A(x, \omega) = (Ax, A\omega)$ for $A \in SO(d)$ and $(x, \omega) \in \mathbb{R}^d \times \mathbb{R}^d$, this is a nice non-trivial geometric problem, which is of interest for itself.

We note that parts of this chapter appeared as preprint in [Rau03a].

4.1 Translation Invariant Function Spaces

Throughout this chapter we will use the notation introduced in Sections 1.5.3 and 1.2. In particular, we assume $\mathcal{G}$ to be a locally compact, σ-compact group. Recall the left and right translation operators $L_y F(x) = F(y^{-1}x)$ and $R_y F(x) = F(xy)$. We also need the involutions $F^{\vee}(x) = F(x^{-1})$, $F^{\triangledown}(x) = \overline{F(x^{-1})}$ and $F^*(x) = \overline{F(x^{-1})}\Delta(x^{-1})$.

Suppose $(Y, \|\cdot|Y\|)$ is a Banach space of functions on $\mathcal{G}$. We call Y **left invariant** if $L_x Y \subset Y$ for all $x \in \mathcal{G}$ and **right invariant** if $R_x Y \subset Y$ for all $x \in \mathcal{G}$. If Y is left and right invariant it is said to be **two-sided invariant**. Also recall the semi-norms p_K on $L^1_{loc}(\mathcal{G})$ defined in (1.1).

Definition 4.1.1. A Banach space Y of complex-valued functions on $\mathcal{G}$ is called a **solid Banach function space** (solid BF-space for short) if the following conditions are satisfied:

(i) Y is continuously embedded into $L^1_{loc}(\mathcal{G})$, i.e., for all compact sets $K \subset \mathcal{G}$ there exists a constant C_K such that $p_K(F) \leq C_K \|F|Y\|$ for all $F \in Y$.

(ii) If $|F(x)| \leq |G(x)|$ a.e. with F measurable and $G \in Y$, then $F \in Y$ and $\|F|Y\| \leq \|G|Y\|$.

Let us prove some basic properties of two-sided invariant solid BF-spaces.

Lemma 4.1.1. *Suppose Y is a two-sided invariant solid BF-space. Then R_x and L_x are bounded operators on Y for all $x \in \mathcal{G}$.*

Proof: Assume that $(F_n)_{n \in \mathbb{N}} \subset Y$ is a sequence converging to $F \in Y$ such that $L_x F_n \to G \in Y$. By the closed graph theorem we need to show that $G = L_x F$. Property (i) of Definition 4.1.1 implies that convergence is also in L^1_{loc}. Thus for any compact set $K \subset \mathcal{G}$ we obtain by left invariance of the Haar measure

$$\begin{aligned} p_K(G - L_x F) &\leq p_K(G - L_x F_n) + p_K(L_x F_n - L_x F) \\ &= p_K(G - L_x F_n) + p_{x^{-1}K}(F_n - F) \to 0 \quad (n \to \infty). \end{aligned}$$

We conclude that $G = L_x F$ almost everywhere. By the solidity condition (ii) of Definition 4.1.1 this implies that $G = L_x F$ in Y. The proof for R_x is essentially the same. □

Lemma 4.1.2. *Suppose Y is a two-sided invariant solid BF-space and let K be an arbitrary compact subset of $\mathcal{G}$. Then Y contains the characteristic function χ_K of K.*

<u>Proof:</u> Since Y is assumed to be non-trivial there is a non-zero function $F \in Y$. Then there exists some $\epsilon > 0$ such that the set $U_\epsilon := \{x \in \mathcal{G}, |F(x)| \geq \epsilon\}$ has positive measure. Otherwise, $\cup_{n\in\mathbb{N}} U_{1/n} = \operatorname{supp} F$ would have measure zero, which is a contradiction to F being non-zero in Y.

Since $\chi_{U_\epsilon} \leq \epsilon|F|$ the solidity condition (ii) implies that the characteristic function χ_{U_ϵ} is contained in Y. Since $\mu(U_\epsilon) > 0$ the interior U_ϵ° is non-empty and by translation invariance of Y we may assume $e \in U_\epsilon^\circ$ without loss of generality. Now let K be an arbitrary compact subset of $\mathcal{G}$. Clearly, we have the covering $K \subset \cup_{x\in K} x U_\epsilon^\circ$ and by compactness of K there exists a finite number of points $x_i \in K, i = 1, \ldots n$, such that $K \subset \cup_{i=1}^n x_i U_\epsilon^\circ$, which implies the pointwise inequality

$$\chi_K \leq \sum_{i=1}^n \chi_{x_i U_\epsilon^\circ}.$$

By translation invariance the right hand side of this inequality is contained in Y and, hence, also χ_K belongs to Y by the solidity condition (ii). □

With this lemma one realizes that restricting two-sided invariant solid BF-spaces to real-valued functions yields Köthe function spaces on $\mathcal{G}$, see [LT79, Definition 1.b.17], which in turn are special instances of Banach lattices.

In the study of translation invariant function spaces the following definition plays a central role.

Definition 4.1.2. A strictly positive continuous weight function w is called **submultiplicative**, if $w(e) = 1$ and

$$w(xy) \leq w(x)w(y) \qquad \text{for all } x, y \in \mathcal{G}.$$

If w is submultiplicative, then another continuous weight function m is called *w*-**moderate** if

$$m(xyz) \leq w(x)m(y)w(z) \qquad \text{for all } x, y, z \in \mathcal{G}. \tag{4.1}$$

In particular, a submultiplicative weight w is w-moderate. The assumption $w(e) = 1$ is not essential, but convenient in our context. Also the restriction to continuous weight functions is not severe as shows the following theorem due to Feichtinger.

Theorem 4.1.3. *[Fei79, Satz 2.7] Assume $w \in L^1_{loc}(\mathcal{G})$ satisfies $w(xy) \leq w(x)w(y)$ for almost all $x, y \in \mathcal{G}$. Then there exists a continuous submultiplicative function $\tilde{w}$, which is equivalent to w, that is, there are constants $C_1, C_2 > 0$ such that*

$$C_1\tilde{w}(x) \leq w(x) \leq C_2\tilde{w}(x) \qquad a.e.\,.$$

Moreover, if $m \in L^1_{loc}(\mathcal{G})$ satisfies (4.1) for some submultiplicative weight function w, then there exists a continuous w-moderate function $\tilde{m}$, which is equivalent to m.

Bearing Lemma 4.1.1 in mind it makes sense to define the functions

$$u(x) := \|L_x|Y\|, \qquad v(x) := \Delta(x^{-1})\|R_{x^{-1}}|Y\|, \quad x \in \mathcal{G}, \qquad (4.2)$$

where $\| \cdot |Y\|$ denotes the operator norm of a bounded operator, which maps Y into itself. Clearly, it holds $u(e) = v(e) = 1$. By $L_{xy} = L_x L_y$ and $R_{xy} = R_x R_y$ we conclude further that both u and v are submultiplicative (provided they are continuous). In the sequel we will work with function spaces Y that satisfy the assumptions in the following definition.

Definition 4.1.3. A function space Y is said to satisfy property (Y) if it is a two-sided invariant solid BF-space, whose associated functions u, v defined in (4.2) are continuous.

We note that the continuity assumption in the previous definition is for the sake of convenience. By Theorem 4.1.3 it may be replaced by the assumption $u, v \in L^1_{loc}$. (In this case one replaces the functions u, v by their equivalent continuous ones according to Theorem 4.1.3.) Continuity makes it easier to work with sample values of u, v. This will be needed frequently later on.

Let us now consider convolutions.

Lemma 4.1.4. *Suppose Y satisfies (Y) and let $F \in Y$ and $G \in L^1_v(\mathcal{G})$. Then the convolution (compare with (1.8))*

$$F * G = \int_{\mathcal{G}} G(x)\Delta(x^{-1})R_{x^{-1}}F \, dx,$$

to be interpreted as a Bochner integral, is a well-defined element of Y and satisfies

$$\|F * G|Y\| \leq \|F|Y\| \, \|G|L^1_v\|.$$

Proof: Using the monotonicity of Bochner integrals (see e.g. Corollary 1 in [Yos80, Chapter V.5]) we obtain

$$\|F * G\| \leq \int_{\mathcal{G}} |G(x)|\Delta(x^{-1})\|R_{x^{-1}}F|Y\|dx \leq \|F|Y\| \int_{\mathcal{G}} |G(x)|\Delta(x^{-1})\|R_{x^{-1}}|Y\|dx$$
$$= \|F|Y\| \, \|G|L^1_v\|,$$

which also shows that the convolution is well-defined, see e.g. Theorem 1 in [Yos80, Chapter V.5]. $\qquad \square$

Analogously, one shows that the convolution $G * F$ with $G \in L_u^1(\mathcal{G})$ and $F \in Y$ is well-defined and satisfies $\|G * F|Y\| \leq \|G|L_u^1\| \, \|F|Y\|$. This will, however, not be needed in the sequel.

Our main examples of two-sided invariant solid BF-spaces are certain weighted $L_m^p(\mathcal{G})$-spaces. It is clear that for any choice of weight function $m \in L_{loc}^1(\mathcal{G})$ and all $1 \leq p \leq \infty$ the space $L_m^p(\mathcal{G})$ is a solid BF-space. Translation invariance is ensured by moderateness of m as shown in the following theorem.

Lemma 4.1.5. *Suppose that w is submultiplicative and m is a w-moderate weight function on $\mathcal{G}$. Then $L_m^p(\mathcal{G}), 1 \leq p \leq \infty$, is two-sided-invariant with*

$$\|L_x|L_m^p\| \leq w(x), \qquad \|R_x|L_m^p\| \leq w(x^{-1})\Delta(x^{-1})^{1/p}, \quad x \in \mathcal{G}.$$

Proof: Let $1 \leq p < \infty$ and suppose $F \in L_m^p, x \in \mathcal{G}$. A straightforward calculation shows

$$\int_{\mathcal{G}} |R_x F(y)|^p m(y)^p dy = \Delta(x^{-1}) \int_{\mathcal{G}} |F(y)|^p m(yx^{-1})^p dy$$

$$\leq \Delta(x^{-1}) \int_{\mathcal{G}} |F(y)|^p \underbrace{w(e)^p}_{=1} m(y)^p w(x^{-1})^p dy,$$

which implies the estimation for $\|R_x|L_m^p\|$. The assertions for $p = \infty$ and for L_x are proven similarly. $\qquad\square$

The converse of this Lemma is also true, i.e., if L_m^p is two-sided invariant then m is necessarily w-moderate for some submultiplicative function w [Fei79, Satz 2.5].

Corollary 4.1.6. *Suppose w is a submultiplicative weight function. Then $L_w^1(\mathcal{G})$ is a Banach algebra (called Beurling algebra), i.e., $L_w^1(\mathcal{G}) * L_w^1(\mathcal{G}) \subset L_w^1(\mathcal{G})$ and*

$$\|F * G|L_w^1\| \leq \|F|L_w^1\| \, \|G|L_w^1\|.$$

Proof: It follows from Lemma 4.1.5 that the function u associated to $Y = L_w^1$ by (4.2) is dominated by w and, hence, the assertion follows from Lemma 4.1.4. $\qquad\square$

Corollary 4.1.7. *Suppose w is submultiplicative. Then it holds $\|R_x|L_{1/w}^\infty\| \leq w(x)$ and $\|L_x|L_{1/w}^\infty\| \leq w(x^{-1})$, $x \in \mathcal{G}$.*

Proof: By submultiplicativity of w we have

$$\frac{1}{w(xyz)} = \frac{w(x^{-1})w(z^{-1})}{w(x^{-1})w(xyz)w(z^{-1})} \leq \frac{w(x^{-1})w(z^{-1})}{w(y)}.$$

Hence, $1/w$ is $w^{\vee}$- moderate and the claim follows from Lemma 4.1.5. $\quad\square$

Further examples of two-sided invariant solid BF-spaces include general rearrangement invariant function space on $\mathcal{G}$, in particular Lorentz spaces or Orlicz spaces, see [LT79]. Up to now, symmetry groups did not enter the game. So we assume further that $\mathcal{A}$ is a compact automorphism group of $\mathcal{G}$ as in Section 1.2. Recall that $\mathcal{A}$ acts on functions on $\mathcal{G}$ by $U_A : F \mapsto F_A, A \in \mathcal{A}$, where $F_A(x) = F(A^{-1}x)$. A function space Y on $\mathcal{G}$ is called invariant under $\mathcal{A}$ if $U_A Y \subset Y$ for all $A \in \mathcal{A}$. As in Lemma 4.1.1 continuity of the operators U_A follows automatically.

Lemma 4.1.8. *Suppose Y is a solid BF-space, which is invariant under $\mathcal{A}$. Then U_A is a bounded operator on Y for all $A \in \mathcal{A}$. Moreover, if for any $F \in Y$ the mapping $A \mapsto \|F_A|Y\|$ is integrable on $\mathcal{A}$ then there exists an equivalent norm $\|\cdot|Y\|'$ on Y such that*

$$\|F_A|Y\|' = \|F|Y\|' \quad \text{for all } A \in \mathcal{A}, F \in Y.$$

Proof: Boundedness of U_A follows from the closed graph theorem as in the proof of Lemma 4.1.1. We remark that instead of translation invariance of the Haar measure μ its invariance under the action of $\mathcal{A}$ is used (Lemma 1.2.1).

In order to prove the second assertion we first show that $A \mapsto s(A) := \|U_A\|$ is bounded. Since $U_{AB} = U_A U_B$, the function s is submultiplicative and by (local) integrability of $A \mapsto \|F_A|Y\|$ also s is integrable. It follows from Theorem 4.1.3 that s is bounded from above and below, i.e., $C_1 \leq s(A) \leq C_2$ a.e. for constants $C_1, C_2 > 0$. With boundedness of U_A we get the inequality

$$C_2^{-1}\|F|Y\| \leq \|U_{A^{-1}}|Y\|^{-1}\|F|Y\| \leq \|F_A|Y\| \leq \|U_A|Y\|\,\|F|Y\| \leq C_2\|F|Y\|. \quad (4.3)$$

Now define a new norm by

$$\|F|Y\|' := \int_{\mathcal{A}} \|F_A|Y\|dA.$$

which is well-defined by the integrability assumption. It is apparent that $\|\cdot|Y\|'$ is a norm on Y. By (4.3) it is equivalent to $\|\cdot|Y\|$. From the invariance of the Haar measure it follows $\|F_A|Y\|' = \|F|Y\|'$ for all $A \in \mathcal{A}, F \in Y$. $\quad\square$

In view of the previous lemma the next definition is reasonable.

Definition 4.1.4. A function space Y is said to have property (YA) if it has property (Y) (see Definition 4.1.3) and if U_A is an isometry on Y for all $A \in \mathcal{A}$.

If Y has property (YA) then it is easy to see that the functions u, v defined in (4.2) satisfy

$$u(Ax) = u(x), \qquad v(Ax) = v(x) \quad \text{for all } A \in \mathcal{A}, x \in \mathcal{G}.$$

Moreover, if $m(Ax) = m(x)$ for all $A \in \mathcal{A}, x \in \mathcal{G}$ then $Y = L_m^p(\mathcal{G})$ has property (YA). For a space Y satisfying (YA) we may define its subspace of invariant functions by

$$Y_{\mathcal{A}} := \{F \in Y, F_A = F \text{ for all } A \in \mathcal{A}\}.$$

Since $Y_{\mathcal{A}} = \cap_{A \in \mathcal{A}} \ker(\mathrm{Id} - U_A)$ it is closed by continuity of U_A on Y. Moreover, it is non-trivial. Indeed, Y is non-trivial and by property (ii) of Definition 4.1.1 it contains a non-zero positive function G. The function $F(x) = \int_{\mathcal{A}} G(Ax) dA$ is also non-zero and positive and easily checked to be contained in $Y_{\mathcal{A}}$. Clearly, functions in $Y_{\mathcal{A}}$ may be identified with functions on $\mathcal{K} = \mathcal{A}^{\mathcal{G}}$ as in Section 1.2.

Lemma 4.1.9. *For $F \in Y_{\mathcal{A}}$ and $G \in (L_v^1)_{\mathcal{A}}$ the convolution $F * G$ is again contained in $Y_{\mathcal{A}}$.*

Proof: By Lemma 4.1.4 the convolution $F * G$ is a well-defined element of Y. Using the invariance of the Haar measure and of the modular function Δ under $\mathcal{A}$ (Lemma 1.2.1) we obtain

$$(F * G)_A = \int_{\mathcal{G}} G(x)\Delta(x^{-1})(R_{x^{-1}}F)_A dx = \int_{\mathcal{G}} G(x)\Delta(x^{-1})R_{Ax^{-1}}F_A dx$$
$$= \int_{\mathcal{G}} G_A(x)\Delta(x^{-1})R_{x^{-1}}F_A dx = F_A * G_A,$$

where all expression have to be interpreted as vector-valued integrals. This yields the invariance of $F * G$. We remark that alternatively the claim follows from Lemma 1.2.3(a) by taking $\mathbb{S} = C_c(\mathcal{G})$. □

The next result states that also $\mathcal{L}_x$ and $\mathcal{R}_x$ defined in (1.13) and (1.14) are well-defined on Y.

Lemma 4.1.10. *Suppose Y possesses property (YA). Then $\mathcal{L}_x$ and $\mathcal{R}_x$ are bounded operators on Y with*

$$\|\mathcal{L}_x|Y\| \leq \|L_x|Y\|, \qquad and \qquad \|\mathcal{R}_x|Y\| \leq \|R_x|Y\|.$$

Proof: Since $\mathcal{L}_x = P_{\mathcal{A}} L_x$ by Lemma 1.2.4(a), we have $\|\mathcal{L}_x|Y\| \leq \|P_{\mathcal{A}}|Y\| \|L_x|Y\| \leq \|L_x|Y\|$ by Lemma 1.2.2(d). The estimation for $\mathcal{R}_x$ is the same. □

4.2   Invariant Coverings

It will become important in Section 4.6 that it is possible to cover the group $\mathcal{G}$ with invariant sets of the form $\mathcal{A}(x_i U)$ for some relatively compact set U and points $\{x_i\}_{i \in I}$.

In addition the covering has the property that different sets of the covering do not overlap "too much".

First we note that there exist relatively compact neighborhoods V of $e \in \mathcal{G}$, that satisfy $V = V^{-1} = \mathcal{A}(V)$. Indeed, there exists a set U that possesses all of these properties except possibly the invariance $U = \mathcal{A}(U)$. We define $V := \mathcal{A}(U)$. This set is still relatively compact by compactness of $\mathcal{A}$ and has non-void interior since $U \subset \mathcal{A}(U)$. It is easily checked that $\mathcal{A}(V) = V$. The property $V = V^{-1}$ follows from $U = U^{-1}$ and from the fact that $\mathcal{A}$ consists of automorphisms.

Definition 4.2.1. Let $X = (x_i)_{i \in I}$ be some discrete set of points in $\mathcal{G}$ and V a relative compact neighborhood of e in $\mathcal{G}$ satisfying $V = V^{-1} = \mathcal{A}(V)$.

(a) X is called V-**dense** with respect to $\mathcal{A}$ if $\mathcal{G} = \bigcup_{i \in I} \mathcal{A}(x_i V)$.

(b) X is called **relatively separated** with respect to $\mathcal{A}$ if for all compact sets $W \subset \mathcal{G}$ there exists some constant C_W such that

$$\sup_{j \in I} \#\{i \in I, \, \mathcal{A}(x_i W) \cap \mathcal{A}(x_j W) \neq \emptyset\} \leq C_W < \infty.$$

(c) If X is V-dense with respect to $\mathcal{A}$ for some V and, in addition, there exists an open relatively compact set $W = W^{-1} = \mathcal{A}(W)$ such that

$$\mathcal{A}(x_i W) \cap \mathcal{A}(x_j W) = \emptyset \qquad \text{for all } i, j \in I, i \neq j. \qquad (4.4)$$

then X is called **well-spread** with respect to $\mathcal{A}$.

In the following we prove the existence of point sets of the above type for arbitrary locally compact groups and arbitrary compact automorphism groups. This generalizes a classical result in [EG67], see also [MB79, Fei81a].

We start with an auxiliary lemma.

Lemma 4.2.1. *Suppose that* $(x_i)_{i \in I} \subset \mathcal{G}$ *is some point set and* W *some relatively compact neighborhood of* $e \in \mathcal{G}$ *satisfying* $W = W^{-1} = \mathcal{A}(W)$ *such that (4.4) holds. If* $K_1, K_2 \subset \mathcal{G}$ *are relatively compact sets then we have*

$$\sup_{y \in \mathcal{G}} \#\{i \in I, \, \mathcal{A}(y K_1) \cap \mathcal{A}(x_i K_2) \neq \emptyset\} \leq \frac{|\mathcal{A}(K_1 K_2^{-1})W|}{|W|} < \infty.$$

Proof: Suppose that $z \in \mathcal{A}(y K_1) \cap \mathcal{A}(x_i K_2) \neq \emptyset$ with $y \in \mathcal{G}$ for some $i \in I$. Then $z = A_1(y)k_1 = A_2(x_i)k_2$ with $A_1, A_2 \in \mathcal{A}$ and $k_j \in \mathcal{A}(K_j), j = 1, 2$. Denoting $A_{i,y} = A_1^{-1} A_2$ we immediately deduce $A_{i,y}(x_i) \in y \mathcal{A}(K_1 K_2^{-1})$ and hence $\mathcal{A}_{i,y}(x_i)W \subset y \mathcal{A}(K_1 K_2^{-1})W$. The assumption $\mathcal{A}(x_i)W \cap \mathcal{A}(x_j)W = \emptyset$ for all $i, j \in I, i \neq j$ implies in particular $x_i W \cap x_j W = \emptyset$. Furthermore, the number of non-overlapping sets of the form xW

that fit into $y\mathcal{A}(K_1 K_2^{-1})W$ is obviously bounded by $|\mathcal{A}(K_1 K_2^{-1})W|/|W|$. Altogether we obtain

$$\#\{i \in I, \, \mathcal{A}(yK_1) \cap \mathcal{A}(x_i K_2) \neq \emptyset\} \leq \#\{i \in I, \, A_{i,y}(x_i)W \subset y\mathcal{A}(K_1 K_2^{-1})W\}$$
$$\leq \frac{|\mathcal{A}(K_1 K_2^{-1})W|}{|W|}. \qquad \square$$

In particular, the previous lemma implies that a well-spread set is relatively separated. We remark that one could replace the assumption (4.4) in the definition of well-spreadness by the slightly less restrictive condition that X be relatively separated. Actually, this corresponds to the original notion of well-spread points in [FG89a]. However, in order to avoid additional technicalities at some points in the sequel we use the definition as stated.

Theorem 4.2.2. Let $\mathcal{A}$ be a compact automorphism group of some locally compact group $\mathcal{G}$ and let $V = V^{-1} = \mathcal{A}(V)$ be a relatively compact neighborhood of $e \in \mathcal{G}$. Then there exists a set of points $(x_i)_{i \in I} \subset \mathcal{G}$ that is V-dense and well-spread with respect to $\mathcal{A}$. If $\mathcal{G}$ is σ-compact then the index set I is countable.

Proof: Choose a neighborhood $W = W^{-1} = \mathcal{A}(W)$ of e that satisfies $W^2 \subset V$. The existence of such a set W follows for instance from [MZ55, §1.15, Theorem] together with the invariance $V = \mathcal{A}(V)$. By Zorn's Lemma there exists a maximal set of points $(x_i)_{i \in I}$ such that

$$\mathcal{A}(x_i W) \cap \mathcal{A}(x_j W) = \emptyset \qquad \text{for all } i, j \text{ with } i \neq j. \tag{4.5}$$

We claim that $\mathcal{G} = \bigcup_{i \in I} \mathcal{A}(x_i W^2)$ implying $\mathcal{G} = \bigcup_{i \in I} \mathcal{A}(x_i V)$. Suppose the contrary, i.e., there exists $y \in \mathcal{G}$ such that y is contained in none of the sets $\mathcal{A}(x_i W^2)$, $i \in I$. This means that $y \neq A(x_i w_1 w_2)$ for all $A \in \mathcal{A}$ and all $w_1, w_2 \in W$. The latter is equivalent to $y(Aw_2)^{-1} \neq A(x_i w_1)$. Since $W = W^{-1} = \mathcal{A}(W)$ this implies that $\mathcal{A}(yW) \cap \mathcal{A}(x_i W) = \emptyset$. Thus we arrived at a contradiction to the maximality of the set $(x_i)_{i \in I}$. Property (4.5) is nothing else than (4.4). It is clear that the index set I is countable whenever $\mathcal{G}$ is σ-compact. This finishes the proof.

Since we find it instructive we add a constructive proof of (4.5) in case that $\mathcal{G} = \bigcup_{n=1}^{\infty} V^n$, that is, $\mathcal{G}$ is compactly generated with generator $\overline{V}$. We choose $x_1 := e$. Now form $K^{(2)} := \overline{V^2} \setminus V$. If $K^{(2)} = \emptyset$ (only possible if $\mathcal{G}$ is compact) then we are ready, since then $\mathcal{G} = V$. Otherwise we choose $x_2 \in K^{(2)}$ and form $K^{(3)} := \overline{V^2} \setminus (V \cup \mathcal{A}(x_2 V))$. If $K^{(3)} \neq \emptyset$ choose $x_3 \in K^{(3)}$. Continuing in this way one obtains $\overline{V^2} \subset \bigcup_{i=1}^{N_2} \mathcal{A}(x_i V)$ with $x_j \notin \bigcup_{i=1}^{j-1} \mathcal{A}(x_i V)$. Inductively, we obtain a covering $\mathcal{G} = \bigcup_{i=1}^{\infty} \mathcal{A}(x_i V)$. If $\mathcal{G}$ is compact then the covering is finite. It is easy to check that the point set constructed in this way also satisfies (4.5). $\qquad \square$

As another important tool we will need an adaption of the concept of bounded uniform partitions of unity (BUPU) as introduced by Feichtinger in [Fei81a]. In our context all functions belonging to the partition of unity must be invariant under $\mathcal{A}$.

Definition 4.2.2. Suppose $U = U^{-1} = \mathcal{A}(U)$ is a relatively compact neighborhood of $e \in \mathcal{G}$. A collection of functions $\Psi = (\psi_i)_{i \in I}, \psi_i \in C_0(\mathcal{G})$, is called $\mathcal{A}$-invariant bounded uniform partition of unity of size U (for short U-$\mathcal{A}$-IBUPU) if the following conditions are satisfied:

(1) $0 \leq \psi_i(x) \leq 1$ for all $i \in I$, $x \in \mathcal{G}$,
(2) $\sum_{i \in I} \psi_i(x) \equiv 1$,
(3) $\psi_i(Ax) = \psi_i(x)$ for all $x \in \mathcal{G}, A \in \mathcal{A}, i \in I$,
(4) there exists a family $(x_i)_{i \in I}$, which is well-spread with respect to $\mathcal{A}$, such that $\operatorname{supp} \psi_i \subset \mathcal{A}(x_i U)$.

It follows from Lemma 4.2.1 that

$$\sup_{j \in I} \#\{i \in I, \operatorname{supp} \psi_i \cap \operatorname{supp} \psi_j \neq \emptyset\} \leq C < \infty.$$

If the automorphism group is trivial ($\mathcal{A} = \{e\}$) then we get back Feichtinger's original definition of BUPUs. Existence of IBUPUs is settled in the following theorem.

Theorem 4.2.3. *Let $\mathcal{G}$ be a locally compact group, $\mathcal{A}$ be a compact automorphism group of $\mathcal{G}$ and $U = U^{-1} = \mathcal{A}(U)$ be a relatively compact neighborhood of $e \in \mathcal{G}$. Then there exists a U-$\mathcal{A}$-IBUPU in the sense of Definition 4.2.2.*

Proof: We proceed similarly as in the proof of Theorem 1 in [Fei81a]. Choose $V = V^{-1} = \mathcal{A}(V)$ such that $V^2 \subset U$ and $X = (x_i)_{i \in I}$ according to Theorem 4.2.2 (where we construct X with respect to V). For every $i \in I$ we let $\phi_i \in C_c(\mathcal{G})$ such that $\phi_i(x) = 1$ for $x \in \mathcal{A}(x_i V)$, $\operatorname{supp} \phi_i \subset \mathcal{A}(x_i U)$, $0 \leq \phi_i(x) \leq 1$ for all $x \in \mathcal{G}$ and $\phi_i(Ax) = \phi_i(x)$ for all $A \in \mathcal{A}, x \in \mathcal{G}$. Since the sets $\operatorname{supp} \phi_i$ cover $\mathcal{G}$ we have by Lemma 4.2.1

$$1 \leq \Phi(x) := \sum_{i \in I} \phi_i(x) \leq C < \infty.$$

Now set $\psi_i(x) := \phi_i(x)/\Phi(x) \in C_c(\mathcal{G})$ yielding $\sum_{i \in I} \psi_i(x) \equiv 1$ and $\operatorname{supp} \psi_i = \operatorname{supp} \phi_i \subset \mathcal{A}(x_i U)$. Invariance under $\mathcal{A}$ of the functions ψ_i is clear. $\square$

4.3 Sequence Spaces

Another important ingredient in the discretization machinery to be developed in Section 4.6 consists of sequence spaces which are associated to a function space Y and to an invariant covering of $\mathcal{G}$.

For a family $X = (x_i)_{i \in I}$, which is well-spread with respect to $\mathcal{A}$, a relatively compact set $U = \mathcal{A}(U)$ with nonvoid interior and a space Y that satisfies (YA), we define the two sequence spaces

$$Y_{\mathcal{A}}^{\flat} := Y_{\mathcal{A}}^{\flat}(X) := \{(\lambda_i)_{i \in I}, \sum_{i \in I} |\lambda_i| \chi_{\mathcal{A}(x_i U)} \in Y\},$$

$$Y_{\mathcal{A}}^{\natural} := Y_{\mathcal{A}}^{\natural}(X) := \{(\lambda_i)_{i \in I}, \sum_{i \in I} |\lambda_i| |\mathcal{A}(x_i U)|^{-1} \chi_{\mathcal{A}(x_i U)} \in Y\}$$

with natural norms

$$\|(\lambda_i)_{i \in I} |Y_{\mathcal{A}}^{\flat}\| := \| \sum_{i \in I} |\lambda_i| \chi_{\mathcal{A}(x_i U)} |Y\|,$$

$$\|(\lambda_i)_{i \in I} |Y_{\mathcal{A}}^{\natural}\| := \| \sum_{i \in I} |\lambda_i| |\mathcal{A}(x_i U)|^{-1} \chi_{\mathcal{A}(x_i U)} |Y\|.$$

Hereby, $|\mathcal{A}(x_i U)|$ denotes the Haar measure of the set $\mathcal{A}(x_i U)$. In case that $\mathcal{A} = \{e\}$ is trivial we get the sequence space $Y_d := Y_{\{e\}}^{\natural} = Y_{\{e\}}^{\flat}$, which was used in [FG88, FG89a, FG89b, Grö91]. Note that Lemma 4.1.2 implies that the finite sequences are contained in $Y_{\mathcal{A}}^{\flat}$ and in $Y_{\mathcal{A}}^{\natural}$. Let us state some further basic properties of these spaces (compare also [FG89a, Lemma 3.5]).

Lemma 4.3.1. *(a) The spaces $Y_{\mathcal{A}}^{\flat}$ and $Y_{\mathcal{A}}^{\natural}$ do not depend on the choice of the set $U = \mathcal{A}(U)$ and different sets U define equivalent norms.*

(b) $Y_{\mathcal{A}}^{\flat}$ and $Y_{\mathcal{A}}^{\natural}$ are solid Banach sequence spaces.

(c) If the functions with compact support are dense in Y then the finite sequences are dense in $Y_{\mathcal{A}}^{\flat}$ and in $Y_{\mathcal{A}}^{\natural}$.

(d) If $\mathcal{A}$ is a finite group then $Y_{\mathcal{A}}^{\flat} = Y_{\mathcal{A}}^{\natural}$ with equivalent norms.

(e) If $Y = L_m^p(\mathcal{G}), 1 \leq p \leq \infty$, with invariant moderate weight function m then $Y_{\mathcal{A}}^{\flat}(X) = l_{\nu_p}^p(I)$ and $Y_{\mathcal{A}}^{\natural}(X) = l_{m_p}^p(I)$ (with equivalent norms) where

$$\nu_p(i) := m(x_i) |\mathcal{A}(x_i U)|^{\frac{1}{p}}, \qquad m_p(i) := m(x_i) |\mathcal{A}(x_i U)|^{\frac{1}{p} - 1}.$$

We have in particular $\nu_\infty(i) = m_1(i) = m(x_i)$.

Proof: (a) Suppose that $V = \mathcal{A}(V)$ is another relatively compact set with nonvoid interior. By compactness there exists a finite number of points $y_j \in \mathcal{G}, j = 1, \ldots, n$, such that $V \subset \cup_{j=1}^n U y_j$. Since $V = \mathcal{A}(V)$ and $U = \mathcal{A}(U)$ it holds

$$\mathcal{A}(x_i V) = \mathcal{A}(x_i) V \subset \cup_{j=1}^n \mathcal{A}(x_i) U y_j = \cup_{j=1}^n \mathcal{A}(x_i U) y_j$$

yielding

$$\sum_{i\in I}|\lambda_i|\chi_{\mathcal{A}(x_iV)} \leq \sum_{j=1}^{n}\sum_{i\in I}|\lambda_i|\chi_{\mathcal{A}(x_iU)y_j} = \sum_{j=1}^{n}R_{y_j^{-1}}\left(\sum_{i\in I}|\lambda_i|\chi_{\mathcal{A}(x_iU)}\right).$$

By solidity and right translation invariance of Y this implies

$$\|\sum_{i\in I}|\lambda_i|\chi_{\mathcal{A}(x_iV)}|Y\| \leq \left(\sum_{j=1}^{n}\|R_{y_j^{-1}}|Y\|\right)\|\sum_{i\in I}|\lambda_i|\chi_{\mathcal{A}(x_iU)}|Y\|.$$

Exchanging the roles of U and V yields a converse inequality. The proof for $Y_{\mathcal{A}}^{\natural}$ is the same.

(b) Suppose that $(\Lambda^n)_{n\in\mathbb{N}}, \Lambda^n = (\lambda_i^n)_{i\in I}$ is a Cauchy sequence of elements in $Y_{\mathcal{A}}^{\flat}$. This means that the functions $F_n = \sum_{i\in I}\lambda_i^n\chi_{\mathcal{A}(x_iU)}$ form a Cauchy sequence in $Y_{\mathcal{A}}$. By completeness of $Y_{\mathcal{A}}$ the limit $F = \lim_{n\to\infty}F_n$ exists and is an element of $Y_{\mathcal{A}}$. By (a) and (4.4) the sets $\mathcal{A}(x_iU)$ are mutually disjoint without loss of generality. The continuous embedding of Y into L_{loc}^1 implies

$$p_{\mathcal{A}(x_iU)}(F - F_n) = \int_{\mathcal{A}(x_iU)}|F(x) - \lambda_i^n|dx \to 0 \quad (n \to \infty) \quad \text{for all } i \in I.$$

This means that $F = \sum_{i\in I}\lambda_i\chi_{\mathcal{A}(x_iU)}$ with $\lambda_i = \lim_{n\to\infty}\lambda_i^n$ for all $i \in I$. Thus, $\Lambda = (\lambda_i)_{i\in I}$ is the limit of $(\Lambda^n)_{n\in\mathbb{N}}$ in $Y_{\mathcal{A}}$, which shows that $Y_{\mathcal{A}}^{\flat}$ is complete. The solidity of $Y_{\mathcal{A}}^{\flat}$ follows immediately from the solidity of Y. The proof for $Y_{\mathcal{A}}^{\natural}$ is the same.

(c) Suppose $(\lambda_i)_{i\in I} \in Y_{\mathcal{A}}^{\flat}$ and denote $F = \sum_{i\in I}\lambda_i\chi_{\mathcal{A}(x_iU)} \in Y$. By assumption there exists for any $\epsilon > 0$ a function $G \in Y$ with compact support such that $\|F - G|Y\| \leq \epsilon$. The set $J := \{i \in I, \mathcal{A}(x_iU) \cap \text{supp}\, G \neq \emptyset\}$ is finite since $(x_i)_{i\in I}$ satisfies (4.4). The function $F_J := \sum_{i\in J}\lambda_i\chi_{\mathcal{A}(x_iU)}$ coincides with F on $V := \cup_{i\in J}\mathcal{A}(x_iU) \supset \text{supp}\, G$ and with G outside V. Thus, we have $|F - F_J| \leq |F - G|$ a.e., which implies by solidity that $\|F - F_J|Y\| \leq \epsilon$. Let Λ_J denote the (finite) sequence, which coincides with Λ on J and is zero outside J. We conclude $\|\Lambda - \Lambda_J|Y_{\mathcal{A}}^{\flat}\| = \|F - F_J|Y\| \leq \epsilon$. The proof for $Y_{\mathcal{A}}^{\natural}$ is completely analogous.

(d) If $\mathcal{A}$ consists of n elements (n finite), we have $|U| \leq |\mathcal{A}(x_iU)| \leq n|U|$ by left invariance of the Haar measure, and $Y_{\mathcal{A}}^{\flat} = Y_{\mathcal{A}}^{\natural}$ together with equivalence of norms is an immediate consequence.

(e) Without loss of generality we may assume that the sets $\mathcal{A}(x_iU)$ are mutually disjoint. For $Y = L_m^p(\mathcal{G}), 1 \leq p < \infty$, we obtain

$$\|(\lambda_i)_{i\in I}|Y_{\mathcal{A}}^{\flat}\| = \left(\sum_{i\in I}|\lambda_i|^p\int_{\mathcal{G}}\chi_{\mathcal{A}(x_iU)}m(x)^pdx\right)^{1/p}.$$

Since m is w-moderate and invariant we have $m(x) \leq m(x_i) \max_{u \in U} w(u) = C_1 m(x_i)$ for all $x \in \mathcal{A}(x_i U)$. Since w is continuous and U is relatively compact the constant C_1 is finite. The same considerations show that $m(x_i) \leq C_2 m(x)$ for all $x \in \mathcal{A}(x_i U)$ with some constant $C_2 > 0$. Hence, we obtain

$$C_2^{-1} \left(\sum_{i \in I} |\lambda_i|^p m(x_i)^p |\mathcal{A}(x_i U)| \right)^{1/p} \leq \|(\lambda_i)_{i \in I} | Y_{\mathcal{A}}^{\flat}\|$$

$$\leq C_1 \left(\sum_{i \in I} |\lambda_i|^p m(x_i)^p |\mathcal{A}(x_i U)| \right)^{1/p}.$$

This implies also the assertion for $Y_{\mathcal{A}}^{\natural}$. The case $p = \infty$ is deduced similarly.   □

Let us now derive a different characterization of $Y_{\mathcal{A}}^{\natural}$. For a positive (non-zero) window function $k \in (C_c)_{\mathcal{A}}(\mathcal{G})$ we define the function

$$m_k(x, z) := \int_{\mathcal{A}} k(z^{-1} A(x)) dA = \mathcal{L}_z k(x) = \mathcal{L}_x k^{\vee}(z). \tag{4.6}$$

Since k is assumed to be invariant, m_k is invariant in both variables. If $\operatorname{supp} k \subset U$ then we have $\operatorname{supp} m_k(\cdot, z) \subset \mathcal{A}(zU)$ and $\operatorname{supp} m_k(x, \cdot) \subset \mathcal{A}(xU^{-1})$. Furthermore, if $k = k^{\vee}$ then m is symmetric, i.e., $m_k(x, z) = m_k(z, x)$ for all $x, z \in \mathcal{G}$.

If $k = \chi_U$ is the characteristic function of some set $U = \mathcal{A}(U)$ then $m_{\chi_U} =: m_U$ has a geometrical interpretation, i.e., $m_U(x, z)$ is the size of the set

$$K_U(x, z) := \{A \in \mathcal{A} \mid z^{-1} A(x) \in U\}$$

(measured with the Haar measure of $\mathcal{A}$), which can be interpreted as the normalized 'surface measure' of $\mathcal{A}x \cap zU$ in the orbit ('surface') $\mathcal{A}x$. We provide a technical lemma concerning the function m_U.

Lemma 4.3.2. *Let $U = U^{-1} = \mathcal{A}U$ and $Q = Q^{-1} = \mathcal{A}Q$ be relatively compact subsets of $\mathcal{G}$ with non-void interior. Then*

$$m_U(x, z) \leq m_{U^3 Q}(y, z) \quad \text{for all } y \in \mathcal{A}(zUQ), x, z \in \mathcal{G}. \tag{4.7}$$

Proof: If $x \notin \operatorname{supp} m_U(\cdot, z) \subset \mathcal{A}(zU)$ there is nothing to prove. Because of the $\mathcal{A}$-invariance of m_U and $m_{U^3 Q}$ it suffices to prove that $m_U(x, z) \leq m_{U^3 Q}(y, z)$ if $x \in zU$, $y \in zUQ$. The latter means $x = zu_x$ and $y = zu_y q$ for some elements $u_x, u_y \in U, q \in Q$. Hence, $x = yq^{-1} u_y^{-1} u_x =: yq^{-1} v \in yQU^2$. Now suppose $A \in K_U(x, z)$, i.e., $z^{-1} A(x) \in U$ implying $z^{-1} A(yq^{-1} v) \in U$. This gives $z^{-1} A(y) \in UA(v^{-1}) A(q) \subset U^3 Q$ since $\mathcal{A}U = U$ and $\mathcal{A}Q = Q$ by assumption. Hence, $K_U(x, z) \subset K_{U^3 Q}(y, z)$ and $m_U(x, z) \leq m_{U^3 Q}(y, z)$.   □

Now we are ready to prove the announced characterization.

Lemma 4.3.3. *There are constants $C_1, C_2 > 0$ such that*

$$C_1 \|(\lambda_i)_{i \in I}|Y_{\mathcal{A}}^{\natural}\| \leq \|\sum_{i \in I} |\lambda_i| m_k(x_i, \cdot)|Y_{\mathcal{A}}\| \leq C_2 \|(\lambda_i)_{i \in I}|Y_{\mathcal{A}}^{\natural}\|, \qquad (4.8)$$

i.e., the expression in the middle defines an equivalent norm on $Y_{\mathcal{A}}^{\natural}$.

Proof: We claim that it suffices to proof (4.8) for characteristic functions $k = \chi_U$ for a relatively compact neighborhood U of $e \in \mathcal{G}$ satisfying $U = \mathcal{A}(U) = U^{-1}$. Indeed, if k is an arbitrary non-zero and positive function in $(C_c)_{\mathcal{A}}(\mathcal{G})$ then there exists a neighborhood $U = U^{-1} = \mathcal{A}(U) \subset \mathcal{G}$ of e and constants $C_1, C_2 > 0$ such that

$$C_1 \chi_U(x) \leq (L_y k)(x) \leq C_2 \chi_{\operatorname{supp} L_y k} \qquad \text{for all } x \in \mathcal{G}$$

for some suitable $y \in \mathcal{G}$. The set $V := \mathcal{A}(\operatorname{supp}(L_y k) \cup (\operatorname{supp}(L_y k))^{-1})$ is a relatively compact neighborhood of e satisfying $V = V^{-1} = \mathcal{A}(V)$ and $\chi_{\operatorname{supp} L_y k} \leq \chi_V$. This implies $C_1 m_U(x, z) \leq m_{L_y k}(x, z) \leq C_2 m_V(x, z)$ for all $x, z \in \mathcal{G}$. Since $m_{L_y k}(x, z) = m_k(x, zy)$ and Y is right translation invariant this shows the claim.

So we assume that $U = U^{-1} = \mathcal{A}(U)$ is a relatively compact neighborhood of e. By invariance of the Haar measure under left translation and under the action of $\mathcal{A}$ (Lemma 1.2.1) we obtain

$$
\begin{aligned}
|U| &= \int_{\mathcal{G}} \chi_{x_i U}(x) dx = \int_{\mathcal{A}} \int_{\mathcal{G}} \chi_{\mathcal{A}(x_i)U}(x) \chi_{\mathcal{A}(x_i U)}(x) dx dA \\
&= \int_{\mathcal{G}} \int_{\mathcal{A}} \chi_{U^{-1}}(x^{-1} A(x_i)) dA \chi_{\mathcal{A}(x_i U)}(x) dx = \int_{\mathcal{A}(x_i U)} m_U(x_i, x) dx \\
&= \int_{\mathcal{A}(x_i U)} m_U(x, x_i) dx \leq \int_{\mathcal{A}(x_i U)} m_{U^4}(y, x_i) dx = |\mathcal{A}(x_i U)| m_{U^4}(x_i, y)
\end{aligned}
$$

for all $y \in \mathcal{A}(x_i U^2)$ by choosing $Q = U$ in inequality (4.7). Thus we have

$$|U| \chi_{\mathcal{A}(x_i U)}(y) \leq |U| \chi_{\mathcal{A}(x_i U^2)}(y) \leq |\mathcal{A}(x_i U)| m_{U^4}(x_i, y) \quad \text{for all } y \in \mathcal{G}.$$

To obtain a reversed inequality we choose again $Q = U$. For all $x \in \mathcal{G}$, Lemma 4.3.2 yields

$$
\begin{aligned}
|\mathcal{A}(x_i U^2)| m_U(x_i, x) &= \int_{\mathcal{A}(x_i U^2)} m_U(x, x_i) dy \leq \int_{\mathcal{A}(x_i U^2)} m_{U^4}(y, x_i) dy \\
&= \int_{\mathcal{A}(x_i U^2)} \int_{\mathcal{A}} \chi_{x_i U^4}(A(y)) dA dy \leq \int_{\mathcal{A}} \int_{\mathcal{A}(x_i U^4)} \chi_{x_i U^4}(y) dy dA = |U^4|. \qquad (4.9)
\end{aligned}
$$

Hereby, we used the invariance of the Haar measure under $\mathcal{A}$ once more. Since $\operatorname{supp} m_U(x_i, \cdot) \subset \mathcal{A}(x_i U)$ we obtain

$$|\mathcal{A}(x_i U^2)| m_U(x_i, y) \leq |U^4| \chi_{\mathcal{A}(x_i U)}(y) \quad \text{for all } y \in \mathcal{G}.$$

By solidity of Y and since the definition of $Y_{\mathcal{A}}^{\natural}$ does not depend on the choice of the set U by Lemma 4.3.1(a) we finally get inequality (4.8).   □

Let us finally state an important inclusion relation.

Lemma 4.3.4. *Let $\tilde{u}(i) := u(x_i) = \|L_{x_i}|Y\|$ and $r(i) := |\mathcal{A}(x_iU)|u(x_i^{-1})$ for some relatively compact neighborhood $U = U^{-1} = \mathcal{A}(U)$ of $e \in \mathcal{G}$. Then we have the continuous embeddings*

$$l_{\tilde{u}}^1 \subset Y_{\mathcal{A}}^{\natural} \subset l_{1/r}^{\infty}.$$

Proof: Observe that

$$\left\| \sum_{i \in I} |\lambda_i| m_k(x_i, \cdot) |Y \right\| \leq \sum_{i \in I} |\lambda_i| \| m_k(x_i, \cdot) |Y\|.$$

Furthermore, it follows from Lemma 4.1.10 that

$$\| m_k(x_i, \cdot) |Y_{\mathcal{A}}\| = \|\mathcal{L}_{x_i} k |Y_{\mathcal{A}}\| \leq \|\mathcal{L}_{x_i} |Y_{\mathcal{A}}\| \, \|k |Y_{\mathcal{A}}\| \leq C u(x_i).$$

By Lemma 4.3.3 this gives the continuous embedding $l_{\tilde{u}}^1 \subset Y_{\mathcal{A}}^{\natural}$.

For the second embedding observe that by solidity and left translation invariance of Y we obtain

$$\|\chi_U|Y\| = \|L_{x_i^{-1}} \chi_{x_iU}|Y\| \leq u(x_i^{-1}) \|\chi_{x_iU}|Y\| \leq u(x_i^{-1}) \|\chi_{\mathcal{A}(x_iU)}|Y\|.$$

This gives

$$|\lambda_i| |\mathcal{A}(x_iU)|^{-1} \|\chi_U|Y\| \leq u(x_i^{-1}) \| |\lambda_i| |\mathcal{A}(x_iU)|^{-1} \chi_{\mathcal{A}(x_iU)}|Y\|$$
$$\leq u(x_i^{-1}) \| \sum_{j \in I} |\lambda_j| |\mathcal{A}(x_jU)|^{-1} \chi_{\mathcal{A}(x_jU)}|Y\| = u(x_i^{-1}) \|(\lambda_j)_{j \in I} |Y_{\mathcal{A}}^{\natural}\|$$

and thus, $Y_{\mathcal{A}}^{\natural} \subset l_{1/r}^{\infty}$.   □

4.4 Wiener Amalgam Spaces

As another tool we will need Wiener amalgam spaces. The idea of these spaces is to measure at the same time local and global properties of a function. For the definition we take a two-sided invariant solid BF-space Y and another two-sided invariant Banach space B of functions or measures on $\mathcal{G}$. For our purpose it suffices to assume that B contains $C_c(\mathcal{G})$. Our two main examples will be $B = C_0(\mathcal{G})$ and $B = M(\mathcal{G})$. Using a non-zero *window function* $k \in C_c(\mathcal{G})$ (usually a function that satisfies $0 \leq k(x) \leq 1$ and

$k(x) = 1$ for x in some compact neighborhood of the identity) we define the **control function** by

$$K(F, k, B)(x) := \|(L_x k)F|B\|, \quad x \in \mathcal{G}, \tag{4.10}$$

where F is locally contained in B, in symbols $F \in B_{loc}$ (see Section 1.1). The **Wiener amalgam space** $W(B, Y)$ is now defined by

$$W(B, Y) := \{F \in B_{loc}, \ K(F, k, B) \in Y\}$$

with norm

$$\|F|W(B, Y)\| := \|K(F, k, B)|Y\|.$$

It was shown in [Fei83b] that these spaces are two-sided invariant Banach spaces which do not depend on the particular choice of the window function k. Moreover, different functions k define equivalent norms. It follows that we can also take characteristic functions $k = \chi_U$ of relatively compact sets U with non-void interior as window functions (at least for $W(C_0, Y)$ and $W(M, Y)$). For the various properties of Wiener amalgam spaces we refer to [Fei81a, Fei83b, Fei81b, Fei90, FG89a, FG89b, Grö01].

Replacing the left translation L_x with the right translation R_x in the definition (4.10) of the control function leads to right Wiener amalgam spaces $W^R(B, Y)$.

Lemma 4.4.1. *A function F is contained in $W^R(C_0, Y)$ if and only if $F^\vee$ is contained in $W(C_0, Y)$.*

Proof: Suppose $k = k^\vee \in C_c(\mathcal{G})$ is the window function used for the definition of $W(C_0, Y)$ and $W^R(C_0, Y)$. The control function of $F^\vee$ is given by

$$K(F^\vee, k, C_0)(x) = \|(L_x k)F^\vee\|_\infty = \|k(L_{x^{-1}}F^\vee)\|_\infty = \|k(R_{x^{-1}}F)^\vee\|_\infty$$
$$= \|k^\vee(R_{x^{-1}}F)\|_\infty = \|(R_x k)F\|_\infty = K^R(F, k, C_0)(x).$$

This immediately yields the claim. □

We state some convolution properties that will be essential for our purpose.

Proposition 4.4.2. *Suppose Y is a two-sided invariant solid BF-space with associated weight functions u, v defined in (4.2).*

(a) (Proposition 3.10 in [FG89a]) With $v^*(x) = v(x^{-1})\Delta^{-1}(x) = \|R_x|Y\|$ it holds $W(M, Y) * W^R(C_0, L^1_{v^*}) \subset Y$ and

$$\|\mu * G|Y\| \le C\|\mu|W(M, Y)\| \, \|G|W^R(C_0, L^1_{v^*})\|$$

for some constant $C > 0$ and all $\mu \in W(M, Y), G \in W^R(C_0, L^1_{v^*})$.

(b) (Theorem 7.1(b) in [FG89b]) It holds $Y * W(C_0, L_v^1) \subset W(C_0, Y)$ and for some constant $D > 0$ we have

$$\|F * G | W(C_0, Y)\| \leq D \|F|Y\| \, \|G|W(C_0, L_v^1)\| \quad \text{for all } F \in Y, G \in W(C_0, L_v^1).$$

Recall that $\mathcal{G}$ is said to be an IN-group if there exists a compact neighborhood Q of e such that $xQ = Qx$ for all $x \in \mathcal{G}$. In particular, the Heisenberg group is an IN-group (choose $Q = U \times U \times \mathbb{T} \subset \mathbb{H}_d$ with some compact neighborhood U of $0 \in \mathbb{R}^d$). For such groups, we have the following convolution property.

Theorem 4.4.3. *[Fei83b, Theorem 3] Suppose $\mathcal{G}$ is an IN-group. Further assume that (Y_1, Y_2, Y_3) and (B_1, B_2, B_3) are Banach convolution triples, i.e., Banach spaces of functions (measures) on $\mathcal{G}$ satisfying $Y_1 * Y_2 \subset Y_3$ and $B_1 * B_2 \subset B_3$ (with corresponding norm estimates). Then also $(W(B_1, Y_1), W(B_2, Y_2), W(B_3, Y_3))$ is a Banach convolution triple.*

The following lemma contains some embeddings involving Wiener amalgams.

Lemma 4.4.4. *[FG89a, Lemma 3.9] Suppose Y satisfies (Y) and denote by u the function associated to Y by (4.2). Then we have the continuous embeddings*

(a) $W(C_0, Y) \subset Y \subset W(L^1, Y)$,

(b) $Y \subset W(L^1, L_{1/u^\vee}^\infty)$.

There is a relation between Wiener amalgams and the sequence spaces introduced in the previous section. Denote $Y_d := Y_{\{e\}}^\natural = Y_{\{e\}}^\flat$ where $\mathcal{A} = \{e\}$ is the trivial automorphism group. Recall that a BUPU is an IBUPU corresponding to the trivial group $\mathcal{A} = \{e\}$, see Definition 4.2.2.

Theorem 4.4.5. *[Fei83b, Theorem 2] Suppose $\Psi = (\psi_i)_{i \in I}$ is a BUPU. Then $F \in W(B, Y)$ if and only if $(\|F\psi_i|B\|)_{i \in I} \in Y_d$ (with equivalent norms).*

We note that Theorem 4.4.5 is still valid when replacing $W(B, Y)$ by $W^R(B, Y)$ and the BUPU Ψ by a "right" BUPU $\Psi^R = (\psi_j^R)_{j \in J}$, where we require that supp $\psi_j^R \subset U z_j^{-1}$. As always throughout this chapter we further assume that $\mathcal{A}$ acts isometrically on Y and B. Clearly, $\mathcal{A}$ acts then also isometrically on $W(B, Y)$ and we may define the closed subspace

$$W_{\mathcal{A}}(B, Y) := \{F \in W(B, Y), F_A = F \text{ for all } A \in \mathcal{A}\},$$

and analogously for the right Wiener amalgams.

Corollary 4.4.6. *Proposition 4.4.2 remains valid when replacing each of the spaces by its corresponding subspace of measures, which are invariant under $\mathcal{A}$.*

Proof: Observe, that the elements of any of the spaces in 4.4.2 can be realized as Radon measures on $\mathcal{G}$, i.e., as functionals on $C_c(\mathcal{G})$. By Lemma 1.2.3(a) the convolution of two invariant Radon measures is again invariant (provided the convolution exists). $\square$

One might ask whether a similar result to (4.4.5) holds also in presence of invariance under $\mathcal{A}$. Unfortunately, this question cannot be answered easily for general spaces B. (Actually, this is one of the reason why we have to deal with the two sequence spaces $Y_{\mathcal{A}}^{\flat}$ and $Y_{\mathcal{A}}^{\natural}$.) However, for the special case $B = M(\mathcal{G})$ we have the following result relating $W_{\mathcal{A}}(M,Y)$ and $Y_{\mathcal{A}}^{\natural}$. Recall the definition (1.15) of the invariant Dirac measures $\epsilon_{\mathcal{A}x}$.

Lemma 4.4.7. *For some well-spread family $X = (x_i)_{i \in I}$ the measure*

$$\mu_\Lambda := \sum_{i \in I} \lambda_i \epsilon_{\mathcal{A}x_i}$$

is contained in $W_{\mathcal{A}}(M,Y)$ if and only if $\Lambda = (\lambda_i)_{i \in I}$ is contained in $Y_{\mathcal{A}}^{\natural}(X)$ and there are constants $C_1, C_2 > 0$ such that

$$C_1 \|\Lambda| Y_{\mathcal{A}}^{\natural}(X)\| \leq \|\mu_\Lambda | W_{\mathcal{A}}(M,Y)\| \leq C_2 \|\Lambda| Y_{\mathcal{A}}^{\natural}(X)\|.$$

Proof: Observe, that the control function of $\epsilon_{\mathcal{A}x}$ is actually given by the function m_k defined in (4.6), i.e.,

$$m_k(x,z) = \|(L_z k)\epsilon_{\mathcal{A}x}|M\| = K(\epsilon_{\mathcal{A}x}, k, M)(z).$$

Clearly, the supports of the measures $L_z k \epsilon_{\mathcal{A}x_i}, i \in I$, do not overlap for any $z \in \mathcal{G}$. Hence, for the control function applied to μ_Λ it holds

$$K(\mu_\Lambda, k, M)(z) = \|\sum_{i \in I} \lambda_i L_z k \epsilon_{\mathcal{A}x_i}|M\| = \sum_{i \in I} |\lambda_i| m_k(x_i, z).$$

Thus, the assertion follows immediately from Lemma 4.3.3. $\square$

Another important observation is that sampling functions in $W_{\mathcal{A}}^{R}(C_0, L_w^1)$ is a "nice" operation.

Lemma 4.4.8. *If $G \in W_{\mathcal{A}}^{R}(C_0, L_w^1)$ with submultiplicative invariant weight w and $(x_i)_{i \in I}$ is well-spread with respect to $\mathcal{A}$ then $(\mathcal{L}_{x_i} G(x))_{i \in I} \in l_r^1$ for all $x \in \mathcal{G}$ where $r(i) := w(x_i)|\mathcal{A}(x_i U)|$ for some relatively compact neighborhood $U = U^{-1} = \mathcal{A}(U)$ of $e \in \mathcal{G}$.*

Proof: It follows from Theorem 4.4.5 (and the remarks after it) that $G \in W^R(C_0, L^1_w)$ has a decomposition $G = \sum_{j \in J} R_{z_j} G_j$ with $\operatorname{supp} G_j \subset Q = Q^{-1} = \mathcal{A}(Q)$ for some compact Q and

$$\sum_{j \in J} \|G_j\|_\infty w(z_j) \leq C\|G|W^R(C_0, L^1_w)\|.$$

By definition of m_Q we have

$$|\mathcal{L}_{x_i} G_j(x)| = |\epsilon_{\mathcal{A}x_i} * (\chi_Q G_j)(x)| \leq \|G_j\|_\infty m_Q(x_i, x).$$

Hence, we obtain the estimation

$$\sum_{i \in I} |\mathcal{L}_{x_i} G(x)| w(x_i) |\mathcal{A}(x_i U)| \leq \sum_{i \in I} \sum_{j \in J} |\epsilon_{\mathcal{A}(x_i)} * R_{z_j} G_j(x)| w(x_i) |\mathcal{A}(x_i U)|$$

$$\leq \sum_{j \in J} \sum_{i \in I_{x,j}} \|G_j\|_\infty m_Q(x_i, x z_j) w(x_i) |\mathcal{A}(x_i U)|.$$

The inner sum runs over the finite index set

$$I_{x,j} = \{i \in I, x_i \in \mathcal{A}(x z_j Q)\}.$$

Since $(x_i)_{i \in I}$ is well-spread it holds $|I_{x,j}| \leq C_Q < \infty$ uniformly for all x, j. For each $i \in I_{x,j}$ we may write $x_i = A_i(x z_j q_i)$ for some $q_i \in Q, A_i \in \mathcal{A}$, hence by invariance and submultiplicativity $w(x_i) \leq w(x) w(z_j) w(q_i)$. Further, it follows from (4.9) that $m_Q(x_i, x z_n) \leq C'|\mathcal{A}(x_i U)|^{-1}$ for some suitable constant $C' > 0$. Thus, we finally obtain

$$\sum_{i \in I} |\mathcal{L}_{x_i} G(x)| w(x_i) |\mathcal{A}(x_i U)| \leq w(x) C' C_Q \sup_{q \in Q} w(q) \sum_{j \in J} \|G_j\|_\infty w(z_j) < \infty. \qquad (4.11)$$

This finishes the proof. $\square$

Note that (4.11) implies that the function $x \mapsto \sum_{i \in I} \mathcal{L}_{x_i} G(x) w(x_i) |\mathcal{A}(x_i U)|$ is contained in $L^\infty_{1/w}(\mathcal{G})$. The following inequalities will be essential in later estimations.

Lemma 4.4.9. *Suppose that* $F \in W_{\mathcal{A}}(C_0, Y)$ *and* $\Psi = (\psi_i)_{i \in I}$ *is some* U-$\mathcal{A}$-*IBUPU with corresponding well-spread set* $X = (x_i)_{i \in I}$*. Then*

$$\left\| \sum_{i \in I} F(x_i) \psi_i | W_{\mathcal{A}}(C_0, Y) \right\| \leq \gamma(U) \|F|W_{\mathcal{A}}(C_0, Y)\|$$

and

$$\|(F(x_i))_{i \in I}|Y^\flat_{\mathcal{A}}\| \leq \gamma(U) C \|F|W_{\mathcal{A}}(C_0, Y)\| \qquad (4.12)$$

for constants $\gamma(U), C > 0$. *If* U *varies through a family of subsets of some compact* $U_0 \subset \mathcal{G}$ *then* $\gamma(U)$ *is uniformly bounded by some constant* γ_0.

Proof: We proceed similarly as in [Grö91, Lemma 4.4]. Without loss of generality we assume that a characteristic function χ_Q for some relatively compact neighborhood $Q = Q^{-1} = \mathcal{A}(Q)$ of $e \in \mathcal{G}$ is taken for the definition of the norm of $W(C_0, Y)$. We obtain for the control function

$$K\left(\sum_{i \in I} |F(x_i)| \psi_i, \chi_Q, C_0\right)(x) = \|(L_x \chi_Q) \sum_{i \in I} |F(x_i)| \psi_i\|_\infty =: H(x).$$

The sum in the last expression runs only over the finite index set

$$I_x := \{i \in I, xQ \cap \mathcal{A}(x_iU) \neq \emptyset\} = \{i \in I, \mathcal{A}(x_i) \cap xQU^{-1} \neq \emptyset\}.$$

Since F is $\mathcal{A}$ invariant and since $(\psi_i)_{i \in I}$ is a partition of unity we therefore have

$$H(x) \leq \|(L_x \chi_{QU^{-1}})F\|_\infty = K(F, \chi_{QU^{-1}}, C_0)(x).$$

Since different window functions define equivalent norms on $W(C_0, Y)$ [Fei83b] there exists a constant $\gamma(U)$ such that

$$\|K(F, \chi_{QU^{-1}}, C_0)|Y\| \leq \gamma(U)\|K(F, \chi_Q, C_0)|Y\|. \tag{4.13}$$

We finally obtain

$$\|\sum_{i \in I} |F(x_i)| \psi_i | W_{\mathcal{A}}(C_0, Y)\| = \|K(\sum_{i \in I} |F(x_i)| \psi_i, \chi_Q, C_0)|Y_{\mathcal{A}}\|$$
$$\leq \|K(F, \chi_{QU^{-1}}, C_0)|Y_{\mathcal{A}}\| \leq \gamma(U)\|K(F, \chi_Q, C_0)|Y_{\mathcal{A}}\| = \gamma(U)\|F|W_{\mathcal{A}}(C_0, Y)\|.$$

To prove inequality (4.12) one proceeds analogously using

$$\|(F(x_i))_{i \in I}|Y_{\mathcal{A}}^\flat\| \leq \|\sum_{i \in I} F(x_i) \chi_{\mathcal{A}(x_iU)}|W_{\mathcal{A}}(C_0, Y)\|,$$

which is easily seen with the finite overlap property of the well-spread family $(x_i)_{x \in I}$ (Lemma 4.2.1).

In order to show the assertion on $\gamma(U)$ we need to give a prove of (4.13) that provides an estimation of the constant $\gamma(U)$ (which is actually hard to extract from the proof in [Fei83b]). Since QU^{-1} is relatively compact there exists a covering $QU^{-1} \subset \bigcup_{k=1}^n z_kQ$ for some points $z_k \in \mathcal{G}$. If $V = V^{-1}$ is such that $V^2 \subset Q$ then the points $z_k, k = 1, \ldots, n$, can be chosen such that

$$n \leq \frac{|QU^{-1}V|}{|V|}. \tag{4.14}$$

Indeed, choose a maximal set of points $z_k \in QU^{-1}$, $k = 1, \ldots, n$, such that the sets $z_kV \subset QU^{-1}V$ are mutually disjoint. Then the maximal number n is given by (4.14) and the sets $z_k V^2$ (and hence also the sets $z_k Q$) cover QU^{-1}. We obtain the estimation

$$K(F, \chi_{QU^{-1}}, C_0)(x) = \|(L_x \chi_{QU^{-1}})F\|_\infty \leq \| \sum_{k=1}^{n}(L_x \chi_{z_k Q})F\|_\infty$$

$$\leq \sum_{k=1}^{n} \|(L_{xz_k} \chi_Q)F\|_\infty = \sum_{k=1}^{n} R_{z_k} K(F, \chi_Q, C_0)(x),$$

and hence

$$\|K(F, \chi_{QU^{-1}}, C_0)|Y\| \leq \sum_{k=1}^{n} \|R_{z_k} K(F, \chi_Q, C_0)|Y\| \leq \sum_{k=1}^{n} w(z_k)\|K(F, \chi_Q, C_0)|Y\|.$$

Thus, it holds

$$\gamma(U) \leq \sum_{k=1}^{n} w(z_k) \leq n \sup_{z \in QU^{-1}} w(z) \leq \frac{|QU^{-1}V|}{|V|} \sup_{z \in QU^{-1}} w(z).$$

If U runs through a family of subsets of some U_0 then $\gamma(U)$ is clearly bounded. $\square$

Wiener amalgam spaces are also a natural tool to handle maximal functions of the following type.

Definition 4.4.1. If $U \subset \mathcal{G}$ is a relatively compact neighborhood of e then

$$G_U^{\#}(x) := \sup_{u \in U} |G(ux) - G(x)|$$

is the U-oscillation of G.

It is easily seen that $G_U^{\#}$ is invariant under $\mathcal{A}$ whenever G is $\mathcal{A}$-invariant and $U = \mathcal{A}U$.

Lemma 4.4.10. *([Grö91, Lemma 4.6])*

(a) *A function G is contained in $W^R(L^\infty, L_w^1)$ if and only if $G \in L_w^1$ and $G_U^{\#} \in L_w^1$ for some (and hence for all) open relatively compact neighborhood U of e.*

(b) *If, in addition, G is continuous, i.e., $G \in W^R(C_0, L_w^1)$, then*

$$\lim_{U \to \{e\}} \|G_U^{\#}|L_w^1\| = 0. \tag{4.15}$$

(c) *If $y \in xU$, then $|L_y G - L_x G| \leq L_y G_U^{\#}$ holds pointwise.*

Corollary 4.4.11. *If G is $\mathcal{A}$-invariant and $y \in \mathcal{A}(xU)$ then $|\mathcal{L}_y G - \mathcal{L}_x G| \leq \mathcal{L}_y G_U^{\#}$ holds pointwise.*

Proof: Since $y \mapsto \mathcal{L}_y G$ is invariant under $\mathcal{A}$ it is enough to consider $y \in xU$. In this case it holds by Lemma 4.4.10(c)

$$|\mathcal{L}_y G - \mathcal{L}_x G|(z) = \left| \int_{\mathcal{A}} G(y^{-1}Az) - G(x^{-1}Az)dA \right|$$
$$\leq \int_{\mathcal{A}} |L_y G(Az) - L_x G(Az)|dA \leq \int_{\mathcal{A}} L_y G_U^{\#}(Az)dA = \mathcal{L}_y G_U^{\#}(z). \qquad \square$$

4.5 Coorbit Spaces

After having discussed the necessary technical background we are ready to introduce the coorbit spaces and to prove some of their basic properties. These Banach spaces provide the natural setting to obtain discretizations of the form $\{\widetilde{\pi}(x_i)g, i \in I\}, g \in \mathcal{H}_{\mathcal{A}}$, of the continuous transforms treated in Chapter 2.

Again we assume that π is an irreducible unitary representation of $\mathcal{G}$ on $\mathcal{H}$ and that σ is a unitary representation of $\mathcal{A}$ on $\mathcal{H}$ such that (2.5) holds, i.e., $\pi(A(x))\sigma(A) = \sigma(A)\pi(x)$. We also assume the space $\mathcal{H}_{\mathcal{A}}$ to be non-trivial. Additionally, we require throughout the rest of this chapter that there exists a non-zero $g \in \mathcal{H}_{\mathcal{A}}$ such that $V_g g \in L^1(\mathcal{G})$. This means in particular that π is **integrable** [FG88, FG89a]. Since $|\langle g, \pi(x)g \rangle|^2 \leq \|g\|^2 |\langle g, \pi(x)g \rangle|$ we observe that π is also square-integrable.

Definition 4.5.1. Let w be a submultiplicative invariant weight function satisfying $w(x) \geq 1$ for all $x \in \mathcal{G}$. The space of analyzing vectors associated to w is defined by

$$\mathbb{A}_w := \{g \in \mathcal{H}, V_g g \in L^1_w(\mathcal{G})\}$$

and its subspace of invariant elements by

$$\mathbb{A}_w^{\mathcal{A}} := \mathbb{A}_w \cap \mathcal{H}_{\mathcal{A}} = \{g \in \mathcal{H}_{\mathcal{A}}, \widetilde{V}_g g \in L^1_w(\mathcal{G})\}.$$

In the sequel we will only consider those weights w for which $\mathbb{A}_w^{\mathcal{A}}$ is non-trivial. By the integrability assumption on π this is at least the case for $w \equiv 1$. Observe also that $\mathbb{A}_w^{\mathcal{A}}$ is a subset of the admissible invariant elements $\mathcal{D}_{\mathcal{A}}(K)$, where K denotes the operator in the Duflo-Moore theorem 1.5.3.

Lemma 4.5.1. *Suppose w is a weight function as in Definition 4.5.1 such that $\mathbb{A}_w^{\mathcal{A}} \neq \{0\}$. Then $\mathbb{A}_w$ is dense in $\mathcal{H}$ and $\mathbb{A}_w^{\mathcal{A}}$ is dense in $\mathcal{H}_{\mathcal{A}}$.*

Proof: Suppose $0 \neq g \in \mathbb{A}_w^{\mathcal{A}}$. By definition of the generalized left and right translation (1.13), (1.14) we obtain

$$\widetilde{V}_{\widetilde{\pi}(x)g}(\widetilde{\pi}(x)g)(y) = \langle \widetilde{\pi}(x)g, \widetilde{\pi}(y)\widetilde{\pi}(x)g \rangle = \epsilon_{\mathcal{A}x^{-1}} * \epsilon_{\mathcal{A}y} * \epsilon_{\mathcal{A}x}(\widetilde{V}_g g) = \mathcal{L}_x \mathcal{R}_x \widetilde{V}_g g(y).$$

Since L^1_w is two-sided invariant by Lemma 4.1.5 we conclude from Lemma 4.1.10 that $\widetilde{\pi}(x)g \in \mathbb{A}_w^{\mathcal{A}}$ for all $x \in \mathcal{G}$. With Lemma 1.4.10 we conclude that span$\{\widetilde{\pi}(x)g, x \in \mathcal{G}\}$ is dense in $\mathcal{H}_{\mathcal{A}}$. Thus, $\mathbb{A}_w^{\mathcal{A}}$ is dense in $\mathcal{H}_{\mathcal{A}}$. Setting $\mathcal{A} = \{e\}$ yields the assertion for $\mathcal{H}$. $\qquad \square$

4.5.1 The Space $(\mathcal{H}_w^1)_{\mathcal{A}}$

Before introducing coorbit spaces in their full generality, we first have to consider a special case.

Definition 4.5.2. Let w be a weight function as in Definition 4.5.1 such that $\mathbb{A}_w^{\mathcal{A}}$ is non-trivial. With a fixed non-zero vector $g \in \mathbb{A}_w^{\mathcal{A}}$ the space $\mathcal{H}_w^1$ is defined by

$$\mathcal{H}_w^1 := \{f \in \mathcal{H}, \, V_g f \in L_w^1\}$$

with norm

$$\|f|\mathcal{H}_w^1\| := \|V_g f|L_w^1\|.$$

Its subspace of invariant elements is denoted by

$$(\mathcal{H}_w^1)_{\mathcal{A}} := \mathcal{H}_{\mathcal{A}} \cap \mathcal{H}_w^1 = \{f \in \mathcal{H}_{\mathcal{A}}, \, \widetilde{V}_g f \in L_w^1\}.$$

It follows from the injectivity of V_g that $\| \cdot |\mathcal{H}_w^1\|$ is indeed a norm and not only a semi-norm.

In the following we will formulate most statements only for $(\mathcal{H}_w^1)_{\mathcal{A}}$. Choosing the trivial group $\mathcal{A} = \{e\}$ gives, of course, the corresponding statements for $\mathcal{H}_w^1$.

Lemma 4.5.2. *(a)* $(\mathcal{H}_w^1)_{\mathcal{A}}$ *is a $\widetilde{\pi}$-invariant Banach space with* $\|\widetilde{\pi}(x)|\mathcal{H}_w^1\| \leq w(x)$.

(b) The definition of $(\mathcal{H}_w^1)_{\mathcal{A}}$ does not depend on the choice of $g \in \mathbb{A}_w^{\mathcal{A}}$.

(c) $(\mathcal{H}_w^1)_{\mathcal{A}}$ is continuously embedded and dense in $\mathcal{H}_{\mathcal{A}}$.

(d) With $w^{\bullet}(x) := w(x) + \Delta(x^{-1})w(x^{-1})$ it holds $\mathbb{A}_w^{\mathcal{A}} = \mathbb{A}_{w^{\bullet}}^{\mathcal{A}} = (\mathcal{H}_{w^{\bullet}}^1)_{\mathcal{A}}$.

Proof: (a) Without loss of generality we may assume $\|Kg\| = 1$. Let $(f_n)_{n \in N}$ be a Cauchy sequence in $(\mathcal{H}_w^1)_{\mathcal{A}}$ implying that $V_g f_n$ is a Cauchy sequence in $(L_w^1)_{\mathcal{A}}$. By completeness of $(L_w^1)_{\mathcal{A}}$ there exists a limit $F \in (L_w^1)_{\mathcal{A}}$. Since by Corollary 4.1.6 $(L_w^1)_{\mathcal{A}}$ is a Banach-algebra and $V_g g \in (L_w^1)_{\mathcal{A}}$ by assumption, the reproducing formula (1.82) yields $F = \lim V_g f_n = \lim V_g f_n * V_g g = F * V_g g$ in $(L_w^1)_{\mathcal{A}}$. Moreover, $w \geq 1$ implies $L_w^1 \subset L^1$ and hence, $L_w^1 * L^{\infty} \subset L^{\infty}$. Since $V_g g \in C^b(\mathcal{G})$ this means that $F \in L^1(\mathcal{G}) \cap L^{\infty}(\mathcal{G}) \subset L^2(\mathcal{G})$. Thus by Corollary 1.5.5(d) we have $F = V_g f$ for some $f \in \mathcal{H}_{\mathcal{A}}$, which is also contained in $(\mathcal{H}_w^1)_{\mathcal{A}}$ since $F \in L_w^1$. Hence, $(\mathcal{H}_w^1)_{\mathcal{A}}$ is complete. By Theorem 2.2.5 we have $V_g(\widetilde{\pi}(x)f) = \mathcal{L}_x V_g f$ and by Lemma 4.1.5 and Lemma 4.1.10 the latter is contained in L_w^1 whenever $f \in (\mathcal{H}_w^1)_{\mathcal{A}}$. We conclude

$$\|\widetilde{\pi}(x)|\mathcal{H}_w^1\| \leq \|\mathcal{L}_x|L_w^1\| \leq \|L_x|L_w^1\| \leq w(x).$$

(b) Let $0 \neq g, g' \in \mathbb{A}_w^{\mathcal{A}}$. The inversion formula for V_g implies $g' = \int_{\mathcal{G}} V_g g'(y)\pi(y)g \, dy$ and hence

$$V_{g'}f(x) = \int_{\mathcal{G}} \overline{V_g g'(y)}\langle f, \pi(xy)g \rangle dy = V_g f * (V_g g')^{\nabla}(x) \quad \text{for all } f \in \mathcal{H}. \quad (4.16)$$

Further note that $(V_g g)^\nabla = V_g g$ implies $\mathbb{A}_w^{\mathcal{A}} = \mathbb{A}_{w^\bullet}^{\mathcal{A}}$ with $w^\bullet(x) := w(x) + w(x^{-1})\Delta(x^{-1})$. Clearly, also $w^\bullet$ is submultiplicative. Now choose a vector $h \in \mathbb{A}_w^{\mathcal{A}} \subset \mathcal{D}_{\mathcal{A}}(K)$ such that Kh is not orthogonal to Kg and Kg'. With Corollary 1.5.5(b) we obtain

$$0 \neq \langle Kg', Kh \rangle \langle Kh, Kg \rangle V_g g' = V_{g'} g' * V_h h * V_g g$$

and, hence, $V_g g' \in L_{w^\bullet}^1$. Since $L_{w^\bullet}^1$ is invariant under the map $F \mapsto F^\nabla$ also $(V_g g')^\nabla$ is contained in $L_{w^\bullet}^1$. By (4.16) we conclude that $V_{g'} f \in L_w^1$ if $V_g f \in L_w^1$. Exchanging the roles of g and g' shows that also the converse is true and, hence, $(\mathcal{H}_w^1)_{\mathcal{A}}$ is independent of the choice of $g \in \mathbb{A}_w^{\mathcal{A}}$ with equivalent norms for different choices.

(c) The above arguments show also that $\mathbb{A}_w^{\mathcal{A}} \subset (\mathcal{H}_w^1)_{\mathcal{A}}$. By Lemma 4.5.1 we conclude that $(\mathcal{H}_w^1)_{\mathcal{A}}$ is dense in $\mathcal{H}_{\mathcal{A}}$. Moreover, if $f \in (\mathcal{H}_w^1)_{\mathcal{A}}$ then

$$\|f|\mathcal{H}_{\mathcal{A}}\|^2 = \|Kg|\mathcal{H}\|^{-2} \|V_g f|L^2\|^2 \leq \|Kg|\mathcal{H}\|^{-2} \|V_g f|L^\infty\| \, \|V_g f|L^1\|$$
$$\leq \|Kg|\mathcal{H}\|^{-2} \|g|\mathcal{H}\| \, \|f|\mathcal{H}_{\mathcal{A}}\| \, \|f|(\mathcal{H}_w^1)_{\mathcal{A}}\|.$$

This shows the continuity of the embedding $(\mathcal{H}_w^1)_{\mathcal{A}} \subset \mathcal{H}_{\mathcal{A}}$.

(d) The relation $\mathbb{A}_w^{\mathcal{A}} = \mathbb{A}_{w^\bullet}^{\mathcal{A}}$ was already shown in (b). Let $0 \neq g \in \mathbb{A}_w^{\mathcal{A}} = \mathbb{A}_{w^\bullet}^{\mathcal{A}}$ and assume that g' is another element of $\mathbb{A}_w^{\mathcal{A}}$. The proof of (b) showed that $V_g g' \in L_{w^\bullet}^1$, i.e., $g' \in (\mathcal{H}_{w^\bullet}^1)_{\mathcal{A}}$. Conversely, if $g' \in (\mathcal{H}_{w^\bullet}^1)$ then (4.16) specialized to $f = g'$ yields $V_{g'} g' = V_g g' * (V_g g')^\nabla \in L_{w^\bullet}^1$ because $L_{w^\bullet}^1$ is a Banach algebra, which is invariant under ∇. Thus, we have $g' \in \mathbb{A}_{w^\bullet}^{\mathcal{A}} = \mathbb{A}_w^{\mathcal{A}}$. □

Further, we will need the anti-dual $(\mathcal{H}_w^1)^\daleth$, that is, the space of all continuous conjugate-linear functionals on $(\mathcal{H}_w^1)$. We extend the bracket $\langle \cdot, \cdot \rangle$ to $(\mathcal{H}_w^1)^\daleth \times \mathcal{H}_w^1$ by means of $\langle f, g \rangle = f(g)$. We remark that by taking the anti-dual instead of the usual dual we can formally use the bracket in the same way as for the Hilbert space $\mathcal{H}$ and all formulas carry over without change. Note that the anti-dual can always be identified with the dual via the mapping $J : (\mathcal{H}_w^1)' \to (\mathcal{H}_w^1)^\daleth$, $J(f)(h) = \overline{f(h)}, h \in \mathcal{H}_w^1$. We also extend the bracket of $L^2(\mathcal{G})$ by $\langle F, G \rangle = \int_{\mathcal{G}} F(x)\overline{G(x)}dx$ for $F \in L_{1/w}^\infty(\mathcal{G}), G \in L_w^1(\mathcal{G})$.

In the presence of invariance also the anti-dual $((\mathcal{H}_w^1)_{\mathcal{A}})^\daleth$ becomes relevant. Define a map $\tilde{} : ((\mathcal{H}_w^1)_{\mathcal{A}})^\daleth \to (\mathcal{H}_w^1)^\daleth$ by $\tilde{f}(g) := f(Q_{\mathcal{A}}g), g \in \mathcal{H}_w^1$, where $Q_{\mathcal{A}}g = \int_{\mathcal{A}} g_A dA$ is the projection operator defined in (2.6). As in Lemma 1.2.2(g) one proves that the map $\tilde{}$ establishes an isometric isomorphism between $((\mathcal{H}_w^1)_{\mathcal{A}})^\daleth$ and $((\mathcal{H}_w^1)^\daleth)_{\mathcal{A}}$, the space of all functionals f in $(\mathcal{H}_w^1)^\daleth$ that satisfy $f(g_A) = f(g)$ for all $A \in \mathcal{A}$ and $g \in \mathcal{H}_w^1$. We may therefore unambiguously write $(\mathcal{H}_w^1)_{\mathcal{A}}^\daleth$.

Let us recall the concept of **Gelfand triple** as described in [GV64], [FN03, p.102]. Suppose B is a Banach space continuously and densely embedded into some Hilbert space $\mathcal{H}$ and denote by B' the dual of B. Then $(B, \mathcal{H}, B')$ is called a Gelfand triple. The Hilbert space $\mathcal{H}$ is then weak-* densely embedded into B', see also [Sch71, p.125].

Returning to our specific situation, $((\mathcal{H}_w^1)_\mathcal{A}, \mathcal{H}_\mathcal{A}, (\mathcal{H}_w^1)_\mathcal{A}^\daleth)$ is in fact a Gelfand triple by Lemma 4.5.2. In particular, we note that

$$(\mathcal{H}_w^1)_\mathcal{A} \subset \mathcal{H}_\mathcal{A} \subset (\mathcal{H}_w^1)_\mathcal{A}^\daleth$$

and $\mathcal{H}_\mathcal{A}$ is weak-$*$ dense in $(\mathcal{H}_w^1)_\mathcal{A}^\daleth$. A more precise statement of this fact will be shown in Lemma 4.5.8(b) below.

By Lemma 4.5.2 all elements $\widetilde{\pi}(x)g$ are contained in $(\mathcal{H}_w^1)_\mathcal{A}$ for $g \in (\mathcal{H}_w^1)_\mathcal{A}$. Hence, the action of $\widetilde{\pi}$ on $(\mathcal{H}_w^1)_\mathcal{A}$ can be extended to $(\mathcal{H}_w^1)_\mathcal{A}^\daleth$ by the usual rule $(\widetilde{\pi}(x)f)(g) = f(\widetilde{\pi}(x^{-1})g)$ for $f \in (\mathcal{H}_w^1)_\mathcal{A}^\daleth, g \in (\mathcal{H}_w^1)_\mathcal{A}$ and it is reasonable to extend also the voice transform to $(\mathcal{H}_w^1)_\mathcal{A}^\daleth$ by

$$\widetilde{V}_g f(x) := \langle f, \widetilde{\pi}(x)g \rangle = f(\widetilde{\pi}(x)g), \quad f \in (\mathcal{H}_w^1)_\mathcal{A}^\daleth, g \in \mathbb{A}_w^\mathcal{A}. \tag{4.17}$$

Let us show some basic properties of $\widetilde{V}_g$ on $(\mathcal{H}_w^1)_\mathcal{A}^\daleth$.

Lemma 4.5.3. For $g \in \mathbb{A}_w^\mathcal{A}, f \in (\mathcal{H}_w^1)_\mathcal{A}^\daleth$ the function $\widetilde{V}_g f$ is continuous.

Proof: It holds

$$|\widetilde{V}_g f(x) - \widetilde{V}_g f(x_0)| = |\langle f, (\widetilde{\pi}(x) - \widetilde{\pi}(x_0))g \rangle| \leq \|f|(\mathcal{H}_w^1)_\mathcal{A}^\daleth\| \, \|(\widetilde{\pi}(x) - \widetilde{\pi}(x_0))g|(\mathcal{H}_w^1)_\mathcal{A}\|$$
$$= \|f|(\mathcal{H}_w^1)_\mathcal{A}^\daleth\| \, \|\mathcal{L}_x \widetilde{V}_g g - \mathcal{L}_{x_0} \widetilde{V}_g g|L_w^1\|.$$

The mapping $x \mapsto L_x F$ is continuous from $\mathcal{G}$ into L_w^1 for any fixed $F \in L_w^1$ and, hence, the same holds true for $\mathcal{L}_x = P_\mathcal{A} L_x$ by Lemma 1.2.2(d). This proves the claim. $\square$

Lemma 4.5.4. Suppose $g \in \mathbb{A}_w^\mathcal{A} \setminus \{0\}$. Then $\|\widetilde{V}_g f|L_{1/w}^\infty\|$ defines an equivalent norm on $(\mathcal{H}_w^1)_\mathcal{A}^\daleth$. Consequently, $\widetilde{V}_g$ is injective on $(\mathcal{H}_w^1)_\mathcal{A}^\daleth$.

Proof: Together with Lemma 4.5.2(a) we obtain for $f \in (\mathcal{H}_w^1)_\mathcal{A}$:

$$|\widetilde{V}_g f(x)| = |\langle f, \widetilde{\pi}(x)g \rangle| \leq \|f|(\mathcal{H}_w^1)_\mathcal{A}^\daleth\| \, \|\widetilde{\pi}(x)|(\mathcal{H}_w^1)_\mathcal{A}\| \, \|g|(\mathcal{H}_w^1)_\mathcal{A}\|$$
$$\leq \|f|(\mathcal{H}_w^1)_\mathcal{A}^\daleth\| \|g|(\mathcal{H}_w^1)_\mathcal{A}\| \, w(x).$$

For a converse inequality recall the adjoint operator $\widetilde{V}_g^* : (L^2)_\mathcal{A} \to \mathcal{H}_\mathcal{A}$ computed in Lemma 2.2.7(c). Denote $R := \widetilde{V}_g(\mathcal{H}_w^1)_\mathcal{A} \subset (L_w^1)_\mathcal{A}$ the range of $\widetilde{V}_g$ restricted to $(\mathcal{H}_w^1)_\mathcal{A}$. By the inversion formula in Lemma 2.2.7(d) $\widetilde{V}_g^*$ is an isometric isomorphism between R and $(\mathcal{H}_w^1)_\mathcal{A}$. We obtain

$$\|f|(\mathcal{H}_w^1)_\mathcal{A}^\daleth\| = \sup_{\|h|(\mathcal{H}_w^1)_\mathcal{A}\| \leq 1} |\langle f, h \rangle| = \sup_{H \in R, \|H|L_w^1\| \leq 1} |\langle f, \widetilde{V}_g^* H \rangle|$$
$$\leq \sup_{H \in (L_w^1)_\mathcal{A}, \|H|L_w^1\| \leq 1} |\langle \widetilde{V}_g f, H \rangle| = \|\widetilde{V}_g f|L_{1/w}^\infty\|$$

and the proof is completed. □

It is important that the linear span of $\{\widetilde{\pi}(x)g, x \in \mathcal{G}\}, g \in (\mathcal{H}^1_w)_\mathcal{A}$, is not only dense in $\mathcal{H}_\mathcal{A}$ but also in $(\mathcal{H}^1_w)_\mathcal{A}$ (with respect to its norm). This follows easily from the next result. We remark that our method of proving the density differs from the original one in [FG88].

Theorem 4.5.5. *Let* $g \in \mathbb{A}^\mathcal{A}_w \setminus \{0\}$. *Then* $(\mathcal{H}^1_w)_\mathcal{A}$ *is characterized by the vector space of all uniform expansions of the form*

$$f = \sum_{i \in I} c_i \widetilde{\pi}(x_i) g \qquad (4.18)$$

where $\{x_i, i \in I\}$ *is an arbitrary countable subset of* $\mathcal{G}$ *and*

$$\sum_{i \in I} |c_i| w(x_i) < \infty.$$

The expression

$$\|f|(\mathcal{H}^1_w)_\mathcal{A}\|' := \inf \sum_{i \in I} |c_i| w(x_i),$$

where the infimum is taken over all representations (4.18) of f, *defines an equivalent norm on* $(\mathcal{H}^1_w)_\mathcal{A}$.

Proof: We proceed similarly as in [Bon86], see also [Grö01, Theorem 12.1.8]. Let $l^1_w(\mathcal{G})$ denote the Banach space of all complex-valued functions on $\mathcal{G}$ with norm

$$\|c|l^1_w(\mathcal{G})\| = \sum_{x \in \mathcal{G}} |c(x)| w(x).$$

Clearly, an element $c \in l^1_w(\mathcal{G})$ vanishes everywhere except on a countable set $\{x_i, i \in I\} \subset \mathcal{G}$. Further, define an operator on $l^1_w(\mathcal{G})$ by

$$D_g c := \sum_{x \in \mathcal{G}} c(x) \widetilde{\pi}(x) g.$$

Using Lemma 4.5.2 we obtain

$$\|D_g c|(\mathcal{H}^1_w)_\mathcal{A}\| \leq \sum_{x \in \mathcal{G}} |c(x)| \|\widetilde{\pi}(x)g|(\mathcal{H}^1_w)_\mathcal{A}\| \leq \sum_{x \in \mathcal{G}} |c(x)| w(x) \|g|(\mathcal{H}^1_w)_\mathcal{A}\|$$
$$\leq \|c|l^1_w(\mathcal{G})\| \, \|g|(\mathcal{H}^1_w)_\mathcal{A}\|.$$

The range of D_g is hence contained in $(\mathcal{H}^1_w)_\mathcal{A}$.

To show that D_g is surjective onto $(\mathcal{H}_w^1)_{\mathcal{A}}$ we use the closed range theorem. So let us compute the adjoint operator $D_g^* : (\mathcal{H}_w^1)_{\mathcal{A}}^{\urcorner} \to l_{1/w}^{\infty}(\mathcal{G})$. If $f \in (\mathcal{H}_w^1)_{\mathcal{A}}^{\urcorner}$ and $c \in l_w^1(\mathcal{G})$ then

$$\langle D_g^* f, c \rangle = \langle f, D_g c \rangle = \sum_{x \in \mathcal{G}} \overline{c(x)} \langle f, \widetilde{\pi}(x)g \rangle.$$

Consequently, we have $D_g^* f(x) = \widetilde{V}_g f(x)$ and

$$\|D_g^* f | l_{1/w}^{\infty}(\mathcal{G})\| = \|\widetilde{V}_g f | L_{1/w}^{\infty}\|.$$

Since by Lemma 4.5.4 the latter term is an equivalent norm on $(\mathcal{H}_w^1)_{\mathcal{A}}^{\urcorner}$ the mapping D_g^* is injective and has closed range. Thus by the closed range theorem (see e.g. Lemma IV.5.3 in [Wer95]), D_g is surjective onto $(\mathcal{H}_w^1)_{\mathcal{A}}$.

Let $\mathcal{N}$ denote the kernel of D_g. Then the operator

$$\tilde{D}_g : l_w^1(\mathcal{G})/\mathcal{N} \to (\mathcal{H}_w^1)_{\mathcal{A}}, \quad \tilde{D}_g(c + \mathcal{N}) = D_g c$$

is an isomorphism by surjectivity of D_g. Hence, by the open mapping theorem the norms $\|c + \mathcal{N} | l_w^1(\mathcal{G})/\mathcal{N}\| = \inf_{n \in \mathcal{N}} \|c + n | l_w^1(\mathcal{G})\|$ and $\|\tilde{D}_g(c + \mathcal{N}) | (\mathcal{H}_w^1)_{\mathcal{A}}\| = \|D_g c | (\mathcal{H}_w^1)_{\mathcal{A}}\|$ are equivalent. If $f = D_g c$ then by definition

$$\|f | (\mathcal{H}_w^1)_{\mathcal{A}}\|' = \|c + \mathcal{N} | l_w^1(\mathcal{G})/\mathcal{N}\|,$$

which proves the assertion. $\square$

We immediately conclude the following.

Corollary 4.5.6. *The linear span of* $\{\widetilde{\pi}(x)g, x \in \mathcal{G}\}$ *is dense in* $(\mathcal{H}_w^1)_{\mathcal{A}}$ *for all non-zero* $g \in \mathbb{A}_w^{\mathcal{A}}$.

Corollary 4.5.7. *Suppose B is a $\widetilde{\pi}$-invariant Banach space with $B \cap \mathbb{A}_w^{\mathcal{A}} \neq \{0\}$ satisfying $\|\widetilde{\pi}(x)|B\| \leq w(x)$. Then $(\mathcal{H}_w^1)_{\mathcal{A}}$ is continuously embedded into B, i.e., $(\mathcal{H}_w^1)_{\mathcal{A}}$ is the minimal Banach space with the mentioned properties.*

Proof: Let $0 \neq g \in \mathbb{A}_w^{\mathcal{A}} \cap B$ and $f = \sum_{i \in I} c_i \widetilde{\pi}(x_i)g$ with $\sum_{i \in I} |c_i| w(x_i) < \infty$, i.e., $f \in (\mathcal{H}_w^1)_{\mathcal{A}}$ by Theorem 4.5.5. Using the $\widetilde{\pi}$-invariance of B we obtain

$$\|f | B\| \leq \sum_{i \in I} |c_i| \|\widetilde{\pi}(x_i)g | B\| \leq \|g | B\| \sum_{i \in I} |c_i| w(x_i) < \infty$$

implying $f \in B$. Taking the infimum over all possible expansions of f yields $\|f | B\| \leq C \|f | (\mathcal{H}_w^1)_{\mathcal{A}}\|$ by Theorem 4.5.5, which shows the continuity of the embedding $(\mathcal{H}_w^1)_{\mathcal{A}} \subset B$. $\square$

The following lemma collects some properties of the extension of $\widetilde{V}_g$ to $(\mathcal{H}_w^1)_{\mathcal{A}}^{\urcorner}$ (compare with [FG89a, Theorem 1]).

Lemma 4.5.8. *Let $g \in \mathbb{A}_w^{\mathcal{A}}$ with $\|Kg|\mathcal{H}\| = 1$.*

(a) *For $f \in (\mathcal{H}_w^1)_{\mathcal{A}}^\urcorner$ it holds $\widetilde{V}_g(\widetilde{\pi}(x)f) = \mathcal{L}_x \widetilde{V}_g f$ for all $x \in \mathcal{G}$, i.e. the covariance property extends to $(\mathcal{H}_w^1)_{\mathcal{A}}^\urcorner$.*

(b) *A bounded net (f_α) in $(\mathcal{H}_w^1)_{\mathcal{A}}^\urcorner$ is weak-$*$ convergent to an element $f \in (\mathcal{H}_w^1)_{\mathcal{A}}^\urcorner$ if and only if $\widetilde{V}_g f_\alpha$ converges pointwise to $\widetilde{V}_g f$. Moreover, if $f \in (\mathcal{H}_w^1)_{\mathcal{A}}^\urcorner$ then there exists a sequence $(f_n)_{n \in \mathbb{N}}$ of elements $f_n \in \mathcal{H}_{\mathcal{A}}$ with $\|f_n|(\mathcal{H}_w^1)^\urcorner\| \leq C\|f|(\mathcal{H}_w^1)^\urcorner\|$ for all $n \in \mathbb{N}$ such that (f_n) is weak-$*$ convergent to f.*

(c) *The reproducing formula extends to $(\mathcal{H}_w^1)_{\mathcal{A}}^\urcorner$, i.e.,*

$$\widetilde{V}_g f = \widetilde{V}_g f * \widetilde{V}_g g \qquad \text{for all } f \in (\mathcal{H}_w^1)_{\mathcal{A}}^\urcorner.$$

(d) *Conversely, if $F \in (L_{1/w}^\infty)_{\mathcal{A}}$ satisfies the reproducing formula $F = F * \widetilde{V}_g g$ then there exists some $f \in (\mathcal{H}_w^1)_{\mathcal{A}}^\urcorner$ such that $F = \widetilde{V}_g f$.*

Proof: (a) By definition (4.17) of the voice transform on $(\mathcal{H}_w^1)_{\mathcal{A}}^\urcorner$ we have

$$\widetilde{V}_g(\widetilde{\pi}(x)f)(y) = f(\widetilde{\pi}(x^{-1})\widetilde{\pi}(y)g) = f(\widetilde{\pi}(\epsilon_{\mathcal{A}x^{-1}} * \epsilon_{\mathcal{A}y})g) = \epsilon_{\mathcal{A}x^{-1}} * \epsilon_{\mathcal{A}y}(\widetilde{V}_g f) = \mathcal{L}_x \widetilde{V}_g f(y).$$

Hereby, also Theorem 2.2.4 and Lemma 1.2.4(a) were used. We note that $\epsilon_{\mathcal{A}x^{-1}} * \epsilon_{\mathcal{A}y}(\widetilde{V}_g f)$ is well-defined since $\widetilde{V}_g f$ is continuous (Lemma 4.5.3) and supp $\epsilon_{\mathcal{A}x^{-1}} * \epsilon_{\mathcal{A}y}$ is compact.

(b) Clearly, if (f_α) is weak-$*$ convergent to f then a fortiori $\widetilde{V}_g f_\alpha(x) = \langle f_\alpha, \widetilde{\pi}(x)g \rangle$ converges to $\widetilde{V}_g f(x)$ for all $x \in \mathcal{G}$. Conversely, suppose $\widetilde{V}_g f_\alpha$ converges pointwise to $\widetilde{V}_g f$ and let $h \in (\mathcal{H}_w^1)_{\mathcal{A}}$. Since span$\{\widetilde{\pi}(x)g, x \in \mathcal{G}\}$ is dense in $(\mathcal{H}_w^1)_{\mathcal{A}}$ (Corollary 4.5.6), there exists for all $\epsilon > 0$ an element of the form $\sum_{i=1}^n c_i \widetilde{\pi}(x_i)g$ such that $\|h - \sum_{i=1}^n c_i \widetilde{\pi}(x_i)g|(\mathcal{H}_w^1)_{\mathcal{A}}\| < \epsilon$. We obtain

$$|f(h) - f_\alpha(h)| \leq |f(h) - f(\sum_{i=1}^n c_i \widetilde{\pi}(x_i)g)| + |f(\sum_{i=1}^n c_i \widetilde{\pi}(x_i)g) - f_\alpha(\sum_{i=1}^n c_i \widetilde{\pi}(x_i)g)|$$

$$+ |f_\alpha(h) - f_\alpha(\sum_{i=1}^n c_i \widetilde{\pi}(x_i)g)|$$

$$\leq \|f|(\mathcal{H}_w^1)_{\mathcal{A}}^\urcorner\| \|h - \sum_{i=1}^n c_i \widetilde{\pi}(x_i)g)|(\mathcal{H}_w^1)_{\mathcal{A}}\| + \sum_{i=1}^n |c_i||\widetilde{V}_g f(x_i) - \widetilde{V}_g f_\alpha(x_i)|$$

$$+ \|f_\alpha|(\mathcal{H}_w^1)_{\mathcal{A}}^\urcorner\| \|h - \sum_{i=1}^n c_i \widetilde{\pi}(x_i)|(\mathcal{H}_w^1)_{\mathcal{A}}\|$$

$$\leq \epsilon(\|f|(\mathcal{H}_w^1)_{\mathcal{A}}^\urcorner\| + \sup_\alpha \|f_\alpha|(\mathcal{H}_w^1)_{\mathcal{A}}^\urcorner\|) + \sum_{i=1}^n |c_i||\widetilde{V}_g f(x_i) - \widetilde{V}_g f_\alpha(x_i)|.$$

By boundedness of the net (f_α) and pointwise convergence of $\widetilde{V}_g f_\alpha$ we conclude that $f_\alpha(h)$ converges to $f(h)$ and since $h \in (\mathcal{H}_w^1)_{\mathcal{A}}$ was arbitrary this shows the weak-$*$ convergence of (f_α).

Let us now show the second claim. By σ-compactness of $\mathcal{G}$ there exists a sequence $(U_n)_{n\in\mathbb{N}}$ of compact sets that cover $\mathcal{G}$ and that are nested, i.e., $\mathcal{G} = \cup_{n=1}^\infty U_n$ and $U_n \subset U_{n+1}$ for all $n \in \mathbb{N}$. Since $\mathcal{G} = \mathcal{A}(\mathcal{G}) = \cup_{n=1}^\infty \mathcal{A}(U_n)$ we may assume the sets U_n to be invariant under $\mathcal{A}$ without loss of generality. Now let $F_n := \chi_{U_n} \widetilde{V}_g f$ with $g \in \mathbb{A}_w^{\mathcal{A}}$ such that $\|Kg|\mathcal{H}\| = 1$. By compactness of U_n and continuity of $\widetilde{V}_g f$ each F_n is contained in $L_{\mathcal{A}}^2(\mathcal{G})$. Moreover, F_n is invariant under $\mathcal{A}$. By Theorem 2.2.7(c) the element

$$f_n := \widetilde{V}_g^* F_n = \int_{U_n} \widetilde{V}_g f(x) \pi(x) g \, dx$$

belongs to $\mathcal{H}_{\mathcal{A}}$. Formally, the adjoint $\widetilde{V}_g^* : (L_{1/w}^\infty)_{\mathcal{A}} \to (\mathcal{H}_w^1)_{\mathcal{A}}^\daleth$ is given by the same formula as $\widetilde{V}_g^* : L_{\mathcal{A}}^2 \to \mathcal{H}_{\mathcal{A}}$, i.e., for $h \in (\mathcal{H}_w^1)_{\mathcal{A}}$ and $F \in (L_{1/w}^\infty)_{\mathcal{A}}$ we have

$$\langle \widetilde{V}_g^* F, h \rangle = \langle F, \widetilde{V}_g h \rangle = \int_{\mathcal{G}} F(x) \overline{\widetilde{V}_g h(x)} \, dx.$$

Setting $F = F_n$ we obtain

$$\lim_{n\to\infty} \langle f_n, h \rangle = \lim_{n\to\infty} \int_{\mathcal{G}} F_n(x) \overline{\widetilde{V}_g h(x)} \, dx = \int_{\mathcal{G}} \widetilde{V}_g f(x) \overline{\widetilde{V}_g h(x)} \, dx = \langle f, \widetilde{V}_g^* \widetilde{V}_g h \rangle = \langle f, h \rangle.$$

Hereby, we used Theorem 2.2.7(d) in the last equality. Since $h \in (\mathcal{H}_w^1)_{\mathcal{A}}$ is arbitrary we conclude that $(f_n)_{n\in\mathbb{N}}$ is weak-$*$ convergent to f. The norm-boundedness follows from

$$\|f_n|(\mathcal{H}_w^1)_{\mathcal{A}}^\daleth\| = \sup_{h\in(\mathcal{H}_w^1)_{\mathcal{A}}, \|h|(\mathcal{H}_w^1)_{\mathcal{A}}\|\leq 1} |\langle V_g^* F_n, h \rangle| \leq \sup_{H\in L_w^1, \|H|L_w^1\|\leq 1} \left| \int_{\mathcal{G}} F_n(x) H(x) dx \right|$$

$$= \|F_n|L_{1/w}^\infty\| \leq \|\widetilde{V}_g f|L_{1/w}^\infty\| \leq C\|f|(\mathcal{H}_w^1)_{\mathcal{A}}^\daleth\|.$$

(c) Let $f \in (\mathcal{H}_w^1)_{\mathcal{A}}^\daleth$. By (b) there exists a sequence $(f_n)_{n\in\mathbb{N}} \subset \mathcal{H}_{\mathcal{A}}$ that is weak-$*$ convergent to f and uniformly bounded in $(\mathcal{H}_w^1)_{\mathcal{A}}^\daleth$. By Lemma 4.5.4 it holds $|\widetilde{V}_g f_n(x)| \leq Cw(x)$ uniformly in n. Since $\widetilde{V}_g g \in L_w^1$ we therefore conclude that for all $x \in \mathcal{G}$ the function $y \mapsto C \, |V_g g(y^{-1}x)| w(y)$ is integrable and dominates $y \mapsto |\widetilde{V}_g f_n(y) \widetilde{V}_g g(y^{-1}x)|$ for all $n \in \mathbb{N}$. Therefore, the reproducing formula (1.82) for $\mathcal{H}$ together with Lebesgue's dominated convergence theorem yields

$$\widetilde{V}_g f(x) = \lim_{n\to\infty} \widetilde{V}_g f_n(x) = \lim_{n\to\infty} \widetilde{V}_g f_n * \widetilde{V}_g g(x) = \widetilde{V}_g f * \widetilde{V}_g g(x), \quad x \in \mathcal{G},$$

by pointwise convergence of $\widetilde{V}_g f_n$ to $\widetilde{V}_g f$, see (b).

(d) Let $\widetilde{V}_g^* : (L_{1/w}^\infty)_{\mathcal{A}} \to (\mathcal{H}_w^1)_{\mathcal{A}}^{\daleth}$ denote the adjoint of $\widetilde{V}_g : (\mathcal{H}_w^1)_{\mathcal{A}} \to (L_w^1)_{\mathcal{A}}$. For $F \in (L_{1/w}^\infty)_{\mathcal{A}}$ it holds

$$\widetilde{V}_g(\widetilde{V}_g^* F)(x) = \langle \widetilde{V}_g^* F, \widetilde{\pi}(x)g \rangle = \langle F, \widetilde{V}_g(\widetilde{\pi}(x)g) \rangle = \int_{\mathcal{G}} F(y)\overline{\mathcal{L}_x \widetilde{V}_g g(y)} dy = F * \widetilde{V}_g g(x)$$

as a consequence of $\widetilde{V}_g g^\nabla = \widetilde{V}_g g$. Hence, if $F = F * \widetilde{V}_g g$ then we have $F = \widetilde{V}_g f$ with $f = \widetilde{V}_g^* F \in (\mathcal{H}_w^1)_{\mathcal{A}}^{\daleth}$. $\qquad\square$

4.5.2 General Coorbits

Now we are prepared to introduce the coorbit spaces in full generality. Hereby, we make use of the two-sided invariant solid BF-spaces treated in Section 4.1.

Definition 4.5.3. Suppose Y satisfies property (YA). We associate to Y the weight function

$$w(x) := \max\{\|L_x|Y\|, \|L_{x^{-1}}|Y\|, \|R_x|Y\|, \Delta(x^{-1})\|R_{x^{-1}}|Y\|\}. \tag{4.19}$$

We assume further that the corresponding set of analyzing vectors $\mathbb{A}_w^{\mathcal{A}}$ is non-trivial. Then for a fixed non-zero vector $g \in \mathbb{A}_w^{\mathcal{A}}$ the coorbit of Y is defined by

$$\mathsf{Co}Y := \{f \in (\mathcal{H}_w^1)^{\daleth}, V_g f \in Y\}$$

with natural norm

$$\|f|\mathsf{Co}Y\| := \|V_g f|Y\|.$$

Its subspace of invariant elements is denoted by

$$\mathsf{Co}Y_{\mathcal{A}} := \mathsf{Co}Y \cap (\mathcal{H}_w^1)_{\mathcal{A}}^{\daleth} = \{f \in (\mathcal{H}_w^1)_{\mathcal{A}}^{\daleth}, \widetilde{V}_g f \in Y_{\mathcal{A}}\}.$$

We note that the weight function w in (4.19) is submultiplicative and invariant under $\mathcal{A}$. Since $u(x) = \|L_x|Y\|$ is submultiplicative it holds $1 = u(e) \le u(x)u(x^{-1})$ and hence, $w(x) \ge 1$. By definition, both functions u, v defined in (4.2) are dominated by w. In particular, we conclude from Lemma 4.1.4 and Lemma 4.1.9 that $Y_{\mathcal{A}} * (L_w^1)_{\mathcal{A}} \subset Y_{\mathcal{A}}$ with

$$\|F * G|Y_{\mathcal{A}}\| \le \|F|Y_{\mathcal{A}}\| \|G|(L_w^1)_{\mathcal{A}}\| \qquad \text{for all } F \in Y_{\mathcal{A}}, G \in (L_w^1)_{\mathcal{A}}. \tag{4.20}$$

Observe also that $\|\cdot|\mathsf{Co}Y\|$ is indeed a norm not just a semi-norm. In fact, since Y is a translation invariant solid BF-space, $\|V_g f|Y\| = 0$ implies with continuity of $V_g f$ (Lemma 4.5.3) that $V_g f = 0$. By injectivity of V_g on $(\mathcal{H}_w^1)_{\mathcal{A}}^{\daleth}$ (Lemma 4.5.4) this means that $f = 0$.

We will only state results for $\mathrm{CoY}_\mathcal{A}$ in the sequel. The corresponding results for CoY can be seen as the special case $\mathcal{A} = \{e\}$. An unspecified weight w will always denote the one associated to Y as in Definition 4.5.3.

In order to study the properties of $\mathrm{CoY}_\mathcal{A}$ in more detail we need to introduce the "better" space of analyzing vectors

$$\mathbb{B}_w^\mathcal{A} := \{g \in \mathbb{A}_w^\mathcal{A}, \ \widetilde{V}_g g \in W_\mathcal{A}^R(C_0, L_w^1)\}.$$

Lemma 4.5.9. *If $\mathbb{A}_w^\mathcal{A}$ is non-trivial then $\mathbb{B}_w^\mathcal{A}$ is dense in $(\mathcal{H}_w^1)_\mathcal{A}$ and in $\mathcal{H}_\mathcal{A}$ (with respect to the corresponding norm topologies).*

Proof: Assume $0 \neq g \in \mathbb{A}_w^\mathcal{A}$ and $0 \neq \phi \in (C_c)_\mathcal{A}(\mathcal{G})$. We form the element

$$h := \pi(\phi)g = \widetilde{V}_g^* \phi = \int_\mathcal{G} \phi(x)\pi(x)g \, dx$$

contained in $\mathcal{H}_\mathcal{A}$. We claim that it is possible to choose ϕ such that $h \neq 0$. Indeed, let $0 \neq f \in \mathcal{H}_\mathcal{A}$. By density of $(C_c)_\mathcal{A}$ in $L_\mathcal{A}^2$ we may choose a sequence $(\phi_n) \subset (C_c)_\mathcal{A}$ that approximates $\widetilde{V}_g f$. This implies that $\widetilde{V}_g^* \phi_n$ approximates $\widetilde{V}_g^* \widetilde{V}_g f = f$ and there exists an n such that $\widetilde{V}_g^* \phi_n \neq 0$. We obtain

$$V_h h(x) = \int_\mathcal{G} \int_\mathcal{G} \phi(y)\overline{\phi(z)}\langle \pi(y)g, \pi(x)\pi(z)g\rangle dydz = (\phi * V_g g * \phi^\nabla)(x). \qquad (4.21)$$

As ϕ is compactly supported both ϕ and ϕ^∇ are contained in $W_\mathcal{A}(C_0, L_w^1) \subset L_w^1$ (Lemma 4.4.4(a)). Applying Proposition 4.4.2(b) twice (specialized to $Y = L_w^1$) yields $V_h h \in W_\mathcal{A}(C_0, L_w^1)$. Hereby, we note that the function v associated to $Y = L_w^1$ by (4.2) coincides with w by Lemma 4.1.5.

Choosing an approximate identity $(\phi_n) \subset (C_c)_\mathcal{A}(\mathcal{G})$ we obtain (using another $g' \in \mathbb{A}_w^\mathcal{A}$ for the norm in $(\mathcal{H}_w^1)_\mathcal{A}$)

$$\|\pi(\phi_n)g - g|(\mathcal{H}_w^1)_\mathcal{A}\| \leq \|\widetilde{V}_{g'}(\pi(\phi_n)g - g)|L_w^1\| = \|\phi_n * \widetilde{V}_{g'}g - \widetilde{V}_{g'}g|L_w^1\| \to 0$$

as n tends to ∞. Thus by density of $\mathbb{A}_w^\mathcal{A}$ in $(\mathcal{H}_w^1)_\mathcal{A}$ (Corollary 4.5.6), $\mathbb{B}_w^\mathcal{A}$ is dense in $(\mathcal{H}_w^1)_\mathcal{A}$. Since $(\mathcal{H}_w^1)_\mathcal{A}$ is continuously embedded into $\mathcal{H}_\mathcal{A}$ this implies that $\mathbb{B}_w^\mathcal{A}$ is dense also in $\mathcal{H}_\mathcal{A}$. □

For IN-groups more can be said about $\mathbb{B}_w^\mathcal{A}$, see also [FG89b, Lemma 7.2].

Lemma 4.5.10. *If $\mathcal{G}$ is an IN-group then $\mathbb{B}_w^\mathcal{A} = \mathbb{A}_w^\mathcal{A}$.*

Proof: Let $g_0 \in \mathbb{B}_w^\mathcal{A} \setminus \{0\}$ and $g \in \mathbb{A}_w^\mathcal{A}$. Without loss of generality we may assume $\|Kg|\mathcal{H}\| = \|Kg_0|\mathcal{H}\| = 1$. The inversion formula reads $g = \int_\mathcal{G} \widetilde{V}_{g_0}g(x)\pi(x)g_0 dx$ and specializing (4.21) to $\phi = \widetilde{V}_{g_0}g$ yields

$$\widetilde{V}_g g = \widetilde{V}_{g_0}g * \widetilde{V}_{g_0}g_0 * (\widetilde{V}_{g_0}g)^\nabla \in (L_w^1)_\mathcal{A} * W_\mathcal{A}(C_0, L_w^1) * (L_w^1)_\mathcal{A}.$$

The space $W_{\mathcal{A}}(C_0, L_w^1)$ coincides with $W_{\mathcal{A}}(C^b, L_w^1)$. By Lemma 4.4.4(a) the transform $V_g g$ is thus contained in $W_{\mathcal{A}}(L^1, L_w^1) * W_{\mathcal{A}}(C^b, L_w^1) * W_{\mathcal{A}}(L^1, L_w^1)$. It holds $L^1 * C^b \subset C^b$ and L_w^1 is a Banach algebra, i.e., (L^1, C^b, C^b) and (L_w^1, L_w^1, L_w^1) are Banach convolution triples. Hence, Theorem 4.4.3 applies and we conclude that $V_g g \in W_{\mathcal{A}}(C^b, L_w^1) = W(C_0, L_w^1)$, i.e., $g \in \mathbb{B}_w^{\mathcal{A}}$. The reversed inclusion $\mathbb{B}_w^{\mathcal{A}} \subset \mathbb{A}_w^{\mathcal{A}}$ is clear by definition. $\qquad\square$

Let us now turn our attention back to the study of $\mathsf{Co}Y_{\mathcal{A}}$.

Proposition 4.5.11. *Given* $0 \neq g \in \mathbb{A}_w^{\mathcal{A}}$ *such that* $\|Kg|\mathcal{H}\| = 1$, *a function* $F \in Y_{\mathcal{A}}$ *is of the form* $\widetilde{V}_g f$ *for some* $f \in \mathsf{Co}Y_{\mathcal{A}}$ *if and only if* F *satisfies the reproducing formula* $F = F * \widetilde{V}_g g$.

Proof: If $f \in \mathsf{Co}Y_{\mathcal{A}} \subset (\mathcal{H}_w^1)_{\mathcal{A}}^{\urcorner}$ then the reproducing formula $\widetilde{V}_g f = \widetilde{V}_g f * \widetilde{V}_g g$ holds by Lemma 4.5.8(c) and is valid also in $Y_{\mathcal{A}}$ by (4.20).

Conversely, assume $F \in Y_{\mathcal{A}}$ with $F * \widetilde{V}_g g = F$. Provided $F \in L_{1/w}^\infty$ we may conclude from Lemma 4.5.8(d) the existence of some $f \in (\mathcal{H}_w^1)_{\mathcal{A}}^{\urcorner}$ such that $F = \widetilde{V}_g f$ and the assumption on F implies $f \in \mathsf{Co}Y_{\mathcal{A}}$. Thus, it remains to show $F \in L_{1/w}^\infty$.

For a normalized $g_0 \in \mathbb{B}_w^{\mathcal{A}}$ (which exists by Lemma 4.5.9) the inversion formula yields

$$g = \int_{\mathcal{G}} V_{g_0} g(x) \pi(x) g_0 dx$$

and the proof of Lemma 4.5.2 already showed that $V_{g_0} g \in L_{w^\bullet}^1$ with $w^\bullet(x) = w(x) + w(x^{-1})\Delta(x^{-1})$. Denoting $\phi = V_{g_0} g$ we obtain $V_g g = \phi * V_{g_0} g_0 * \phi^\nabla$ as in (4.21) and, hence,

$$F = F * V_g g = F * \phi * V_{g_0} g_0 * \phi^\nabla.$$

By (4.20) together with Lemma 4.4.4(b), we have

$$F * \phi \in Y \subset W(L_1, L_{1/u^\vee}^\infty) \subset W(L^1, L_{1/w}^\infty),$$

where the latter inclusion follows from $u^\vee \leq w$ (by definition of w). Further note that for the function v_0 associated to $Y_0 = L_{1/w}^\infty$ as in (4.2), it holds $v_0 \leq w^*$ by Corollary 4.1.7. Hence, $W(C_0, L_w^1) \subset W(C_0, L_{v_0^*}^1)$ and $V_{g_0} g_0$ is contained in the latter. Identifying $L^1(\mathcal{G})$ with a subspace of $M(\mathcal{G})$ and replacing Y by $L_{1/w}^\infty$ in Proposition 4.4.2(a) yields

$$F * \phi * V_{g_0} g_0 \in W(L^1, L_{1/w}^\infty) * W(C_0, L_{v_0^*}^1) \subset L_{1/w}^\infty.$$

Noting further that $\phi^\nabla \in L_{w^\bullet}^1$ we finally obtain $F \in L_{1/w}^\infty * L_{w^\bullet}^1 \subset L_{1/w}^\infty * L_{v_0^*}^1 \subset L_{1/w}^\infty$ by Lemma 4.1.4. $\qquad\square$

Corollary 4.5.12. *Let* $g \in \mathbb{A}_w^{\mathcal{A}}$ *with* $\|Kg|\mathcal{H}\| = 1$.

(a) The transform $\widetilde{V}_g : \text{Co}Y_{\mathcal{A}} \to Y_{\mathcal{A}}$ establishes an isometric isomorphism between $\text{Co}Y_{\mathcal{A}}$ and the closed subspace $Y_{\mathcal{A}} * \widetilde{V}_g g$ of $Y_{\mathcal{A}}$. Furthermore, the mapping $F \mapsto F * \widetilde{V}_g g$ defines a bounded projection from $Y_{\mathcal{A}}$ onto $Y_{\mathcal{A}} * \widetilde{V}_g g$.

(b) Every function $F = F * \widetilde{V}_g g \in Y_{\mathcal{A}}$ is continuous, automatically belongs to $L^\infty_{1/w}$ and $F(x) = \langle F, L_x \widetilde{V}_g g \rangle = \langle F, \mathcal{L}_x \widetilde{V}_g g \rangle$, $x \in \mathcal{G}$.

Proof: The statement in (a) follows immediately from Proposition 4.5.11.
(b) It was already shown in the proof of Proposition 4.5.11 that $F = F * \widetilde{V}_g g \in Y$ belongs to $L^\infty_{1/w}$ and by Lemma 4.5.3 F is continuous. Moreover, the relation $\widetilde{V}_g g = (\widetilde{V}_g g)^\nabla$ yields

$$F(x) = F * \widetilde{V}_g g(x) = \int_{\mathcal{G}} F(y) \overline{\widetilde{V}_g g(x^{-1}y)} dy = \langle F, L_x \widetilde{V}_g g \rangle = \langle F, \mathcal{L}_x \widetilde{V}_g g \rangle,$$

where the last equality follows from invariance of F and $\widetilde{V}_g g$ under $\mathcal{A}$. ☐

Now we are finally prepared to show basic properties of $\text{Co}Y_{\mathcal{A}}$.

Theorem 4.5.13. (a) $\text{Co}Y_{\mathcal{A}}$ is a $\widetilde{\pi}$-invariant Banach space with $\|\widetilde{\pi}(x)|\text{Co}Y_{\mathcal{A}}\| \leq w(x)$.

(b) The definition of $\text{Co}Y_{\mathcal{A}}$ does not depend on the choice of $g \in \mathbb{A}^{\mathcal{A}}_w \setminus \{0\}$.

(c) The definition of $\text{Co}Y_{\mathcal{A}}$ is independent of the reservoir $(\mathcal{H}^1_w)^\daleth_{\mathcal{A}}$ in the following sense: Assume that $S_{\mathcal{A}} \subset (H^1_w)_{\mathcal{A}}$ is a non-trivial locally convex vector space, which is invariant under $\widetilde{\pi}$. Assume further that there exists a non-zero vector $g \in S_{\mathcal{A}} \cap \mathbb{A}^{\mathcal{A}}_w$ for which the reproducing formula $\widetilde{V}_g f = \widetilde{V}_g f * \widetilde{V}_g g$ holds for all $f \in S^\daleth_{\mathcal{A}}$ (the topological anti-dual of $S_{\mathcal{A}}$). Then we have

$$\text{Co}Y_{\mathcal{A}} = \{f \in (\mathcal{H}^1_w)^\daleth_{\mathcal{A}}, \widetilde{V}_g f \in Y_{\mathcal{A}}\} = \{f \in S^\daleth_{\mathcal{A}}, \widetilde{V}_g f \in Y_{\mathcal{A}}\}.$$

(d) We have the continuous embeddings

$$(\mathcal{H}^1_w)_{\mathcal{A}} \subset \text{Co}Y_{\mathcal{A}} \subset (\mathcal{H}^1_w)^\daleth_{\mathcal{A}}.$$

Proof: (a) Suppose that (f_n) is a Cauchy sequence in $\text{Co}Y_{\mathcal{A}}$ which means that $\widetilde{V}_g f_n$ is a Cauchy sequence in $Y_{\mathcal{A}}$. Since $Y_{\mathcal{A}}$ is complete there exists a limit $F \in Y_{\mathcal{A}}$. Moreover, the reproducing formula in Proposition 4.5.11 and the continuity of the operator $F \mapsto F * \widetilde{V}_g g$ on $Y_{\mathcal{A}}$ (4.20) imply

$$F = \lim_{n\to\infty} \widetilde{V}_g f_n = \lim_{n\to\infty} \widetilde{V}_g f_n * \widetilde{V}_g g = F * \widetilde{V}_g g.$$

Using Proposition 4.5.11 once more we conclude that $F = \widetilde{V}_g f$ for some element $f \in \mathsf{Co}Y_{\mathcal{A}}$. Thus, $\mathsf{Co}Y_{\mathcal{A}}$ is complete.

Lemma 4.5.8(a), Lemma 4.1.10 and the definition of w yield

$$\|\widetilde{\pi}(x)f|\mathsf{Co}Y_{\mathcal{A}}\| = \|\mathcal{L}_x\widetilde{V}_g f|Y_{\mathcal{A}}\| \leq \|\mathcal{L}_x|Y_{\mathcal{A}}\| \, \|f|\mathsf{Co}Y_{\mathcal{A}}\| \leq w(x)\|f|\mathsf{Co}Y_{\mathcal{A}}\|$$

proving $\|\widetilde{\pi}(x)|\mathsf{Co}Y_{\mathcal{A}}\| \leq w(x)$.

(b) If g' is another non-zero vector in $\mathbb{A}_w^{\mathcal{A}}$ then the proof of Lemma 4.5.2 already showed $\widetilde{V}_{g'}f = \widetilde{V}_g f * (\widetilde{V}_{g'}g)^{\nabla}$ and $\widetilde{V}_{g'}g \in L_{w^\bullet}^1$. Thus by (4.20), we have $\widetilde{V}_{g'}f \in Y$ if and only if $\widetilde{V}_g f \in Y$ with equivalence of norms. (The "only if" part follows from exchanging the roles of g and g'.)

(c) Since $S_{\mathcal{A}}$ is non-trivial and $\widetilde{\pi}$-invariant it is dense in $\mathcal{H}_{\mathcal{A}}$ (Lemma 1.4.10) and, hence, $S_{\mathcal{A}} \subset (\mathcal{H}_w^1)_{\mathcal{A}} \subset \mathcal{H}_{\mathcal{A}} \subset (\mathcal{H}_w^1)_{\mathcal{A}}^{\daleth} \subset S_{\mathcal{A}}^{\daleth}$. Let $0 \neq g \in \mathbb{A}_w^{\mathcal{A}} \cap S_{\mathcal{A}}$ such that the reproducing formula is valid on $S_{\mathcal{A}}^{\daleth}$. By $\widetilde{\pi}$-invariance the transform $\widetilde{V}_g f(x) = f(\widetilde{\pi}(x)g)$ is well-defined on $S_{\mathcal{A}}^{\daleth}$. Assuming $\widetilde{V}_g f \in Y$ for some $f \in S_{\mathcal{A}}^{\daleth}$ we have to verify that f already belongs to $(\mathcal{H}_w^1)_{\mathcal{A}}^{\daleth}$, which is the same as $\widetilde{V}_g f \in L_{1/w}^\infty$ by Lemma 4.5.4. Implicitly, this was already shown in the proof of Proposition 4.5.11 (provided the reproducing formula holds). Indeed the embedding $Y \subset W(L^1, L_{1/w}^\infty)$ is independent of the reservoir $S_{\mathcal{A}}^{\daleth}$ and, hence, following the further arguments in Proposition 4.5.11 we see that $\widetilde{V}_g f \in L_{1/w}^\infty$.

(d) The continuous embedding $(\mathcal{H}_w^1)_{\mathcal{A}} \subset \mathsf{Co}Y_{\mathcal{A}}$ follows from part (a) and Corollary 4.5.7. For the continuity of the embedding $\mathsf{Co}Y_{\mathcal{A}} \subset (\mathcal{H}_w^1)_{\mathcal{A}}^{\daleth}$ we first note that $Y \subset W(L^1, L_{1/u^\vee}^\infty) \subset W(L^1, L_{1/w}^\infty)$ by Lemma 4.4.4 and by definition of w. With Corollary 4.1.7 we conclude that the function v_0 associated to $Y_0 = L_{1/w}^\infty$ satisfies $v_0^* \leq w$. Hence, $W(C_0, L_w^1) \subset W(C_0, L_{v_0^*}^1)$ and for $g \in \mathbb{B}_w^{\mathcal{A}}$ the voice transform $V_g g$ is contained in $W_{\mathcal{A}}(C_0, L_{v_0^*}^1)$. Proposition 4.4.2(a) finally yields for all $f \in \mathsf{Co}Y_{\mathcal{A}}$:

$$\begin{aligned}
\|f|(\mathcal{H}_w^1)_{\mathcal{A}}^{\daleth}\| &\leq C\|\widetilde{V}_g f|L_{1/w}^\infty\| = C\|\widetilde{V}_g f * \widetilde{V}_g g|L_{1/w}^\infty\| \\
&\leq C'\|\widetilde{V}_g f|W_{\mathcal{A}}(L^1, L_{1/w}^\infty)\| \, \|\widetilde{V}_g g|W_{\mathcal{A}}(C_0, L_{v_0^*}^1)\| \leq C''\|f|\mathsf{Co}Y_{\mathcal{A}}\|. \qquad \square
\end{aligned}$$

We remark that part (c) of the previous theorem applies in particular to the spaces $S_{\mathcal{A}} = (\mathcal{H}_{w_2}^1)_{\mathcal{A}}$ with $w \leq Cw_2$. This corresponds to the original formulation in [FG89a, Theorem 4.2].

Corollary 4.5.14. *It holds*

  (a) $\mathsf{Co}(L_{1/w}^\infty)_{\mathcal{A}} = (\mathcal{H}_w^1)_{\mathcal{A}}^{\daleth}$,

  (b) $\mathsf{Co}(L_w^1)_{\mathcal{A}} = (\mathcal{H}_w^1)_{\mathcal{A}}$,

  (c) $\mathsf{Co}L_{\mathcal{A}}^2 = \mathcal{H}_{\mathcal{A}}$.

Proof: The statement in (a) follows from Lemma 4.5.4 and Theorem 4.5.13(c) by noting that the weight w' associated to $Y = L_{1/w}^\infty$ satisfies $w' \geq w$, see Corollary 4.1.7.

(b) The inclusion $(\mathcal{H}_w^1)_\mathcal{A} \subset \mathrm{Co}(L_w^1)_\mathcal{A}$ is clear. It remains to show that $\widetilde{V}_g f \in L_w^1$ for some $f \in (\mathcal{H}_w^1)_\mathcal{A}^\daleth$ already implies $f \in (\mathcal{H}_w^1)_\mathcal{A}$, in particular $f \in \mathcal{H}_\mathcal{A}$. The reproducing formula for $(\mathcal{H}_w^1)_\mathcal{A}^\daleth$ (Lemma 4.5.8(c)) yields $\widetilde{V}_g f = \widetilde{V}_g f * \widetilde{V}_g g$. Since $\widetilde{V}_g g \in L_\mathcal{A}^\infty$ and $w \geq 1$ it holds $\widetilde{V}_g f \in (L_w^1)_\mathcal{A} * L_\mathcal{A}^\infty \subset L_\mathcal{A}^\infty$ and, hence, $\widetilde{V}_g f \subset (L_w^1)_\mathcal{A} \cap L_\mathcal{A}^\infty \subset L_\mathcal{A}^2$. It follows from Corollary 1.5.5(d) that there exists some $f' \in \mathcal{H}_\mathcal{A}$ such that $\widetilde{V}_g f = \widetilde{V}_g f'$ and by injectivity of $\widetilde{V}_g$ we have $f = f'$. This means that $f \in (\mathcal{H}_w^1)_\mathcal{A}$. The assertion (c) is proved similarly as (b). $\qquad \square$

Let us finally provide a result that will be useful in the next section.

Theorem 4.5.15. *Suppose* $g \in \mathbb{B}_w^\mathcal{A}$. *Then* $\widetilde{V}_g f \in W_\mathcal{A}(C_0, Y)$ *for all* $f \in \mathrm{Co}Y_\mathcal{A}$. *If* $X = (x_i)_{i \in I}$ *is a* U-*dense well-spread family with respect to* $\mathcal{A}$ *then*

$$\|(\widetilde{V}_g f(x_i))_{i \in I} | Y_\mathcal{A}^\flat(X)\| \leq \gamma(U) C \|f | \mathrm{Co}Y_\mathcal{A}\|,$$

where the constant C *depends only on* g *and* $\gamma(U)$ *is the one in Lemma 4.4.9.*

Proof: Without loss of generality we may assume $\|Kg\| = 1$. By Proposition 4.5.11 we have $\widetilde{V}_g f = \widetilde{V}_g f * \widetilde{V}_g g$. Proposition 4.4.2(b) together with the definition of w yields

$$\|\widetilde{V}_g f | W_\mathcal{A}(C_0, Y)\| = \|\widetilde{V}_g f * \widetilde{V}_g g | W_\mathcal{A}(C_0, Y)\| \leq D \|\widetilde{V}_g f | Y\| \, \|\widetilde{V}_g g | W_\mathcal{A}(C_0, L_w^1)\|.$$

Since $\widetilde{V}_g g = (\widetilde{V}_g g)^\nabla$ Lemma 4.4.1 yields $\|\widetilde{V}_g g | W_\mathcal{A}(C_0, L_w^1)\| = \|\widetilde{V}_g g | W_\mathcal{A}^R(C_0, L_w^1)\|$. Finally, Lemma 4.4.9 leads to

$$\|(\widetilde{V}_g f(x_i))_{i \in I} | Y_\mathcal{A}^\flat(X)\| \leq \gamma(U) C' \|\widetilde{V}_g f | W_\mathcal{A}(C_0, Y)\|$$
$$\leq \gamma(U) \underbrace{DC' \|\widetilde{V}_g g | W_\mathcal{A}^R(C_0, L_w^1)\|}_{=C} \|f | \mathrm{Co}Y_\mathcal{A}\|. \qquad \square$$

4.6   Discretizations

In this section we will derive discretizations of the continuous transform $\widetilde{V}_g$ on the coorbit spaces $\mathrm{Co}Y_\mathcal{A}$, in particular on $\mathcal{H}_\mathcal{A}$.

In general, it is too restrictive to require $\{\widetilde{\pi}(x_i)g\}_{i \in I}$ to be an (unconditional) Schauder basis of $\mathrm{Co}Y_\mathcal{A}$ [LT77]. Instead it is more suitable to work with the concept of (Banach) frames and atomic decompositions. For Hilbert spaces, frames were introduced by Duffin and Schaefer in the study of non-harmonic Fourier series [DS52]. Daubechies and her coworkers further popularized them in the mid-80's when wavelet theory began to evolve [DGM86].

Definition 4.6.1. A family of elements $\{f_i\}_{i \in I}$ in some Hilbert space $\mathcal{H}$ is called a (Hilbert) **frame** if there exist constants $C_1, C_2 > 0$ such that

$$C_1 \|f\|^2 \leq \sum_{i \in I} |\langle f, f_i \rangle|^2 \leq C_2 \|f\|^2 \quad \text{for all } f \in \mathcal{H}. \tag{4.22}$$

Obviously, an orthonormal (or Riesz) basis is a frame. However, also the union of a finite number of orthonormal bases forms a frame, which shows that a frame may possess some redundancy. This redundancy, however, allows in some cases to avoid certain shortcomings of bases. A remarkable illustration of this fact is provided by the Balian-Low theorem [Chr03, Dau92].

Associated to a frame $\{f_i\}_{i \in I}$ is the frame operator

$$Sf := \sum_{i \in I} \langle f, f_i \rangle f_i,$$

which turns out to be bounded, positive and boundedly invertible. The elements $\{S^{-1} f_i\}_{i \in I}$ again form a frame – called the canonical dual frame – and we have the reconstruction formulae

$$f = S^{-1} S f = \sum_{i \in I} \langle f, f_i \rangle S^{-1} f_i = S S^{-1} f = \sum_{i \in I} \langle f, S^{-1} f_i \rangle f_i. \tag{4.23}$$

So on one hand $f \in \mathcal{H}$ can be reconstructed from the sequence of coefficients $(\langle f, f_i \rangle)_{i \in I}$ and on the other hand we may expand f into a linear combination of the frame elements $\{f_i\}_{i \in I}$. Moreover, the lower bound in the frame inequality (4.22) makes sure that reconstruction is a bounded operator from $l^2(I)$ into $\mathcal{H}$. In other words, we have stability. For more detailed information concerning frames we refer to [Chr03].

Gröchenig extended the concept of frames to Banach spaces in the following way [Grö91].

Definition 4.6.2. Suppose B is a Banach space. A family $\{h_i\}_{i \in I} \subset B'$ is called a **Banach frame** for B if there exists a Banach sequence space $B^\flat = B^\flat(I)$ and a linear bounded reconstruction operator $\Theta : B^\flat \to B$ such that

(i) $(h_i(f))_{i \in I} \in B^\flat$ for all $f \in B$ and there exist constants $C_1, C_2 > 0$ such that

$$C_1 \|f|B\| \leq \|(h_i(f))_{i \in I}|B^\flat\| \leq C_2 \|f|B\|,$$

(ii) $\Theta\left((h_i(f))_{i \in I}\right) = f$ for all $f \in B$.

Identifying $\mathcal{H}$ with its dual $\mathcal{H}'$ a Hilbert frame $\{f_i\}_{i \in}$ for $\mathcal{H}$ is obviously a Banach frame with corresponding sequence space $l^2(I)$ and with reconstruction operator given by one of the series expansions in (4.23). Thus for Hilbert spaces, (ii) is a consequence of (i). For general Banach spaces, however, this is false, see also [Chr03, Chapter 17].

It is important to note that the corresponding sequence space $B^\flat(I)$ should be "nice" (e.g. some $l^p(I)$ space) in order that a Banach frame be useful for practical purposes. Indeed, it was proven in [CCS02] that any total sequence in B' (a sequence $\{h_i\}_{i \in I}$ for which $h_i(f) = 0$ for all $i \in I$ implies $f = 0$) is a Banach frame for B with respect to *some* sequence space. This Banach space, however, might possibly be very hard to handle. In particular, there is an example of a sequence $\{h_i\}_{i \in I}$, which forms a Banach frame for some Hilbert space $\mathcal{H}$, but not a (Hilbert) frame in the sense of definition (4.6.1), see [Chr03, p.396].

Banach frames aim at reconstructing an element $f \in B$ by means of the coefficients $(h_i(f))_{i \in I}$. In general however, the reconstruction process is not performed by a series expansion with respect to some elements $\{g_i\}_{i \in I} \subset B$, as it is the case for Hilbert frames (4.23). So when desiring series expansions another concept is needed.

Definition 4.6.3. A family $\{g_i\}_{i \in I}$ in a Banach space B is called an **atomic decomposition** of B if there exists a family of bounded linear functionals $\{\lambda_i\}_{i \in I} \subset B'$ (not necessarily unique) and a Banach sequence space $B^\natural = B^\natural(I)$ such that

(i) $(\lambda_i(f))_{i \in I} \in B^\natural$ for all $f \in B$ and there exists a constant $C_1 > 0$ such that

$$\|(\lambda_i(f))_{i \in I}|B^\natural\| \leq C_1 \|f|B\|,$$

(ii) $f = \sum_{i \in I} \lambda_i(f) g_i$ for all $f \in B$ (with convergence in some suitable topology),

(iii) if $(\lambda_i)_{i \in I} \in B^\natural$ then $\sum_{i \in I} \lambda_i g_i \in B$ and there exists a constant $C_2 > 0$ such that

$$\|\sum_{i \in I} \lambda_i g_i | B\| \leq C_2 \|(\lambda_i)_{i \in I}|B^\natural\|.$$

Observe that a Hilbert frame $\{f_i\}_{i \in I}$ is an atomic decomposition with corresponding sequence of functionals $\{\lambda_i\}_{i \in I}$ given by $\lambda_i(f) = \langle f, S^{-1}f_i \rangle$, see (4.23). For general Banach spaces, however, the concepts of Banach frame and atomic decomposition differ from each other, see also [Chr03].

The task of this section is to derive Banach frames and atomic decompositions for the coorbits $\text{Co}Y_\mathcal{A}$ of the form $\{\widetilde{\pi}(x_i)g\}_{i \in I}$. The corresponding sequence spaces will turn out to be $Y_\mathcal{A}^\flat$ and $Y_\mathcal{A}^\natural$, which can be considered as sufficiently "nice".

Let us now state our main theorems, which are analogues of Theorems S,T and U in [Grö91]. The first one establishes the existence of atomic decompositions. Any unspecified weight function w will denote the one associated to Y by (4.19) as usual. Recall also Definition 4.4.1 of the U-oscillation $G_U^\#$.

Theorem 4.6.1. *Suppose* $g \in \mathbb{B}_w^{\mathcal{A}}$ *with* $\|Kg\| = 1$. *Choose further a relatively compact neighborhood* $U = U^{-1} = \mathcal{A}U$ *of* $e \in \mathcal{G}$ *such that*

$$\|(\widetilde{V}_g g)_U^\# | L_w^1 \| < 1. \tag{4.24}$$

(This is possible according to Lemma 4.4.10(b).) Then for any family $X = (x_i)_{i \in I}$, *which is* U-*dense and well-spread with respect to* $\mathcal{A}$, *the family* $\{\widetilde{\pi}(x_i)g\}_{i \in I}$ *is an* atomic decomposition *of* $\mathsf{Co}Y_{\mathcal{A}}$ *with corresponding sequence space* $Y_{\mathcal{A}}^{\natural}(X)$. *Moreover, the sum* $\sum_{i \in I} \lambda_i(f)\widetilde{\pi}(x_i)g$ *(where* $\{\lambda_i\}_{i \in I}$ *denotes a corresponding sequence of functionals on* $\mathsf{Co}Y_{\mathcal{A}}$*) converges in the norm of* $\mathsf{Co}Y_{\mathcal{A}}$ *if the finite sequences are dense in* $\mathsf{Co}Y_{\mathcal{A}}$, *and in the weak-∗ topology of* $(\mathcal{H}_w^1)_{\mathcal{A}}^{\top}$, *in general.*

Our next theorem is concerned with Banach frames.

Theorem 4.6.2. *Suppose* $g \in \mathbb{B}_w^{\mathcal{A}}$ *and choose a relatively compact neighborhood* $U = U^{-1} = \mathcal{A}U$ *of* $e \in \mathcal{G}$ *such that*

$$\|(\widetilde{V}_g g)_U^\# | L_w^1 \| < \|\widetilde{V}_g g | L_w^1 \|^{-1}. \tag{4.25}$$

Then for any U-*dense well-spread family* $X = (x_i)_{i \in I}$ *in* $\mathcal{G}$ *(with respect to* $\mathcal{A}$*) the set* $\{\widetilde{\pi}(x_i)g\}_{i \in I}$ *is a* Banach frame *for* $\mathsf{Co}Y_{\mathcal{A}}$ *with corresponding sequence space* $Y_{\mathcal{A}}^{\natural}(X)$. *Moreover, if the finite sequences are dense in* $Y_{\mathcal{A}}^{\natural}$ *then the reconstruction operator is given by*

$$f = \Theta((\langle f, \widetilde{\pi}(x_i)g \rangle)_{i \in I}) := \sum_{i \in I} \langle f, \widetilde{\pi}(x_i)g \rangle e_i \tag{4.26}$$

for some elements $(e_i)_{i \in I} \subset (\mathcal{H}_w^1)_{\mathcal{A}}$ *and with convergence in the norm of* $\mathsf{Co}Y_{\mathcal{A}}$.

Our third main theorem establishes the existence of a "dual frame".

Theorem 4.6.3. *Suppose* $g \in \mathbb{B}_w^{\mathcal{A}}$ *and choose further a relatively compact neighborhood* $U = U^{-1} = \mathcal{A}U$ *of* $e \in \mathcal{G}$ *such that*

$$\|(\widetilde{V}_g g)_U^\# | L_w^1 \| \, (\|\widetilde{V}_g g | L_w^1 \| + \gamma(U)D\|\widetilde{V}_g g | W_{\mathcal{A}}^R(C_0, L_w^1) \|) < 1, \tag{4.27}$$

where D *is the constant in Proposition 4.4.2(b) and* $\gamma(U)$ *the one in Lemma 4.4.9. Then for any family* $X = (x_i)_{i \in I}$, *which is* U-*dense and well-spread with respect to* $\mathcal{A}$, *the set* $\{\widetilde{\pi}(x_i)g\}_{i \in I}$ *is both an atomic decomposition for* $\mathsf{Co}Y_{\mathcal{A}}$ *with corresponding sequence space* $Y_{\mathcal{A}}^{\natural}(X)$ *and a Banach frame for* $\mathsf{Co}Y_{\mathcal{A}}$ *with corresponding sequence space* $Y_{\mathcal{A}}^{\flat}(X)$. *Moreover, there exists a 'dual frame'* $\{e_i\}_{i \in I} \subset (\mathcal{H}_w^1)_{\mathcal{A}}$ *such that*

(a) the following norms are equivalent

$$\|f | \mathsf{Co}Y_{\mathcal{A}} \| \asymp \|(\langle f, e_i \rangle)_{i \in I} | Y_{\mathcal{A}}^{\natural} \| \asymp \|(\langle f, \widetilde{\pi}(x_i)g \rangle)_{i \in I} | Y_{\mathcal{A}}^{\flat} \|; \tag{4.28}$$

(b) for $f \in \mathsf{Co}Y_{\mathcal{A}}$ it holds

$$f = \sum_{i \in I} \langle f, e_i \rangle \widetilde{\pi}(x_i) g,$$

with norm convergence in $\mathsf{Co}Y_{\mathcal{A}}$, if the finite sequences are dense in $Y_{\mathcal{A}}^{\natural}$ and with convergence in the weak-$*$ topology of $(\mathcal{H}_w^1)_{\mathcal{A}}^{\urcorner}$ otherwise;

(c) if the finite sequences are dense in $Y_{\mathcal{A}}^{\flat}$, then the decomposition

$$f = \sum_{i \in I} \langle f, \widetilde{\pi}(x_i) g \rangle e_i$$

is valid for $f \in \mathsf{Co}Y_{\mathcal{A}}$ with norm convergence.

We remark that in contrast to the classical theorems in [Grö91] we have to work with two (possibly different) types of sequence spaces. Banach frames correspond to the space $Y_{\mathcal{A}}^{\flat}$ and atomic decompositions to $Y_{\mathcal{A}}^{\natural}$. If, however, $\mathcal{A}$ is trivial then we get back the classical theorems, and the sequence spaces coincide, see Lemma 4.3.1(d).

Concerning the Hilbert space $\mathcal{H}_{\mathcal{A}}$ we have the following immediate consequence.

Corollary 4.6.4. *Let $w(x) := \max\{1, \Delta(x)^{-1/2}\}$ and $g \in \mathbb{B}_w^{\mathcal{A}}$ with $\|Kg|\mathcal{H}\| = 1$. Choose a relatively compact neighborhood $U = U^{-1} = \mathcal{A}U$ of $e \in \mathcal{G}$ such that*

$$\|(\widetilde{V}_g g)_U^{\#}|L_w^1\| \;<\; \|\widetilde{V}_g g|L_w^1\|^{-1}.$$

Then for any U-dense well-spread family $X = (x_i)_{i \in I}$ in $\mathcal{G}$ (with respect to $\mathcal{A}$) the family

$$\{\sqrt{|\mathcal{A}(x_i U)|}\,\widetilde{\pi}(x_i) g\}_{i \in I} \tag{4.29}$$

forms a (Hilbert) frame for $\mathcal{H}_{\mathcal{A}}$.

Proof: By Lemma 4.1.5 it holds $\|R_x|L^2\| = \Delta(x)^{-1/2}$ and $\|L_x|L^2\| = 1$ and hence, the corresponding weight function is given by $w(x) = \max\{1, \Delta(x)^{-1/2}\}$. Application of Theorem 4.6.2 yields a Banach frame of the form $\{\widetilde{\pi}(x_i) g\}_{i \in I}$ for $\mathsf{Co}L_{\mathcal{A}}^2 = \mathcal{H}_{\mathcal{A}}$ with corresponding sequence space $(L^2)_{\mathcal{A}}^{\flat} = l_{\nu_2}^2(I)$ with $\nu_2(i) = \sqrt{|\mathcal{A}(x_i U)|}$ (Lemma 4.3.1(e)). Renormalization of the frame elements as in (4.29) yields a Banach frame with corresponding sequence space $l^2(I)$, i.e., a Hilbert frame. $\qquad\qquad \Box$

We will split the proof of Theorems 4.6.1–4.6.3 into several steps. Let us shortly describe the main idea. For some arbitrary U-$\mathcal{A}$-IBUPU $\Psi = (\psi_i)_{i \in I}$ with corresponding well-spread set $X = (x_i)_{i \in I}$ and $g \in \mathbb{B}_w^{\mathcal{A}}$ we introduce the operators

$$T_\Psi F := \sum_{i \in I} \langle F, \psi_i \rangle \mathcal{L}_{x_i} \widetilde{V}_g g,$$

$$S_\Psi F := \sum_{i \in I} F(x_i) \psi_i * \widetilde{V}_g g,$$

$$U_\Psi F := \sum_{i \in I} c_i F(x_i) \mathcal{L}_{x_i} \widetilde{V}_g g,$$

where $c_i = \int_{\mathcal{G}} \psi_i(x) dx$. Intuitively, these are approximations of the convolution operator

$$T : Y_{\mathcal{A}} \to Y_{\mathcal{A}}, \quad TF(x) := F * \widetilde{V}_g g(x) = \int_{\mathcal{G}} F(y) \mathcal{L}_y \widetilde{V}_g g(x) dy.$$

So if U is sufficiently small it seems reasonable that these operators are close to T, which is the identity on $Y_{\mathcal{A}} * \widetilde{V}_g g$ by the reproducing formula in Proposition 4.5.11. Hence, an application of the von Neumann series should imply that T_Ψ, S_Ψ, U_Ψ are invertible on $Y_{\mathcal{A}} * \widetilde{V}_g g$. This means for instance

$$\widetilde{V}_g f = U_\Psi U_\Psi^{-1} \widetilde{V}_g f = \sum_{i \in I} c_i (U_\Psi^{-1} \widetilde{V}_g f)(x_i) \mathcal{L}_{x_i} \widetilde{V}_g g \qquad (4.30)$$

and by injectivity of $\widetilde{V}_g$ and the covariance principle we conclude

$$f = \sum_{i \in I} c_i (U_\Psi^{-1} \widetilde{V}_g f)(x_i) \widetilde{\pi}(x_i) g.$$

This is an expansion of an arbitrary $f \in \mathsf{Co} Y_{\mathcal{A}}$ into elements $\{\widetilde{\pi}(x_i) g\}_{i \in I}$. Thus we have a strong hint that it might be possible to derive atomic decompositions by means of U_Ψ. Reversing the order of U_Ψ and U_Ψ^{-1} in (4.30) leads to a reconstruction of f by means of the coefficients $\langle f, \widetilde{\pi}(x_i) g \rangle$ and it is plausible that one has a Banach frame. Similar arguments apply to the other approximation operators. In the sequel we will make these rough ideas precise.

Let us first consider the operator T_Ψ. We show that T_Ψ is a bounded operator from $Y_{\mathcal{A}}$ to $Y_{\mathcal{A}}$ by splitting it into the analysis operator $F \mapsto (\langle F, \psi_i \rangle)_{i \in I}$ and synthesis operator $(\lambda_i)_{i \in I} \mapsto \sum_{i \in I} \lambda_i \mathcal{L}_{x_i} G$ and treating each part separately.

Proposition 4.6.5. *Let $U = U^{-1} = \mathcal{A}(U)$ be a relatively compact neighborhood of $e \in \mathcal{G}$. For any U-$\mathcal{A}$-IBUPU $(\psi_i)_{i \in I}$ and corresponding well-spread family $X = (x_i)_{i \in I}$ the linear coefficient mapping $F \mapsto (\langle F, \psi_i \rangle)_{i \in I}$ is a bounded operator from $Y_{\mathcal{A}}$ into $Y_{\mathcal{A}}^\flat(X)$, i.e.,*

$$\|(\langle F, \psi_i \rangle)_{i \in I} | Y_{\mathcal{A}}^\flat\| \leq C \|F | Y_{\mathcal{A}}\|,$$

for some constant $C > 0$ independent of F.

Proof: Suppose $F \in Y_{\mathcal{A}}$ and let $V = V^{-1} = \mathcal{A}(V)$ be some relatively compact set with nonvoid interior. We assume that χ_V is taken as window function for the definition of $Y_{\mathcal{A}}^{\natural}$. Since $\operatorname{supp} m_V(\cdot, y) \subset \mathcal{A}(yV)$ the function

$$K(F, y) := \sum_{i \in I} \langle |F|, \psi_i \rangle m_V(x_i, y)$$

is a finite sum over the index set $I_y := \{i, \ x_i \in \mathcal{A}(yV)\}$ for every $y \in \mathcal{G}$. Hence, using Lemma 4.3.2 in the third step we obtain

$$
\begin{aligned}
K(F, y) &= \sum_{i \in I_y} \int_{\mathcal{G}} |F(x)| \psi_i(x) dx m_V(x_i, y) \\
&= \int_{\mathcal{A}(yVU)} |F(x)| \sum_{i \in I_y} \psi_i(x) m_V(x_i, y) dx \leq \int_{\mathcal{A}(yVU)} |F(x)| m_{V^3 U}(x, y) dx \\
&\leq \int_{\mathcal{G}} |F(x)| \int_{\mathcal{A}} \chi_{V^3 U}(y^{-1} \mathcal{A}(x)) dA dx = \int_{\mathcal{A}} \int_{\mathcal{G}} L_y \chi_{V^3 U}(Ax) |F(Ax)| dx dA \\
&= \int_{\mathcal{G}} L_y \chi_{V^3 U}(x) |F(x)| dx = |F| * \chi_{V^3 U}^{\vee}(y).
\end{aligned}
$$

By solidity of Y, Lemma 4.3.3 and (4.20) we finally conclude

$$
\begin{aligned}
\|(\langle F, \psi_i \rangle)_{i \in I} | Y_{\mathcal{A}}^{\natural} \| &\leq C_1^{-1} \| K(F, \cdot) | Y_{\mathcal{A}} \| \leq C_1^{-1} \| |F| * \chi_{V^3 U}^{\vee} | Y_{\mathcal{A}} \| \\
&\leq \underbrace{C_1^{-1} \| \chi_{V^3 U}^{\vee} | L_w^1(\mathcal{G}) \|}_{=C} \| F | Y_{\mathcal{A}} \|.
\end{aligned}
$$

$\square$

Proposition 4.6.6. *Let $X = (x_i)_{i \in I}$ be a well-spread set in $\mathcal{G}$ (with respect to $\mathcal{A}$) and let $G \in W_{\mathcal{A}}^R(C_0, L_w^1)$. Then the mapping*

$$\Lambda = (\lambda_i)_{i \in I} \mapsto \sum_{i \in I} \lambda_i \mathcal{L}_{x_i} G$$

is a bounded, linear operator from $Y_{\mathcal{A}}^{\natural}(X)$ into $Y_{\mathcal{A}}$ satisfying

$$\| \sum_{i \in I} \lambda_i \mathcal{L}_{x_i} G | Y_{\mathcal{A}} \| \leq C \| G | W_{\mathcal{A}}^R(C_0, L_w^1) \| \, \| \Lambda | Y_{\mathcal{A}}^{\natural} \|$$

with some constant C independent of Λ. The sum always converges pointwise, and in the norm of Y if the finite sequences are dense in $Y_{\mathcal{A}}^{\natural}$.

Proof: We let $\mu_{\Lambda} = \sum_{i \in I} \lambda_i \epsilon_{\mathcal{A} x_i}$. By Lemma 4.4.7 this measure is contained in $W_{\mathcal{A}}(M, Y)$. Furthermore, it holds $\sum_{i \in I} \lambda_i \mathcal{L}_{x_i} G = \mu_{\Lambda} * G$. Hence, by Proposition

4.4.2(a), the definition of w and once again by Lemma 4.4.7 we have

$$\|\sum_{i\in I}\lambda_i\mathcal{L}_{x_i}G|Y_{\mathcal{A}}\| \leq C\,\|\mu_\Lambda|W_{\mathcal{A}}(M,Y)\|\,\|G|W_{\mathcal{A}}^R(C_0,L_w^1)\|$$

$$\leq CC_2\,\|\Lambda|Y_{\mathcal{A}}^{\natural}\|\,\|G|W_{\mathcal{A}}^R(C_0,L_w^1)\|.$$

If the finite sequences are dense in $Y_{\mathcal{A}}^{\natural}$ the norm convergence in Y is clear. Since $Y^{\natural}(X) \subset l_{1/r}^\infty$ by Lemma 4.3.4 and $(\mathcal{L}_{x_i}G(x))_{i\in I} \in l_r^1$ by Lemma 4.4.8) where $r(i) = w(x_i)|\mathcal{A}(x_iU)|$ the pointwise convergence follows by l_r^1-$l_{1/r}^\infty$-duality. $\square$

Corollary 4.6.7. *Suppose Ψ is a U-$\mathcal{A}$-IBUPU and $G \in W_{\mathcal{A}}^R(C_0,L_w^1)$. Then T_Ψ is a bounded operator from $Y_{\mathcal{A}}$ into $Y_{\mathcal{A}}$.*

Proof: The assertion follows from Propositions 4.6.5 and 4.6.6. $\square$

For the following we consider families of operators T_Ψ where Ψ runs through a system of IBUPUs. We write $\Psi \to 0$ if for the corresponding neighborhoods U of e it holds $U \to \{e\}$.

Theorem 4.6.8. *Assume that $\Psi = (\psi_i)_{i\in I}$ is a U-$\mathcal{A}$-IBUPU for some set $U = U^{-1} = \mathcal{A}U$ and let $G \in W_{\mathcal{A}}^R(C_0,L_w^1)$. Then it holds*

$$\|T - T_\Psi|Y_{\mathcal{A}}\| \leq \|G_U^{\#}|L_w^1\|,$$

and as consequence of (4.15) $\lim_{\Psi\to 0}\|T - T_\Psi|Y_{\mathcal{A}}\| = 0$.

Proof: We have

$$|TF - T_\Psi F| = \left|\sum_{i\in I}\int_{\mathcal{G}} F(y)\psi_i(y)(\mathcal{L}_yG - \mathcal{L}_{x_i}G)dy\right|$$

$$\leq \sum_{i\in I}\int_{\mathcal{G}} |F(y)|\psi_i(y)|\mathcal{L}_yG - \mathcal{L}_{x_i}G|dy.$$

Since $\operatorname{supp}\psi_i \in \mathcal{A}(x_iU)$ we obtain with Corollary 4.4.11

$$|TF - T_\Psi F| \leq \sum_{i\in I}\int_{\mathcal{G}} |F(y)|\psi_i(y)\mathcal{L}_yG_U^{\#}dy = \int_{\mathcal{G}}|F(y)|\mathcal{L}_yG_U^{\#}dy = |F| * G_U^{\#}$$

and finally by (4.20)

$$\|TF - T_\Psi F|Y_{\mathcal{A}}\| \leq \|F|Y_{\mathcal{A}}\|\,\|G_U^{\#}|L_w^1\|.$$

This gives the estimate for the operator norm. $\square$

Let us now consider the operators S_Ψ and U_Ψ. We first prove their boundedness.

Proposition 4.6.9. *Suppose that* Ψ *is a* U-$\mathcal{A}$-*IBUPU.*

(a) *If* $G \in (L_w^1)_{\mathcal{A}}$ *then* S_Ψ *is a bounded operator from* $W_{\mathcal{A}}(C_0, Y)$ *into* $Y_{\mathcal{A}}$ *and*

$$\|S_\Psi|W_{\mathcal{A}}(C_0, Y) \to Y_{\mathcal{A}}\| \le \gamma(U)\|G|L_w^1\|$$

where $\gamma(U)$ *is the constant from Lemma 4.4.9.*

(b) *If* $G \in W_{\mathcal{A}}^R(C_0, L_w^1)$ *then* U_Ψ *is a bounded operator from* $W_{\mathcal{A}}(C_0, Y)$ *into* $Y_{\mathcal{A}}$ *and*

$$\|U_\Psi|W_{\mathcal{A}}(C_0, Y) \to Y_{\mathcal{A}}\| \le \gamma(U)(\|G|L_w^1\| + \|G_U^{\#}|L_w^1\|). \tag{4.31}$$

Proof: (a) For $F \in W_{\mathcal{A}}(C_0, Y)$ the convolution relation (4.20), the norm estimate $\|F|Y\| \le \|F|W(C_0, Y)\|$ (Lemma 4.4.4(a)) and Lemma 4.4.9 yield

$$\|S_\Psi F|Y_{\mathcal{A}}\| = \|(\sum_{i \in I} F(x_i)\psi_i) * G|Y_{\mathcal{A}}\| \le \|\sum_{i \in I} F(x_i)\psi_i|Y_{\mathcal{A}}\| \, \|G|L_w^1\|$$

$$\le \|\sum_{i \in I} F(x_i)\psi_i|W_{\mathcal{A}}(C_0, Y)\| \, \|G|L_w^1\| \le \gamma(U)\|F|W_{\mathcal{A}}(C_0, Y)\| \, \|G|L_w^1\|. \tag{4.32}$$

(b) Since $\operatorname{supp} \psi_i \subset \mathcal{A}(x_i U)$ we may estimate by Corollary 4.4.11

$$|c_i \mathcal{L}_{x_i} G - \psi_i * G| = \left| \int_{\mathcal{G}} \psi_i(y)(\mathcal{L}_{x_i} G - \mathcal{L}_y G) dy \right| \le \int_{\mathcal{G}} \psi_i(y)\mathcal{L}_y G_U^{\#} dy = \psi_i * G_U^{\#}.$$

Hence, we get

$$\|U_\Psi F - S_\Psi F|Y_{\mathcal{A}}\| = \|\sum_{i \in I} F(x_i)(c_i \mathcal{L}_{x_i} G - \psi_i * G)|Y_{\mathcal{A}}\|$$

$$\le \|(\sum_{i \in I} |F(x_i)|\psi_i) * G_U^{\#}|Y_{\mathcal{A}}\|.$$

As in (4.32) we obtain

$$\|U_\Psi F - S_\Psi F|Y_{\mathcal{A}}\| \le \gamma(U)\|F|W_{\mathcal{A}}(C_0, Y)\| \, \|G_U^{\#}|L_w^1\| \tag{4.33}$$

giving (4.31) by the triangle inequality and by (4.32). $\square$

For the analysis of the operator S_Ψ we need to restrict to the subspace $Y_{\mathcal{A}} * G$, where in the original setting $G = \widetilde{V}_g g$ with $\|Kg\| = 1$ implying $G = G^\nabla = G * G$.

Theorem 4.6.10. *Suppose that* $G \in W_{\mathcal{A}}^R(C_0, L_w^1)$ *with* $G = G^\nabla = G * G$ *and that* Ψ *is a* U-$\mathcal{A}$-*IBUPU. Then*

$$\|T - S_\Psi|Y_{\mathcal{A}} * G\| \le \|G_U^{\#}|L_w^1\| \, \|G|L_w^1\|.$$

In particular, it holds $\lim_{\Psi \to 0} \|T - S_\Psi|Y_{\mathcal{A}} * G\| = 0$.

Proof: Suppose $F \in Y_{\mathcal{A}} * G$. Since $G = G * G$ we have the reproducing formula $F * G = F$. Using the convolution relation (4.20) we obtain

$$\|TF - S_\Psi F | Y_{\mathcal{A}}\| \leq \|F - \sum_{i \in I} F(x_i)\psi_i | Y_{\mathcal{A}}\| \, \|G|L_w^1\|.$$

Since $F \in Y_{\mathcal{A}} * G \subset W_{\mathcal{A}}(C_0, Y)$ (Proposition 4.4.2(b)) the expression on the right hand side is well-defined by Lemma 4.4.9.

Assume $y \in \mathcal{A}(x_i U)$, i.e., $x_i = A(yu^{-1})$ for some $A \in \mathcal{A}, u \in U = \mathcal{A}(U)$. By Corollary 4.5.12(b) and by invariance of F we obtain

$$|F(y) - F(x_i)| \leq \sup_{u \in U} |F(y) - F(yu^{-1})| = \sup_{u \in U} |\langle F, L_y G - L_{yu^{-1}} G \rangle|$$

$$= \sup_{u \in U} \left| \int_{\mathcal{G}} F(z)\overline{(G(y^{-1}z) - G(uy^{-1}z))} dz \right| \leq \int_{\mathcal{G}} G_U^\#(y^{-1}z)|F(z)|dz = |F| * (G_U^\#)^\vee(y),$$

and hence,

$$|F(y) - \sum_{i \in I} F(x_i)\psi_i(y)| \leq \sum_{i \in I} |F(y) - F(x_i)|\psi_i(y) \leq \sum_{i \in I} |F| * (G_U^\#)^\vee(y)\psi_i(y)$$

$$= |F| * (G_U^\#)^\vee(y).$$

Finally, this gives

$$\|TF - S_\Psi F | Y_{\mathcal{A}}\| \leq \||F| * (G_U^\#)^\vee | Y_{\mathcal{A}}\| \, \|G|L_w^1\| \leq \|F|Y_{\mathcal{A}}\| \, \|G_U^\#|L_w^1\| \, \|G|L_w^1\|.$$

The last assertion of the theorem follows with Lemma 4.4.10(b). $\qquad\square$

Theorem 4.6.11. *Suppose that* $G \in W_{\mathcal{A}}^R(C_0, L_w^1)$ *with* $G = G^\nabla = G * G$ *and let* Ψ *be a* U-$\mathcal{A}$-IBUPU. *Then*

$$\|T - U_\Psi | Y_{\mathcal{A}} * G\| \leq \|G_U^\#|L_w^1\| \left(\|G|L_w^1\| + \gamma(U)D\|G|W_{\mathcal{A}}^R(C_0, L_w^1)\| \right)$$

where $\gamma(U)$ *is the constant from Lemma 4.4.9 and* D *is the constant in Proposition 4.4.2(b). In particular, it holds* $\lim_{\Psi \to 0} \|T - U_\Psi | Y_{\mathcal{A}} * G\| = 0$.

Proof: Suppose $F \in Y_{\mathcal{A}} * G$. Using the reproducing formula $F * G = F$, (4.33) and Proposition 4.4.2(b) we obtain

$$\|U_\Psi F - S_\Psi F | Y_{\mathcal{A}}\| \leq \gamma(U)\|F|W_{\mathcal{A}}(C_0, Y)\| \, \|G_U^\#|L_w^1\|$$

$$= \gamma(U)\|F * G|W_{\mathcal{A}}(C_0, Y)\| \, \|G_U^\#|L_w^1\|$$

$$\leq \gamma(U)D\|F|Y_{\mathcal{A}}\| \, \|G|W_{\mathcal{A}}^R(C_0, L_w^1)\| \, \|G_U^\#|L_w^1\|.$$

Together with Theorem 4.6.10 and the triangle inequality we obtain the desired estimation. Since $\gamma(U) \leq \gamma_0$ when U runs through a family of subsets of some U_0 (Lemma 4.4.9) the last assertion follows with Lemma 4.4.10(b). □

Now we are prepared to prove Theorems 4.6.1–4.6.3.

Proof of Theorem 4.6.1: The restriction of the operator $TF := F * \widetilde{V}_g g$ to the closed subspace $Y_{\mathcal{A}} * \widetilde{V}_g g$ is the identity by the reproducing formula (Proposition 4.5.11). By the assumption on $(\widetilde{V}_g g)_U^{\#}$ and Theorem 4.6.8 we have $\|T - T_\Psi|Y_{\mathcal{A}} * \widetilde{V}_g g\| < 1$ and, hence, T_Ψ is invertible on $Y_{\mathcal{A}} * \widetilde{V}_g g$ by means of the von Neumann series. Further, if $f \in \mathsf{Co}Y_{\mathcal{A}}$ then $\widetilde{V}_g f \in Y_{\mathcal{A}} * \widetilde{V}_g g$ and

$$\widetilde{V}_g f \; = \; T_\Psi T_\Psi^{-1} \widetilde{V}_g f \; = \; \sum_{i \in I} \langle T_\Psi^{-1} \widetilde{V}_g f, \psi_i \rangle \mathcal{L}_{x_i} \widetilde{V}_g g.$$

Since $\mathcal{L}_{x_i} \widetilde{V}_g g = \widetilde{V}_g(\widetilde{\pi}(x_i)g)$ and since $\widetilde{V}_g$ is an isometric isomorphism between $\mathsf{Co}Y_{\mathcal{A}}$ and $Y_{\mathcal{A}} * \widetilde{V}_g g$ (Corollary 4.5.12(a)) we obtain

$$f \; = \; \sum_{i \in I} \langle T_\Psi^{-1} \widetilde{V}_g f, \psi_i \rangle \widetilde{\pi}(x_i)g.$$

Set $\lambda_i(f) := \langle T_\Psi^{-1} \widetilde{V}_g f, \psi_i \rangle$. As $T_\Psi^{-1} \widetilde{V}_g f \in Y_{\mathcal{A}} * \widetilde{V}_g g \subset Y_{\mathcal{A}}$ we get with Proposition 4.6.5

$$\|(\lambda_i)_{i \in I}|Y_{\mathcal{A}}^\natural\| \leq C \|T_\Psi^{-1} \widetilde{V}_g f|Y_{\mathcal{A}}\| \leq C \|T_\Psi^{-1}|Y_{\mathcal{A}}\| \, \|f|\mathsf{Co}Y_{\mathcal{A}}\|.$$

For a converse inequality we apply $\widetilde{V}_g$ to the series $\sum_{i \in I} \lambda_i \widetilde{\pi}(x_i)g$ to obtain (at least formally)

$$F(x) \; := \; \widetilde{V}_g \left(\sum_{i \in I} \lambda_i \widetilde{\pi}(x_i)g \right)(x) \; = \; \sum_{i \in I} \lambda_i \mathcal{L}_{x_i} \widetilde{V}_g g(x). \tag{4.34}$$

Since $Y_{\mathcal{A}}^\natural \subset l_{1/\tilde{r}}^\infty$ with $r(i) := w(x_i)|\mathcal{A}(x_i U)|$ by Lemma 4.3.4 and $\widetilde{V}_g g \in W_{\mathcal{A}}^R(C_0, L_w^1)$ the right hand side of (4.34) converges pointwise and defines a function in $L_{1/w}^\infty(\mathcal{G})$ by (4.11). By Lemma 4.5.8(b) the pointwise convergence of the partial sums of F implies the weak-$*$ convergence of $\sum_{i \in I} \lambda_i \widetilde{\pi}(x_i)g =: f$. Once f is identified with an element of $(\mathcal{H}_w^1)_{\mathcal{A}}^\neg$ it belongs to $\mathsf{Co}Y_{\mathcal{A}}$ by Proposition 4.6.6, i.e.,

$$\|f|\mathsf{Co}Y_{\mathcal{A}}\| \; = \; \|\sum_{i \in I} \lambda_i \mathcal{L}_{x_i} \widetilde{V}_g g|Y_{\mathcal{A}}\| \leq C \|\widetilde{V}_g g|W_{\mathcal{A}}^R(C_0, L_w^1)\| \, \|(\lambda_i)_{i \in I}|Y_{\mathcal{A}}^\natural\|. \tag{4.35}$$

Also the stated type of convergence is implied by Proposition 4.6.6. □

Proof of Theorem 4.6.2: By Theorem 4.6.10 condition (4.25) implies that S_Ψ is invertible on $Y_{\mathcal{A}} * \widetilde{V}_g g$ and we obtain

$$\widetilde{V}_g f \; = \; S_\Psi^{-1} S_\Psi \widetilde{V}_g f = S_\Psi^{-1} \left(\sum_{i \in I} \widetilde{V}_g f(x_i) \psi_i * \widetilde{V}_g g \right). \tag{4.36}$$

By the correspondence principle (Corollary 4.5.12(a)) $\widetilde{V}_g$ is invertible on its range $Y_{\mathcal{A}} * \widetilde{V}_g g$ yielding

$$f = \widetilde{V}_g^{-1} S_\Psi^{-1} \left(\sum_{i \in I} \langle f, \widetilde{\pi}(x_i) g \rangle \psi_i * \widetilde{V}_g g \right). \tag{4.37}$$

This is a reconstruction of f from the coefficients $(\langle f, \widetilde{\pi}(x_i) g \rangle)_{i \in I}$. The reconstruction operator may be written as $\Theta = \widetilde{V}_g^{-1} S_\Psi^{-1} T H$ where $H : Y_{\mathcal{A}}^\flat \to Y_{\mathcal{A}}$ is defined by $H((\lambda_i)_{i \in I}) := \sum_{i \in I} \lambda_i \phi_i$. Since $\phi_i \leq \chi_{\mathcal{A}(x_i U)}$ the operator H is bounded by definition of $Y_{\mathcal{A}}^\flat$. Hence, also Θ is bounded as the composition of bounded operators.

Setting $Y = L_{1/w}^\infty$ shows that any $f \in \mathrm{Co}(L_{1/w}^\infty)_{\mathcal{A}} = (\mathcal{H}_w^1)_{\mathcal{A}}^{\urcorner}$ (Corollary 4.5.14(a)) can be reconstructed as in (4.37). Now, if for $f \in (\mathcal{H}_w^1)_{\mathcal{A}}^{\urcorner}$ it holds $(\widetilde{V}_g f(x_i))_{i \in I} \in Y_{\mathcal{A}}^\flat$ then the series in (4.36) converges to a function in $W_{\mathcal{A}}(C_0, Y) * G \subset Y_{\mathcal{A}} * \widetilde{V}_g g$ by Lemma 4.4.9. By invertibility of S_Ψ on $Y_{\mathcal{A}} * \widetilde{V}_g g$ the function $\widetilde{V}_g f$ is therefore contained in $Y_{\mathcal{A}} * \widetilde{V}_g g$, hence $f \in \mathrm{Co} Y_{\mathcal{A}}$. Together with Theorem 4.5.15 this shows $f \in \mathrm{Co} Y_{\mathcal{A}}$ if and only if $(\widetilde{V}_g f(x_i))_{i \in I} = (\langle f, \widetilde{\pi}(x_i) g \rangle)_{i \in I} \in Y_{\mathcal{A}}^\flat(X)$.

From (4.36) we obtain the equivalence of norms,

$$\|f|\mathrm{Co} Y_{\mathcal{A}}\| = \|\widetilde{V}_g f|Y_{\mathcal{A}}\| \leq \|S_\Psi^{-1}|Y_{\mathcal{A}} * \widetilde{V}_g g\| \| \sum_{i \in I} \widetilde{V}_g f(x_i) \psi_i * \widetilde{V}_g g | Y_{\mathcal{A}}\|$$

$$\leq \|S_\Psi^{-1}\| \| \sum_{i \in I} \widetilde{V}_g f(x_i) \psi_i |Y_{\mathcal{A}}\| \|\widetilde{V}_g g|L_w^1\|$$

$$\leq \|S_\Psi^{-1}\| \| \sum_{i \in I} |\widetilde{V}_g f(x_i)| \chi_{\mathcal{A}(x_i U)} |Y_{\mathcal{A}}\| \|\widetilde{V}_g g|L_w^1\|$$

$$= \|S_\Psi^{-1}\| \|\widetilde{V}_g g|L_w^1\| \|(\widetilde{V}_g f(x_i))_{i \in I}|Y_{\mathcal{A}}^\flat(X)\| \leq \gamma(U) C \|S_\Psi^{-1}\| \|\widetilde{V}_g g|L_w^1\| \|f|\mathrm{Co} Y_{\mathcal{A}}\|.$$

Hereby we used (4.20), the definition of $Y_{\mathcal{A}}^\flat(X)$ and Theorem 4.5.15.

In remains to prove (4.26) if the finite sequences are dense in $Y_{\mathcal{A}}^\flat$. Since $\psi_i * \widetilde{V}_g g \in (L_w^1)_{\mathcal{A}} * \widetilde{V}_g g$ and S_Ψ is invertible on $L_w^1 * \widetilde{V}_g g$, there exists a unique element $e_i \in (\mathcal{H}_w^1)_{\mathcal{A}}$ such that

$$\widetilde{V}_g e_i = S_\Psi^{-1}(\psi_i * \widetilde{V}_g g).$$

Assume now that $(\lambda_i)_{i \in I}$ is finitely supported on N, say. In this case we may interchange the application of $V_g^{-1} S_\Psi^{-1}$ with the summation in the definition of the reconstruction operator Θ, see also (4.37), yielding $\Theta((\lambda_i)_{i \in I}) = \sum_{i \in N} \lambda_i e_i$. Now, if the finite sequences are dense in $Y_{\mathcal{A}}^\flat$ then the mapping $(\lambda_i)_{i \in I} \mapsto \sum_{i \in I} \lambda_i e_i$ extends uniquely from the finite sequences to a bounded operator from $Y_{\mathcal{A}}^\flat \to \mathrm{Co} Y_{\mathcal{A}}$ by boundedness of Θ and, hence, it must coincide with Θ. This shows also the norm convergence in $\mathrm{Co} Y_{\mathcal{A}}$. $\square$

Proof of Theorem 4.6.3: Similarly as in the two previous proofs condition (4.27) implies by Theorem 4.6.11 that the operator U_Ψ is invertible on $Y_{\mathcal{A}} * \widetilde{V}_g g$. For $f \in \mathrm{Co} Y_{\mathcal{A}}$

we have

$$\widetilde{V}_g f = U_\Psi U_\Psi^{-1} \widetilde{V}_g f = \sum_{i \in I} (U_\Psi^{-1} \widetilde{V}_g f)(x_i) c_i \mathcal{L}_{x_i} \widetilde{V}_g g \tag{4.38}$$

and

$$\widetilde{V}_g f = U_\Psi^{-1} U_\Psi \widetilde{V}_g f = U_\Psi^{-1} \left(\sum_{i \in I} \widetilde{V}_g f(x_i) c_i \mathcal{L}_{x_i} \widetilde{V}_g g \right). \tag{4.39}$$

Now one proceeds similarly to the proofs of Theorems 4.6.1 and 4.6.2, i.e., (4.38) leads to an atomic decomposition of $\mathsf{Co}Y_{\mathcal{A}}$ and (4.39) leads to Banach frames. However, the norm estimates are slightly different since the numbers c_i are not bounded from above, in general, as it is the case in the classical theory [Grö91].

Starting from (4.38) we define $\lambda_i(f) := c_i (U_\Psi^{-1} \widetilde{V}_g f)(x_i)$ yielding $f = \sum_{i \in I} \lambda_i(f) \widetilde{\pi}(x_i) g$. Since $\operatorname{supp} \psi_i \subset \mathcal{A}(x_i U)$ it holds $c_i \leq |\mathcal{A}(x_i V)|$ if $U \subset V$ and we assume without loss of generality that such a set V is chosen for the definition of $Y_{\mathcal{A}}^\flat$. Moreover, we have $U_\Psi^{-1} \widetilde{V}_g f \in W_{\mathcal{A}}(C_0, Y) \cap Y_{\mathcal{A}} * \widetilde{V}_g g$ by Proposition 4.6.9. Applying Lemma 4.4.9 and Proposition 4.4.2(b) yields

$$\|(\lambda_i(f))_{i \in I} | Y_{\mathcal{A}}^\flat\| \leq \|((U_\Psi^{-1} \widetilde{V}_g f)(x_i))_{i \in I} | Y_{\mathcal{A}}^\flat\| \leq \gamma(U) C \|U_\Psi^{-1} \widetilde{V}_g f | W_{\mathcal{A}}(C_0, Y)\|$$

$$= \gamma(U) C \|(U_\Psi^{-1} \widetilde{V}_g f) * \widetilde{V}_g g | W_{\mathcal{A}}(C_0, Y)\| \leq \gamma(U) C D \|U_\Psi^{-1} \widetilde{V}_g f | Y_{\mathcal{A}}\| \, \|\widetilde{V}_g g | W_{\mathcal{A}}^R(C_0, L_w^1)\|$$

$$\leq C' \|U_\Psi^{-1} | Y_{\mathcal{A}} * \widetilde{V}_g g\| \, \|\widetilde{V}_g g | W_{\mathcal{A}}^R(C_0, L_w^1)\| \, \|f | \mathsf{Co}Y_{\mathcal{A}}\|. \tag{4.40}$$

The converse norm estimate is the same as in (4.35). Thus $\{\widetilde{\pi}(x_i) g\}_{i \in I}$ is an atomic decomposition of $\mathsf{Co}Y_{\mathcal{A}}$.

Beginning with (4.39) the norm estimate in the proof of the Banach frame property goes as follows,

$$\|f | \mathsf{Co}Y_{\mathcal{A}}\| = \|\widetilde{V}_g f | Y_{\mathcal{A}}\| = \|U_\Psi^{-1} (\sum_{i \in I} c_i \widetilde{V}_g f(x_i) \mathcal{L}_{x_i} \widetilde{V}_g g) | Y_{\mathcal{A}}\|$$

$$\leq \|U_\Psi^{-1} | Y_{\mathcal{A}} * \widetilde{V}_g g\| \, \|\sum_{i \in I} c_i \widetilde{V}_g f(x_i) \epsilon_{\mathcal{A}(x_i)} * \widetilde{V}_g g | W_{\mathcal{A}}(C_0, Y)\|$$

$$\leq \|U_\Psi^{-1}\| \, \|\sum_{i \in I} c_i \widetilde{V}_g f(x_i) \epsilon_{\mathcal{A}x_i} | W_{\mathcal{A}}(M, Y)\| \, \|\widetilde{V}_g g | W_{\mathcal{A}}^R(C_0, L_w^1)\|$$

$$\leq C \|U_\Psi^{-1}\| \, \|\widetilde{V}_g g | W_{\mathcal{A}}^R(C_0, L_w^1)\| \, \|(\langle f, \widetilde{\pi}(x_i) g \rangle)_{i \in I} | Y_{\mathcal{A}}^\flat\|$$

$$\leq C' \gamma(U)) \|U_\Psi^{-1}\| \, \|\widetilde{V}_g g | W_{\mathcal{A}}^R(C_0, L_w^1)\|^2 \, \|f | \mathsf{Co}Y_{\mathcal{A}}\|. \tag{4.41}$$

Hereby, we used Proposition 4.4.2(a), Lemma 4.4.7, $c_i \leq |\mathcal{A}(x_i U)|$ and Theorem 4.5.15. It follows from (4.39) that the reconstruction operator is given by $\Theta := U_\Psi^{-1} Q$, where $Q : Y_{\mathcal{A}}^\flat \to Y_{\mathcal{A}}$, $Q((\lambda_i)_{i \in I}) := \sum_{i \in I} c_i \lambda_i \mathcal{L}_{x_i} \widetilde{V}_g g$. By Proposition 4.6.6 and since $c_i \leq |\mathcal{A}(x_i U)|$ the operator Q, and hence also Θ, is bounded. Thus, we showed that $\{\widetilde{\pi}(x_i) g\}_{i \in I}$ forms a Banach frame.

Now set $E_i := c_i U_\Psi^{-1}(\mathcal{L}_{x_i}\widetilde{V}_g g)$. Then we have $E_i \in (L_w^1)_\mathcal{A} * \widetilde{V}_g g$ and $E_i = \widetilde{V}_g(e_i)$ for some unique $e_i \in (\mathcal{H}_w^1)_\mathcal{A}$. As in the last part of the previous proof we conclude from (4.39) that it holds $f = \sum_{i \in I}\langle f, \widetilde{\pi}(x_i)g\rangle e_i$ provided the finite sequences are dense in $Y_\mathcal{A}^\flat$.

We claim that
$$\lambda_i(f) = c_i(U_\Psi^{-1}V_g f)(x_i) = \langle f, e_i\rangle.$$

Together with the correspondence principle this would yield $f = \sum_{i \in I}\langle f, e_i\rangle\widetilde{\pi}(x_i)g$ (with weak-$*$ convergence, and if the finite sequences are dense in $Y_\mathcal{A}^\flat$ with norm convergence).

Since $U_\Psi^{-1}F \in Y_\mathcal{A} * \widetilde{V}_g g$ for $F \in Y_\mathcal{A} * \widetilde{V}_g g$ we have $U_\Psi^{-1}F(x_i) = \langle U_\Psi^{-1}F, \mathcal{L}_{x_i}\widetilde{V}_g g\rangle$ by Corollary 4.5.12(b). It follows that U_Ψ satisfies $\langle U_\Psi F, H\rangle = \langle F, U_\Psi H\rangle$ for all $F \in Y_\mathcal{A} * \widetilde{V}_g g, H \in L_w^1 * \widetilde{V}_g g$:

$$\langle U_\Psi F, H\rangle = \sum_{i \in I}c_i F(x_i)\langle \mathcal{L}_{x_i}\widetilde{V}_g g, H\rangle = \sum_{i \in I}c_i\langle F, \mathcal{L}_{x_i}\widetilde{V}_g g\rangle\langle \mathcal{L}_{x_i}\widetilde{V}_g g, H\rangle$$
$$= \sum_{i \in I}c_i\overline{H(x_i)\langle \mathcal{L}_{x_i}\widetilde{V}_g g, F\rangle} = \langle F, U_\Psi H\rangle.$$

Hence, the same relation applies to $U_\Psi^{-1} = \sum_{n=0}^\infty(\mathrm{Id} - U_\Psi)^n$ and we conclude that $\langle U_\Psi^{-1}F, \mathcal{L}_{x_i}\widetilde{V}_g g\rangle = \langle F, U_\Psi^{-1}\mathcal{L}_{x_i}\widetilde{V}_g g\rangle$. Finally, we obtain

$$c_i(U_\Psi^{-1}\widetilde{V}_g f)(x_i) = \langle \widetilde{V}_g f, c_i U_\Psi^{-1}\mathcal{L}_{x_i}\widetilde{V}_g g\rangle = \langle \widetilde{V}_g f, \widetilde{V}_g e_i\rangle = \langle f, \widetilde{V}_g^*\widetilde{V}_g e_i\rangle = \langle f, e_i\rangle,$$

where we used Theorem 2.2.7(d) in the last equality. By Proposition 4.6.6 and Lemma 4.4.9 we have the norm estimate

$$\|f|\mathsf{Co}Y_\mathcal{A}\| = \Big\|\sum_{i \in I}\langle f, e_i\rangle\mathcal{L}_{x_i}\widetilde{V}_g g|Y_\mathcal{A}\Big\| \leq C\|\widetilde{V}_g g|W_\mathcal{A}^R(C_0, L_w^1)\| \|(\langle f, e_i\rangle)_{i \in I}|Y_\mathcal{A}^\flat\|$$
$$\leq C\|\widetilde{V}_g g|W_\mathcal{A}^R(C_0, L_w^1)\| \|((U_\Psi^{-1}\widetilde{V}_g f)(x_i))_{i \in I}|Y_\mathcal{A}^\flat\|$$
$$\leq C\|U_\Psi^{-1}|Y_\mathcal{A} * \widetilde{V}_g g\| \|\widetilde{V}_g g|W_\mathcal{A}^R(C_0, L_w^1)\|^2 \|f|\mathsf{Co}Y_\mathcal{A}\|.$$

Hereby we used $\|(c_i\lambda_i)_{i \in I}|Y_\mathcal{A}^\flat\| \leq \|(\lambda_i)_{i \in I}|Y_\mathcal{A}^\flat\|$. Thus, we derived the first equivalence in (4.28). The second equivalence of (4.28) was already shown in (4.41). □

4.7 Examples

4.7.1 Radial Modulation Spaces

Let us take the Heisenberg group $\mathcal{G} = \mathbb{H}_d$ and the Schrödinger representation ρ on $L^2(\mathbb{R}^d)$ as in Section 1.5.3. The transform V_g is connected to the STFT by (1.83), in particular $|V_g f(x, \omega, \tau)| = |\operatorname{STFT}_g f(x, \omega)|$ for all $(x, \omega, \tau) \in \mathbb{H}_d$. We will identify the corresponding coorbit spaces with modulation spaces on $\mathbb{R}^d$ as introduced by Feichtinger [Fei83c, Fei83a, Fei90], see also [Grö01, Tri83a].

Let w be a submultiplicative weight function on $\mathbb{H}_d$, that depends only on $\mathbb{R}^d \times \mathbb{R}^d$, in particular

$$w(x + x', \omega + \omega') \le w(x, \omega) w(x', \omega') \quad \text{for all } x, x', \omega, \omega' \in \mathbb{R}^d.$$

For simplicity we assume that w is symmetric, i.e $w(x, \omega) = w(-x, -\omega)$, and has polynomial growth, i.e., $w(x, \omega) \le C(1 + \sqrt{|x|^2 + |\omega|^2})^s$ for some $s \ge 0$. Further, let m be a w-moderate weight function on $\mathbb{H}_d$, that also depends only on $\mathbb{R}^d \times \mathbb{R}^d$, i.e.,

$$m(x + x', \omega + \omega') \le w(x, \omega) m(x', \omega') \quad \text{for all } x, x', \omega, \omega' \in \mathbb{R}^d.$$

Polynomial growth of w implies that also m grows polynomially (with the same s). Of common use are the functions

$$m_s(x, \omega) := (1 + |\omega|)^s, \quad s \in \mathbb{R},$$

depending only on the frequency variable. These functions are submultiplicative if $s \ge 0$ and $m_{|s|}$-moderate, in general [Grö01, Chapter 11.1].

For $1 \le p, q \le \infty$ we define the mixed norm space $L_m^{p,q}$ on $\mathbb{H}_d$ consisting of all measurable functions whose norm

$$\|F|L_m^{p,q}\| := \left(\int_{\mathbb{R}^d} \left(\int_{\mathbb{R}^d} \int_0^1 |F(x, \omega, e^{2\pi i t})|^p m(x, \omega)^p dt dx \right)^{q/p} d\omega \right)^{1/q}$$

is finite (with obvious modification for $p = \infty$ or $q = \infty$). If $p = q$ then we have $L_m^{p,p} = L_s^p(\mathbb{H}_d)$. Clearly, the $L_m^{p,q}$-spaces are solid BF-spaces. In order to define the corresponding coorbit spaces we need to prove the translation invariance.

Lemma 4.7.1. *Suppose m is w-moderate (with symmetric w). Then for $1 \le p, q \le \infty$ the space $L_m^{p,q}$ is two-sided invariant with*

$$\|L_{(x, \omega, \tau)} | L_m^{p,q}\| \le w(x, \omega), \qquad \|R_{(x, \omega, \tau)} | L_m^{p,q}\| \le w(x, \omega).$$

Proof: For $(x', \omega', e^{2\pi i t'}) \in \mathbb{H}_d$ and $1 \le p, q < \infty$ we obtain by invariance of the Lebesgue measure and w-moderateness of m

$$\|R_{(x', \omega', e^{2\pi i t'})} F | L_m^{p,q}\|$$

$$= \left(\int_{\mathbb{R}^d} \left(\int_{\mathbb{R}^d} \int_0^1 |F(x + x', \omega + \omega', e^{2\pi i(t + t' + 1/2(x' \cdot \omega - x \cdot \omega'))})|^p m(x, \omega)^p dt dx \right)^{q/p} d\omega \right)^{1/q}$$

$$\le w(x', \omega') \left(\int_{\mathbb{R}^d} \left(\int_{\mathbb{R}^d} \int_0^1 |F(x, \omega, e^{2\pi i t})|^p m(x, \omega)^p dt dx \right)^{q/p} d\omega \right)^{1/q}$$

$$= w(x', \omega') \|F | L_m^{p,q}\|.$$

The cases $p = \infty$ or $q = \infty$ and also the invariance under $L_{(x',\omega',e^{2\pi i t'})}$ are shown analogously. □

Next we have to show that $\mathbb{A}_w$ is not trivial. If $g \in \mathcal{S}(\mathbb{R}^d)$ then Theorem 11.2.5 in [Grö91] states that for all $n \in \mathbb{N}$ there exists a constant $C_n > 0$ such that

$$|V_g g(x,\omega,\tau)| = |\operatorname{STFT}_g g(x,\omega)| \leq C_n (1 + |x|^2 + |\omega|^2)^{-n}, \quad (x,\omega,\tau) \in \mathbb{H}_d.$$

This implies that $V_g g \in L_w^1$ for all submultiplicative weight functions w with polynomial growth, i.e., $\mathcal{S}(\mathbb{R}^d) \subset \mathbb{A}_w$, in particular $\mathbb{A}_w \neq \{0\}$. Thus, the spaces $\mathcal{H}_w^1 = \operatorname{Co}L_w^1$, $(\mathcal{H}_w^1)^{\neg} = \operatorname{Co}L_{1/w}^\infty$ and $\operatorname{Co}L_m^{p,q}$ are well-defined. Moreover, we have $(\mathcal{H}_w^1)^{\neg} \subset \mathcal{S}'(\mathbb{R}^d)$. It follows from Corollary 11.2.7(a) in [Grö01] that the reproducing formula $V_g f = V_g f * V_g g$ extends to $f \in \mathcal{S}'(\mathbb{R}^d)$ and, consequently, with Theorem 4.5.13(c) we see that the coorbit spaces $\operatorname{Co}L_m^{p,q}$ coincide with the modulation spaces, defined by

$$M_m^{p,q} = \{f \in \mathcal{S}'(\mathbb{R}^d), \|f|M_m^{p,q}\| < \infty\},$$

where

$$\|f|M_m^{p,q}\| = \|f|\operatorname{Co}L_m^{p,q}\| = \|V_g f|L_m^{p,q}\|$$

$$= \left(\int_{\mathbb{R}^d} \left(\int_{\mathbb{R}^d} |\operatorname{STFT}_g f(x,\omega)|^p m(x,\omega)^p dx \right)^{q/p} d\omega \right)^{1/q}.$$

Taking the weight function m_s, the corresponding modulation spaces are commonly denoted by $M_s^{p,q} = M_{m_s}^{p,q}$. Of special interest is the Feichtinger algebra $S_0 = M^{1,1} = M_0^{1,1}$, which is known to be the minimal character invariant Segal algebra on $\mathbb{R}^d$ [Fei81c, FZ98], see also Corollary 4.5.7. Also Sobolev spaces (resp. Bessel potential spaces) are particular instances of modulation spaces, i.e., $M_s^{2,2} = H^s(\mathbb{R}^d)$ [Grö01, Proposition 11.3.1(c)].

Since the Heisenberg group is an IN-group the space $\mathbb{B}_w$ coincides with $\mathbb{A}_w$ by Lemma 4.5.10. Since w is assumed to be symmetric and $\mathbb{H}_d$ is unimodular it holds $w^{\bullet} = 2w$. Thus by Lemma 4.5.2(d), we have $\mathbb{A}_w = \mathcal{H}_w^1 = M_w^{1,1}$. Taking the trivial automorphism group $\mathcal{A} = \{e\}$, Theorems 4.6.1–4.6.3 yield classical (irregular) Gabor frames and Gabor type atomic decompositions of modulation spaces, see [Fei89, FG92c, FG92a, Grö91, Grö01].

In the sequel we are interested in radial functions on $\mathbb{R}^d, d \geq 2$, which leads us to take $\mathcal{A} = SO(d)$ and its natural representation σ on $L^2(\mathbb{R}^d)$ as in Section 2.3.2. The Hilbert space $\mathcal{H}_{\mathcal{A}}$ coincides then with the space $L_{rad}^2(\mathbb{R}^d)$ of radial L^2-functions as desired. In order to define coorbit spaces we require additionally that the weight functions m and w are invariant under $SO(d)$, i.e., $m(Ax, A\omega) = m(x,\omega)$ for all $x, \omega \in \mathbb{R}^d, A \in SO(d)$.

This is the case for m_s, in particular. Clearly, $\mathbb{A}_w \supset \mathcal{S}(\mathbb{R}^d)$ contains radial functions, so that $\mathbb{A}_w^{\mathcal{A}} \supset \mathcal{S}_{rad}(\mathbb{R}^d)$ is non-trivial. Moreover, we have

$$\mathsf{Co}(L_m^{p,q})_{SO(d)} = (M_m^{p,q})_{rad} = \{f \in M_m^{p,q}, f \text{ is radial}\}.$$

Theorems 4.6.1–4.6.3 yield radial Gabor frames and Gabor type atomic decompositions for the radial modulation spaces $(M_m^{p,q})_{rad}$. Let us work this out in detail.

The only ingredient, which requires some effort, is an invariant covering of $\mathbb{H}_d$ as in Theorem 4.2.2. Rather than relying on existence we give an explicit construction of a "regular" covering invariant under $SO(d)$. We remark that it is enough to cover $\mathbb{R}^d \times \mathbb{R}^d$ since then we immediately obtain an invariant covering of $\mathbb{H}_d$ by forming Cartesian products with $\mathbb{T}$ (or a covering of $\mathbb{T}$). Let us recall that $SO(d)$ acts on $\mathbb{R}^d \times \mathbb{R}^d$ as $A(x,\omega) = (Ax, A\omega)$ for $A \in SO(d)$, $x, \omega \in \mathbb{R}^d$. The simultaneous action of $SO(d)$ in both components of $\mathbb{R}^d \times \mathbb{R}^d$ is the main reason for some technical difficulties in the sequel.

Given $a, b > 0$ let $U_{a,b} := B(0,a) \times B(0,b) \subset \mathbb{R}^d \times \mathbb{R}^d$, where $B(x,r)$ denotes the closed ball in $\mathbb{R}^d$ of radius r centered at x. Clearly, $U_{a,b}$ is invariant under $SO(d)$. For $(x,\omega) \in \mathbb{R}^d \times \mathbb{R}^d$ we consider the set

$$\begin{aligned} V_{a,b}(x,\omega) &:= SO(d)((x,\omega) + U_{a,b}) \\ &= \{(A(x+y), A(\omega+\xi)), A \in SO(d), y \in B(0,a), \xi \in B(0,b)\}. \end{aligned}$$

By construction this set is again invariant under $SO(d)$ and by the isomorphism in Lemma 2.3.2(a) it may be regarded as a subset $\tilde{V}_{a,b}(x,\omega)$ of $\mathbb{R}_+ \times \mathbb{R}_+ \times [-1,1]$, i.e.,

$$\begin{aligned} &\tilde{V}_{a,b}(x,\omega) \\ &= \left\{ \left(|A(x+y)|, |A(\omega+\xi)|, \frac{A(x+y) \cdot A(\omega+\xi)}{|A(x+y)||A(\omega+\xi)|} \right), |y| \leq a, |\xi| \leq b, A \in SO(d) \right\} \\ &= \left\{ \left(|x+y|, |\omega+\xi|, \frac{(x+y) \cdot (\omega+\xi)}{|x+y||\omega+\xi|} \right), |y| \leq a, |\xi| \leq b \right\}. \\ &= \{(r,s,\theta), \ r \in [u-a, u+a], s \in [\lambda-b, \lambda+b], \theta \in Q_{a,b}(x,\omega,r,s)\}, \quad (4.42) \end{aligned}$$

where $u = |x|, \lambda = |\omega|$ and

$$\begin{aligned} &Q_{a,b}(x,\omega,r,s) \\ &:= \left\{ \frac{(x+y) \cdot (\omega+\xi)}{rs}, \ (y,\xi) \in B(0,a) \times B(0,b) : |x+y| = r, |\omega+\xi| = s \right\}. \end{aligned}$$

Clearly, $Q_{a,b}(x,\omega,r,s)$ is a closed subinterval of $[-1,1]$. We need to compute (or estimate) its upper and lower end point. The following auxiliary result will be useful for this task.

Lemma 4.7.2. *Let $a > 0$, $x, \omega \in \mathbb{R}^d$ with $u := |x| > a$ and $r > 0$ with $|r - u| \le a$. Then it holds*

$$M_d^+(x, \omega, r, a) := \max\{y \cdot \omega, \ |y| \le a, |x + y| = r\}$$
$$= \begin{cases} \lambda(-u \cos \alpha + r) & \text{if } u^2 - 2ru\cos(\alpha) + r^2 \le a^2, \\ \frac{\lambda}{2u}\left((r^2 - u^2 - a^2)\cos\alpha + \sqrt{4u^2a^2 - (r^2 - u^2 - a^2)^2}\sin\alpha\right) & \text{otherwise,} \end{cases}$$

$$M_d^-(x, \omega, r, a) := \min\{y \cdot \omega, \ |y| \le a, |x + y| = r\}$$
$$= \begin{cases} \lambda(-u \cos \alpha - r) & \text{if } u^2 + 2ru\cos(\alpha) + r^2 \le a^2, \\ \frac{\lambda}{2u}\left((r^2 - u^2 - a^2)\cos\alpha - \sqrt{4u^2a^2 - (r^2 - u^2 - a^2)^2}\sin\alpha\right) & \text{otherwise,} \end{cases}$$

where $\lambda = |\omega|$ and $u\lambda\cos\alpha = x \cdot \omega$.

Proof: A simple calculation shows that $M_d^\pm(Ax, A\omega, r, a) = M_d^\pm(x, \omega, r, a)$ for all $A \in SO(d)$, i.e., $M_d^\pm(x, \omega, r, a)$ depends only on $|x|, |\omega|, x \cdot \omega, r$ and a. We may therefore assume $x = (u, 0, \ldots, 0)^T$ and $\omega = (\lambda\cos\alpha, \lambda\sin\alpha, 0, \ldots, 0)^T$. Let us first treat the case $d = 2$. The task is then to compute the minimum and maximum of

$$F_{\lambda,\alpha}(y_1, y_2) := y_1\lambda\cos\alpha + y_2\lambda\sin\alpha \qquad (4.43)$$

under the constraint $y_1^2 + y_2^2 \le a^2$ and $(u + y_1)^2 + y_2^2 = r^2$. Elementary analysis shows that omitting the first constraint yields

$$\min_{(u+y_1)^2 + y_2^2 = r^2} F_{\lambda,\alpha}(y_1, y_2) = \lambda(-u\cos\alpha - r), \qquad (4.44)$$

$$\max_{(u+y_1)^2 + y_2^2 = r^2} F_{\lambda,\alpha}(y_1, y_2) = \lambda(-u\cos\alpha + r), \qquad (4.45)$$

whereas the minimum and maximum are attained in $(-u - r\cos\alpha, -r\sin\alpha)$ and $(-u + r\cos\alpha, r\sin\alpha)$, respectively. The minimum (4.44) of the unconstrained problem is also a solution of the constrained problem whenever $|(-u - r\cos\alpha, -r\sin\alpha)| \le a$, i.e., if $u^2 + 2ru\cos\alpha + r^2 \le a^2$, and the maximum in (4.45) is valid also for the constrained problem, whenever $u^2 - 2ru\cos\alpha + r^2 \le a^2$.

In the remaining case the maximum (minimum) is taken on the intersection of the circles $\partial B(0, a)$ and $\partial B(-x, r)$, i.e., for

$$y^{(1)} := \frac{1}{2u}\left(r^2 - u^2 - a^2, \sqrt{4u^2a^2 - (r^2 - u^2 - a^2)^2}\right)^T$$
$$\text{and} \quad y^{(2)} := \frac{1}{2u}\left(r^2 - u^2 - a^2, -\sqrt{4u^2a^2 - (r^2 - u^2 - a^2)^2}\right)^T.$$

Inserting these points into (4.43) yields the desired result for $d = 2$.

Now, suppose $d \ge 3$. Then we may write $y = z + \xi$ with $z \in \text{span}\{x, \omega\}$ and $\xi \in (\text{span}\{x, \omega\})^\perp$. Our extremal problem consists in maximizing (minimizing) $y \cdot \omega = z \cdot \omega$

under the constraints $|z|^2 + |\xi|^2 = a^2$ and $r^2 = |x + z + \xi|^2 = |x + z|^2 + |\xi|^2$. This implies that we have the following reduction to the problem for $d = 2$:

$$M_d^+(x, \omega, r, a) = \max_{0 \le t \le a} M_2^+(x, \omega, \sqrt{r^2 - t^2}, \sqrt{a^2 - t^2}),$$

$$M_d^-(x, \omega, r, a) = \min_{0 \le t \le a} M_2^-(x, \omega, \sqrt{r^2 - t^2}, \sqrt{a^2 - t^2}).$$

A closer inspection of the right hand sides shows that the extrema are always taken for $t = 0$, which completes the proof also for the general case. $\qquad\square$

We conclude from Lemma 4.7.2 that $Q_{a,b}(x, \omega, r, s) = [\theta_{a,b}^-(x, \omega, r, s), \theta_{a,b}^+(x, \omega, r, s)]$ with

$$\theta_{a,b}^-(x, \omega, r, s) := \frac{1}{rs} \left(x \cdot \omega + M_d^-(x, \omega, r, a) + M_d^-(\omega, x, s, b) + \zeta^- \right), \qquad (4.46)$$

$$\theta_{a,b}^+(x, \omega, r, s) := \frac{1}{rs} \left(x \cdot \omega + M_d^+(x, \omega, r, a) + M_d^+(\omega, x, s, b) + \zeta^+ \right), \qquad (4.47)$$

with some numbers $\zeta^\pm \in \mathbb{R}$ (possibly depending on all parameters) that satisfy $|\zeta^\pm| \le ab$. If $\alpha = 0$, i.e., if x and ω are collinear then it is easily seen directly with the definition that $1 \in Q_{a,b}(x, \omega, r, s)$, i.e., $\theta_{a,b}^+(x, \omega, r, s) = 1$ for all $r \in [u - a, u + a], s \in [\lambda - b, \lambda + b]$. An analogous statement holds for $\alpha = \pi$.

Lemma 4.7.3. Let $(x, \omega) \in \mathbb{R}^d \times \mathbb{R}^d, a, b > 0$ with $u := |x|, \lambda := |\omega|$ and $u\lambda \cos \alpha = x \cdot \omega$. The Lebesgue measure of the sets $V_{a,b}(x, \omega)$ asymptotically behaves as follows.

(a) If $\alpha - \arcsin(b/\lambda) - \arcsin(a/u) \ge \epsilon > 0$ then there exist constants $C_1, C_2 > 0$ (depending on ϵ and d) such that

$$C_1 |V_{a,b}(x, \omega)| \le ab(u\lambda \sin \alpha)^{d-2}(a\lambda + bu) \le C_2 |V_{a,b}(x, \omega)|.$$

(b) Suppose $\alpha = 0$ or $\alpha = \pi$ and $u \ge u_0 > a, \lambda \ge \lambda_0 > b$. Then there exists constants $C_1', C_2' > 0$ (depending on a, b and d) such that

$$C_1' |V_{a,b}(x, \omega)| \le (u^{d-1} + \lambda^{d-1} + 1) \le C_2' |V_{a,b}(x, \omega)|.$$

Proof: (a) By Lemma 2.3.6 and (4.42) we have

$$|V_{a,b}(x, \omega)| = |S^{d-1}| \, |S^{d-2}| \int_{\lambda-b}^{\lambda+b} \int_{u-a}^{u+a} \int_{-1}^{1} \chi_{Q_{a,b}(x,\omega,r,s)}(z)(1 - z^2)^{(d-3)/2} dz \, r^{d-1} s^{d-1} \, drds.$$

For symmetry reasons we only need to consider the case $\cos \alpha \ge 0$. Our assumption on α implies that $\sin \alpha > \max\{\frac{a}{u}, \frac{b}{\lambda}\}$. This means $u^2 - 2ru \cos \alpha + r^2 > a^2$ and $\lambda^2 - 2s\lambda \cos \alpha + s^2 > b^2$ for all $r \in [u - a, u + a]$ and $s \in [\lambda - b, \lambda + b]$, i.e., in

Lemma 4.7.2 always the second case for $M^+(x, \omega, r, s)$ applies (and for $M^-(x, \omega, r, s)$ anyway since $\cos(\alpha) \geq 0$). The mean value theorem yields

$$
\begin{aligned}
F_{a,b}(x, \omega, r, s) &:= \int_{-1}^{1} \chi_{Q_{a,b}(x,\omega,r,s)}(z)(1 - z^2)^{(d-3)/2} dz \\
&= (\theta_{a,b}^+(x, \omega, r, s) - \theta_{a,b}^-(x, \omega, r, s))\kappa(r, s) \\
&= \frac{\kappa(r, s)}{rs} \left(\sin(\alpha) \left(\frac{\lambda}{u} \sqrt{4u^2 a^2 - (r^2 - u^2 - a^2)^2} + \frac{u}{\lambda} \sqrt{4\lambda^2 b^2 - (s^2 - \lambda^2 - b^2)^2} \right) + \zeta \right)
\end{aligned}
$$

with some $\kappa(r, s)$ satisfying

$$
(1 - \theta_{a,b}^+(x, \omega, r, s)^2)^{(d-3)/2} \leq \kappa(r, s) \leq (1 - \theta_{a,b}^-(x, \omega, r, s)^2)^{(d-3)/2}
$$

and $|\zeta| \leq 2ab$. (Note that $\kappa(r, s) = 1$ in case $d = 3$.) We claim that asymptotically

$$
\int_{u-a}^{u+a} r^{d-2} \sqrt{4u^2 a^2 - (r^2 - u^2 - a^2)^2} dr \sim a^2 \pi u^{d-1} \quad (u \to \infty).
$$

Indeed, by Lebesgue's dominated convergence theorem we obtain

$$
\begin{aligned}
&\lim_{u \to \infty} u^{-(d-1)} \int_{u-a}^{u+a} r^{d-2} \sqrt{4u^2 a^2 - (r^2 - u^2 - a^2)^2} dr \\
&= \lim_{u \to \infty} \int_{-a}^{a} \left(1 + \frac{p(r, u)}{u^{d-2}} \right) \sqrt{4a^2 - (u^{-1} r^2 + 2r + u^{-1} a^2)^2} dr \\
&= \int_{-a}^{a} \sqrt{4a^2 - 4r^2} dr + \underbrace{\lim_{u \to \infty} \int_{-a}^{a} \frac{p(r, u)}{u^{d-2}} \sqrt{4a^2 - 4r^2} dr}_{=0} = a^2 \pi,
\end{aligned}
$$

where $p(r, u)$ is a positive polynomial in r and u of degree at most $d - 3$. Since

$$
\left| \int_{\lambda-b}^{\lambda+b} \int_{u-a}^{u+a} \frac{\zeta}{rs} r^{d-1} s^{d-1} dr dr \right| \leq C(ab)^2 \lambda^{d-2} u^{d-2}
$$

we have asymptotically

$$
\begin{aligned}
|V_{a,b}(x, \omega)| &\asymp C(x, \omega) \sin(\alpha) \left(\frac{\lambda}{u} \int_{\lambda-b}^{\lambda+b} s^{d-2} ds \int_{u-a}^{u+a} r^{d-2} \sqrt{4u^2 a^2 - (r^2 - u^2 - a^2)^2} dr \right. \\
&\qquad \left. + \frac{u}{\lambda} \int_{u-a}^{u+a} r^{d-2} dr \int_{\lambda-b}^{\lambda+b} s^{d-2} \sqrt{4\lambda^2 b^2 - (s^2 - \lambda^2 - b^2)^2} ds \right) \\
&\asymp C(x, \omega) 2\pi ab (u\lambda)^{d-2} (a\lambda + bu) \sin \alpha
\end{aligned}
$$

with $C(x,\omega) = (1 - \theta^2)^{(d-3)/2}$ for some θ satisfying

$$\min_{r\in[u-a,u+a],s\in[\lambda-b,\lambda+b]} \theta^-_{a,b}(x,\omega,r,s) \le \theta \le \max_{r\in[u-a,u+a],s\in[\lambda-b,\lambda+b]} \theta^+_{a,b}(x,\omega,r,s). \qquad (4.48)$$

Directly from the definition of $Q_{a,b}(x,\omega,r,s)$ we see that the right hand side of (4.48) attains the value 1 if and only if the cones $K(x,a) := \{\gamma(x+y), \gamma > 0, |y| \le a\}$ and $K(\omega,b)$ have non-empty intersection. A geometric inspection shows $K(x,a) \cap K(\omega,b) = \emptyset$ if and only if $\arcsin\left(\frac{b}{\lambda}\right) + \arcsin\left(\frac{a}{u}\right) < \alpha$. Moreover, our assumption on α implies

$$\delta(\epsilon) \le 1 - \theta^\pm_{a,b}(x,\omega,r,s)^2 \le 1$$

for some $\delta(\epsilon) > 0$. Furthermore, we have

$$\sin(\epsilon)^2 \le \sin^2(\arcsin(b/\lambda) + \arcsin(a/u) + \epsilon) \le \sin^2(\alpha) \le 1.$$

This yields

$$\delta(\epsilon)^{(d-3)/2} \sin^{d-3}\alpha \le C(x,\omega) \le (\sin^{d-3}\epsilon) \sin^{d-3}\alpha$$

giving

$$|V_{a,b}(x,\omega)| \asymp ab(u\lambda\sin\alpha)^{d-2}(a\lambda + bu).$$

We remark that the comparison of $1 - \theta^\pm_{a,b}(x,\omega,r,s)^2$ with $\sin^2(\alpha)$ is natural since for large $|x|$ and $|\omega|$ we have asymptotically $\theta^\pm_{a,b}(x,\omega,r,s) \sim \cos\alpha$.

(b) Let us now investigate the case $\alpha = 0$. Then clearly $\theta^+_{a,b}(x,\omega,r,s) = 1$ for all $(r,s) \in [u-a,u+a] \times [\lambda-b,\lambda+b]$. The cosine theorem shows that the maximal angle $\beta(x,r,a)$ between $x+y$ and x for some $y \in B(0,a)$ satisfying $|x+y| = r$ is given by

$$\cos\beta(x,r,a) = \frac{r^2 + u^2 - a^2}{2ru}. \qquad (4.49)$$

Since x and ω are collinear by assumption, the maximal angle between points $x \in B(0,a)$ and $y \in B(0,b)$ satisfying $|x+y| = r$ and $|\omega+\xi| = s$ is given by $\beta(x,r,a) + \beta(\omega,s,b)$. We conclude that

$$\theta^-(r,s) = \cos(\beta(x,r,a) + \beta(\omega,s,b))$$
$$= \cos\beta(x,r,a)\cos\beta(\omega,s,b) - \sin\beta(x,r,a)\sin\beta(\omega,s,b)$$
$$= \left(\frac{r^2+u^2-a^2}{2ru}\right)\left(\frac{s^2+\lambda^2-b^2}{2\lambda s}\right) - \sqrt{1 - \left(\frac{r^2+u^2-a^2}{2ru}\right)^2}\sqrt{1 - \left(\frac{s^2+\lambda^2-b^2}{2\lambda b}\right)^2}.$$

Let us now assume $u \ge u_0 > a$ and $\lambda \ge \lambda_0 > b$. This implies by (4.49) that $\beta(x,r,a) + \beta(\omega,s,b) < \pi$ for all $(r,s) \in [u-a,u+a] \times [\lambda-b,\lambda+b]$, which in turn means that $\theta^-(r,s) > 0$ for all valid r,s. A simple substitution shows that

$$\int_{-1}^1 \chi_{Q_{a,b}(x,\omega,r,s)}(t)(1-t^2)^{(d-3)/2}dt = \int_{\theta^-(r,s)}^1 (1-t^2)^{(d-3)/2}dt$$
$$= (1-\theta^-(r,s))^{(d-1)/2}\int_0^1 (t(2-(1-\theta^-(r,s))t))^{(d-3)/2}dt \asymp (1-\theta^-(r,s))^{(d-1)/2},$$

where the last relation follows from

$$0 < C_1 := \int_0^1 (t(2-t))^{(d-3)/2} dt \leq \int_0^1 (t(2 - (1 - \theta^-(r,s))t))^{(d-3)/2} dt$$
$$\leq \int_0^1 (2t)^{(d-3)/2} dt =: C_2.$$

Thus we have

$$|V_{a,b}(x,\omega)| \asymp \int_{u-a}^{u+a} \int_{\lambda-b}^{\lambda+b} (1 - \theta^-(r,s))^{(d-1)/2} r^{d-1} s^{d-1} dr ds$$
$$= \int_{-a}^a \int_{-b}^b \underbrace{(u+r)^{d-1}(\lambda+s)^{d-1}(1 - \theta^-(u+r,\lambda+s))^{(d-1)/2}}_{=: H(u,\lambda,r,s)} dr ds.$$

A straightforward computation shows

$$H(u,\lambda,r,s) = \left((u+r)(\lambda+s) \left(\frac{\lambda+s}{u}(a^2 - r^2) + \frac{u+r}{\lambda}(b^2 - s^2) + \frac{(a^2 - r^2)(b^2 - s^2)}{u\lambda} \right. \right.$$
$$\left. \left. + \sqrt{2\frac{u+r}{u}(a^2 - r^2) + \frac{(a^2 - r^2)^2}{u^2}} \sqrt{2\frac{\lambda+s}{\lambda}(b^2 - s^2) + \frac{(b^2 - s^2)^2}{\lambda^2}} \right) \right)^{\frac{d-1}{2}}.$$

Since $|r| \leq a, |s| \leq b$ and $u \geq u_0 > a, \lambda \geq \lambda_0 > b$ there exist constants $C_1, C_2 > 0$ (depending on a, b, d) such that

$$C_1 H(u,\lambda,r,s) \leq \left(u\lambda \left(\frac{\lambda}{u}(a^2 - r^2) + \frac{u}{\lambda}(b^2 - s^2) + 2\sqrt{(a^2 - r^2)(b^2 - s^2)} \right) \right)^{(d-1)/2}$$
$$\leq C_2 H(u,\lambda,r,s).$$

After integrating we see that there exist constants $C_1', C_2' > 0$ such that

$$C_1'|V_{a,b}(x,\omega)| \leq u^{d-1} + \lambda^{d-1} + 1 \leq C_2'|V_{a,b}(x,\omega)|$$

and the proof is completed. $\qquad\square$

We remark that the cases that are left open in the previous lemma will not be needed. Let us now face the problem of constructing a covering of $\mathbb{R}^d \times \mathbb{R}^d$ as in Theorem 4.2.2. We propose the following construction of a point set in $\mathbb{R}^d \times \mathbb{R}^d$ which is well-spread with respect to $SO(d)$. Given $a, b > 0$ we let

$$u_j := aj, \qquad \lambda_k := bk, \qquad j, k \in \mathbb{N}_0$$

and

$$N(j,k) := \left\lceil \frac{\pi}{4} \left(\arctan \frac{k\left(3 + \frac{3}{2j} - (\frac{3}{4j})^2\right)^{1/2} + j\left(3 + \frac{3}{2k} - (\frac{3}{4k})^2\right)^{1/2}}{(jk + \frac{1}{2}(j+k) - \frac{3}{8}(\frac{j}{k} + \frac{k}{j}) + 1)} \right)^{-1} \right\rceil, \quad j,k \in \mathbb{N},$$

$$N(0,k) := N(j,0) := 0, \qquad j,k \in \mathbb{N}_0.$$

We note that asymptotically these numbers behave like

$$N(j,k) \asymp \frac{\pi}{4\sqrt{3}} \frac{jk}{j+k}. \tag{4.50}$$

We further define

$$\theta_{j,k}^\ell := \cos\alpha_{j,k}^\ell := \sin\frac{\pi\ell}{2N(j,k)}, \quad j,k \in \mathbb{N}, \ell = -N(j,k),\dots,N(j,k),$$

$$\theta_{0,k}^0 := \theta_{j,0}^0 = 1, \quad j,k \in \mathbb{N}_0,$$

and, finally, with some fixed unit vectors η,ζ with $\eta \cdot \zeta = 0$ we let

$$(x_{j,k,\ell},\, \omega_{j,k,\ell}) := (u_j\eta,\, \lambda_k(\cos(\alpha_{j,k}^\ell)\eta + \sin(\alpha_{j,k}^\ell)\zeta)), \qquad (j,k,\ell) \in I,$$

where the index set is given by

$$I := \{(j,k,\ell),\, j,k \in \mathbb{N}_0, \ell = -N(j,k),\dots,N(j,k)\}.$$

We further denote by $X_{a,b}$ the collection of all points $\{(x_{j,k,\ell},\omega_{j,k,\ell}),\, (j,k,\ell) \in I\}$ emphasizing the dependence on the parameters a,b. Having in mind the isomorphism of $\mathbb{H}^{SO(d)}$ to $\mathcal{K}$ in Lemma 2.3.2, the set $X_{a,b}$ corresponds to the points

$$\left(aj,\, bk,\, \sin\frac{\pi\ell}{2N(j,k)}\right) \in \mathbb{R}_+ \times \mathbb{R}_+ \times [-1,1], \quad (j,k,\ell) \in I.$$

Theorem 4.7.4. *For all $a,b > 0$ the point set $X_{a,b}$ is a $U_{a,b}$-dense, well-spread set with respect to $SO(d)$. In particular, it holds*

$$\mathbb{R}^d \times \mathbb{R}^d = \bigcup_{(j,k,\ell)\in I} V_{a,b}(x_{j,k,\ell},\omega_{j,k,\ell}). \tag{4.51}$$

Proof: Let us first prove the covering property (4.51). Observe that we may cover $\mathbb{R}^d \times \mathbb{R}^d$ with Cartesian products of annuli, i.e.,

$$\mathbb{R}^d \times \mathbb{R}^d = \bigcup_{j,k\in\mathbb{N}_0} A(aj,a/2) \times A(bk,b/2), \tag{4.52}$$

where $A(r,b) := \{z \in \mathbb{R}^d, r - b \leq |z| \leq r + b\}$. Thus, it suffices to prove that for any fixed j, k we have a covering of these annuli, i.e.,

$$\bigcup_{\ell=-N(j,k)}^{N(j,k)} V_{a,b}(x_{j,k,\ell}, \omega_{j,k,\ell}) \supset A(aj, a/2) \times A(bk, b/2). \tag{4.53}$$

If $j = 0$ then it follows directly from the definition that

$$V_{a,b}(x_{0,k,0}, \omega_{0,k,0}) = V_{a,b}(0, \omega_{0,k,0}) = B(0,a) \times A(bk, b) \supset A(0, a/2) \times A(bk, b/2), \ k \in \mathbb{N}_0,$$

and similarly for $k = 0, j \in \mathbb{N}_0$. So it remains to consider the case $j, k \geq 1$. Then the covering (4.53) is equivalent to

$$\bigcup_{\ell=-N(j,k)}^{N(j,k)} \tilde{V}_{a,b}(x_{j,k,\ell}, \omega_{j,k,\ell}) \supset [aj - a/2, aj + a/2] \times [bk - b/2, bk + b/2] \times [-1, 1],$$

where $\tilde{V}_{a,b}(x_{j,k,\ell}, \omega_{j,k,\ell})$ are the corresponding subsets of $\mathbb{R}_+ \times \mathbb{R}_+ \times [-1, 1]$. Thus, we need to show that

$$\theta_\ell^+(r, s) := \theta_{a,b}^+(x_{j,k,\ell}, \omega_{j,k,\ell}, r, s) \geq \theta_{a,b}^-(x_{j,k,\ell+1}, \omega_{j,k,\ell+1}, r, s) =: \theta_{\ell+1}^-(r, s)$$

for all $(r, s) \in [aj - a/2, aj + a/2] \times [bk - b/2, bk + b/2]$ and $\ell = -N(j,k) \ldots, N(j,k) - 1$. For symmetry reasons it suffices to consider $\ell \geq 0$ implying that for $M_d^-(x, \omega, r, a)$ always the second case in Lemma 4.7.2 applies. Moreover, if the first case of Lemma 4.7.2 applies for $M_d^+(x, \omega, r, s)$ then the term for the second case is less than $M_d^+(x, \omega, r, a)$. Hence, for any $l = 0, \ldots, N(j,k) - 1$ we have

$$\theta_\ell^+(r, s) - \theta_{\ell+1}^-(r, s)$$
$$\geq \frac{1}{rs} \left((\cos \alpha_{j,k}^\ell - \cos \alpha_{j,k}^{\ell+1}) R_{j,k}(r, s) + (\sin \alpha_{j,k}^\ell + \sin \alpha_{j,k}^{\ell+1}) S_{j,k}(r, s) + \zeta_\ell^+ - \zeta_{\ell+1}^- \right) \tag{4.54}$$

with

$$R_{j,k}^{a,b}(r, s) = R_{j,k}(r, s) := u_j \lambda_k + \frac{\lambda_k}{u_j}(r^2 - u_j^2 - a^2) + \frac{u_j}{\lambda_k}(s^2 - \lambda_k^2 - b^2),$$

$$S_{j,k}^{a,b}(r, s) = S_{j,k}(r, s) := \frac{\lambda_k}{u_j}\sqrt{4u_j^2 a^2 - (r^2 - u_j^2 - a^2)^2} + \frac{u_j}{\lambda_k}\sqrt{4\lambda_k^2 b^2 - (s^2 - \lambda_k^2 - b^2)^2}.$$

Here, the numbers $\zeta_\ell^+, \zeta_{\ell+1}^-$ have the same meaning as in (4.46), (4.47). Using trigonometric identities we further obtain

$$(\cos \alpha_{j,k}^\ell - \cos \alpha_{j,k}^{\ell+1}) R_{j,k}(r, s) + (\sin \alpha_{j,k}^\ell + \sin \alpha_{j,k}^{\ell+1}) S_{j,k}(r, s)$$
$$= \left(\sin \frac{\pi\ell}{2N(j,k)} - \sin \frac{\pi(\ell+1)}{2N(j,k)} \right) R_{j,k}(r, s) + \left(\cos \frac{\pi\ell}{2N(j,k)} + \cos \frac{\pi(\ell+1)}{2N(j,k)} \right) S_{j,k}(r, s)$$
$$= 2\cos \frac{\pi(2\ell+1)}{4N(j,k)} \left(S_{j,k}(r, s) \cos \frac{\pi}{4N(j,k)} - R_{j,k}(r, s) \sin \frac{\pi}{4N(j,k)} \right).$$

Observe that both the maximum of $R_{j,k}(r,s)$ and the minimum of $S_{j,k}(r,s)$ for $(r,s) \in [u_j - a/2, u_j + a/2] \times [\lambda_k + b/2, \lambda_k - b/2]$ are attained at $(r,s) = (u_j + a/2, \lambda_k + b/2)$ and

$$\widetilde{R}_{j,k} := R_{j,k}(u_j + a/2, \lambda_k + b/2) = ab \left(jk + \frac{1}{2}(j+k) - \frac{3}{8}\left(\frac{j}{k} + \frac{k}{j} \right) \right)$$

$$\widetilde{S}_{j,k} := S_{j,k}(u_j + a/2, \lambda_k + b/2) = ab \left(k\sqrt{3 + \frac{3}{2j} - (\frac{3}{4j})^2} + j\sqrt{3 + \frac{3}{2k} - (\frac{3}{4k})^2} \right).$$

Let us now consider ζ_ℓ^+ and $\zeta_{\ell+1}^-$. From the proof of Lemma 4.7.2 we conclude that the maxima (minima) $M_d^\pm(x_{j,k,\ell}, \omega_{j,k,\ell}, r, a), M_d^\pm(\omega_{j,k,\ell}, x_{k,\ell,\omega}, s, b)$ are taken for

$$y_{j,k,\ell}^\pm = \frac{1}{2u_j} \left(r^2 - u_j^2 - a^2 \,,\, \pm\sqrt{4u_j^2 a^2 - (r^2 - u_j^2 - a^2)^2}, 0, \ldots, 0 \right)^T ,$$

$$\xi_{j,k,\ell}^\pm = \frac{1}{2\lambda_k} \left((s^2 - \lambda_k^2 - b^2)\cos\alpha_{j,k} \,,\, \pm\sqrt{4\lambda_k^2 b^2 - (s^2 - \lambda_k^2 - b^2)^2}\sin\alpha_{j,k}, 0, \ldots, 0 \right)^T .$$

This implies that

$$\zeta_\ell^+ \geq y_{j,k,\ell}^+ \cdot \xi_{j,k,\ell}^+ = \overbrace{\frac{1}{4u_j\lambda_k}(r^2 - u_j^2 - a^2)(s^2 - \lambda_k^2 - b^2)\cos\alpha_{j,k}^\ell}^{=:\, K_{j,k}(r,s)}$$
$$+ \underbrace{\frac{1}{4u_j\lambda_k}\sqrt{4u_j^2 a^2 - (r^2 - u_j^2 - a^2)^2}\sqrt{4\lambda_k^2 b^2 - (s^2 - \lambda_k^2 - b^2)^2}\sin\alpha_{j,k}^\ell}_{=:\, L_{j,k}(r,s)}$$

and

$$\zeta_{\ell+1}^- \leq y_{j,k,\ell}^- \cdot \xi_{j,k,\ell}^- = K_{j,k}(r,s)\cos\alpha_{j,k}^{\ell+1} + L_{j,k}(r,s)\sin\alpha_{j,k}^{\ell+1}.$$

As $y_{j,k,\ell}^\pm \in B(0,a)$ and $\xi_{j,k,\ell}^\pm \in B(0,b)$ we have $|K_{j,k}(r,s)|, |L_{j,k}(r,s)| \leq ab$ and $L_{j,k}(r,s)$ is always non-negative. Using trigonometric identities we further obtain

$$\zeta_\ell^+ - \zeta_{\ell+1}^-$$
$$\geq \left(\sin\frac{\pi\ell}{2N(j,k)} - \sin\frac{\pi(\ell+1)}{2N(j,k)} \right) K_{j,k}(r,s) + \left(\cos\frac{\pi\ell}{2N(j,k)} - \cos\frac{\pi(\ell+1)}{2N(j,k)} \right) L_{j,k}(r,s)$$
$$= 2\sin\frac{\pi}{4N(j,k)} \left(-K_{j,k}(r,s)\cos\frac{\pi(2\ell+1)}{4N(j,k)} + L_{j,k}(r,s)\sin\frac{\pi(2\ell+1)}{4N(j,k)} \right).$$

Pasting things together we obtain

$$rs(\theta_\ell^+(r,s) - \theta_{\ell+1}^-(r,s))$$
$$\geq 2\cos\frac{\pi(2\ell+1)}{4N(j,k)}\left(\widetilde{S}_{j,k}\cos\frac{\pi}{4N(j,k)} - (\widetilde{R}_{j,k} + K_{j,k}(r,s))\sin\frac{\pi}{4N(j,k)}\right)$$
$$+ 2L_{j,k}(r,s)\sin\frac{\pi}{4N(j,k)}\sin\frac{\pi(2\ell+1)}{4N(j,k)}$$
$$\geq 2\cos\frac{\pi(2\ell+1)}{4N(j,k)}\left(\widetilde{S}_{j,k}\cos\frac{\pi}{4N(j,k)} - (\widetilde{R}_{j,k} + ab)\sin\frac{\pi}{4N(j,k)}\right).$$

By definition of $N(j,k)$ we have

$$\tan\frac{\pi}{4N(j,k)} \leq \frac{\widetilde{S}_{j,k}}{\widetilde{R}_{j,k} + ab}$$

and, hence, we conclude that $\theta_\ell^+(r,s) \geq \theta_{\ell+1}^-(r,s)$ for all $(r,s) \in [u_j - a/2, u_j + a/2] \times [\lambda_k - b/2, \lambda_k + b/2]$. Hence, the covering property is shown.

In order to show that the point set $(x_{j,k,\ell}, \omega_{j,k,\ell})$ is separated, i.e., that (4.4) holds, we need to prove that whenever a', b' are small enough then the sets $V_{a',b'}(x_{j,k,\ell}, \omega_{j,k,\ell})$ are mutually disjoint, see also Lemma 4.2.1. (Hereby, the constants a, b used for the construction of $x_{j,k,\ell}, \omega_{j,k,\ell}$ remain the same as before).

First, observe that $\tilde{V}_{a',b'}(x_{j,k,\ell}, \omega_{j,k,\ell}) \subset [aj - a', aj + a'] \times [bk - b', bk + b'] \times [-1, 1]$ so that whenever $a' < a/2$ and $b' < b/2$ then $\tilde{V}_{a',b'}(x_{j,k,\ell}, \omega_{j,k,\ell})$ and $\tilde{V}_{a',b'}(x_{j',k',\ell'}, \omega_{j',k',\ell'})$ are disjoint if $k \neq k'$ or $\ell \neq \ell'$. So it remains to prove that two of such sets are disjoint if $j = j'$, $k = k'$ and $\ell \neq \ell'$, which also means that we may restrict to $j, k \geq 1$. Again, it also suffices to consider $\ell = 0, \ldots, N(j,k)$.

Observe that

$$\sin\alpha_{j,k}^{N(j,k)-1} = \sin\frac{\pi}{2N(j,k)} \asymp C\frac{j+k}{jk}.$$

Thus whenever $a', b' > 0$ are small enough then

$$\sin\alpha_{j,k}^{N(j,k)-1} > \max\left\{\frac{a'}{u_j}, \frac{b'}{\lambda_k}\right\} = \max\left\{\frac{a'}{aj}, \frac{b'}{bk}\right\}.$$

This means that for all $l = 0, \ldots, N(j,k) - 1$ the second case in Lemma 4.7.2 applies. Thus, we always have equality in (4.54) and we are left with showing that the right hand side in (4.54) is always negative whenever a', b' is small enough.

The minimum of $R_{j,k}^{a',b'}(r,s)$ for $(r,s) \in [u_j - a', u_j + a'] \times [\lambda_k - b', \lambda + b']$ is attained in $(u_j - a', \lambda_k - b')$ yielding

$$\dot{R}_{j,k}^{a',b'} := R_{j,k}^{a',b'}(u_j - a', \lambda_k - b') = abjk - 2ab'j - 2a'bk.$$

Since the function $r \mapsto (r^2 - u_j^2 - a'^2)$ attains the value 0 for some $r \in [u_j - a', u_j + a']$ we have

$$\dot{S}_{j,k}^{a'b'} := \max_{(r,s)\in[u_j-a',u_j+a']\times[\lambda_k-b',\lambda_k+b']} S_{j,k}^{a',b'}(r,s) = 2a'bk + 2ab'j.$$

Altogether we obtain

$$\theta_{a',b'}^{+}(x_{j,k,\ell},\omega,r,s) - \theta_{a',b'}^{-}(x_{j,k,\ell+1},\omega_{j,k,\ell+1},r,s)$$

$$\leq (\cos\alpha_{j,k}^{\ell} - \cos\alpha_{j,k}^{\ell+1})\dot{R}_{j,k}^{a',b'} + (\sin\alpha_{j,k}^{\ell} + \sin\alpha_{j,k}^{\ell+1})\dot{S}_{j,k}^{a',b'} + 2a'b'$$

$$= \cos\frac{\pi(2\ell+1)}{4N(j,k)}\left(-(abjk - 2ab'j - 2a'bk)\sin\frac{\pi}{4N(j,k)} + (2a'bk + 2ab'j)\cos\frac{\pi}{4N(j,k)}\right)$$

$$+ 2a'b$$

$$\leq \cos\frac{\pi(2\ell+1)}{4N(j,k)}\left(-C\frac{j+k}{jk}(abjk - 2ab'j - 2a'bk) + (2a'bk + 2ab'j)\right) + 2a'b'$$

$$\leq \cos\frac{\pi(2\ell+1)}{4N(j,k)}(\underbrace{(-Cab + 6ab')}_{=:\ \gamma(a',b')}j + \underbrace{(-Cab + 6a'b)}_{=:\ \delta(a',b')}k) + 2a'b'.$$

The inequality in the fourth line is due to the asymptotic behaviour (4.50) of $N(j,k)$. Clearly, it is possible to choose $a', b' > 0$ such that $\gamma(a',b'), \delta(a',b') < 0$. Moreover, the smallest value of the cosine term in the last line is taken for $\ell = N(j,k) - 1$ and again by the asymptotic behaviour of $N(j,k)$ it holds

$$\cos\frac{\pi(2(N(j,k)-1)+1)}{4N(j,k)} = \sin\frac{\pi}{4N(j,k)} \leq C'\frac{j+k}{jk}.$$

This yields

$$\theta_{a',b'}^{+}(x_{j,k,\ell},\omega_{j,k,\ell},r,s) - \theta_{a',b'}^{-}(x_{j,k,\ell+1},\omega_{j,k,\ell+1},r,s)$$

$$\leq C'\frac{j+k}{jk}(\gamma(a',b')j + \delta(a',b')k) + 2a'b'$$

$$= C'\left(\gamma(a',b')\frac{j}{k} + \delta(a',b')\frac{k}{j}\right) + C'(\gamma(a',b') + \delta(a',b')) + 2a'b'$$

$$\leq C'(\gamma(a',b') + \delta(a',b')) + 2a'b'.$$

Since for a', b' small enough the latter expression is negative, this proves our claim. $\square$

We remark that starting from the well-spread set $X_{a,b} \subset \mathbb{R}^d \times \mathbb{R}^d$ we easily obtain a well-spread set in $\mathbb{H}_d$ by taking the points $(x_{j,k,\ell}, \omega_{j,k,\ell}, 1) \in \mathbb{H}_d$.

Let us now consider the sequence spaces associated to $X_{a,b}$ and $L_s^{p,q}$. We denote by $\chi_{j,k,\ell}$ the characteristic function of the set $V_{a,b}(x_{j,k,\ell}, \omega_{j,k,\ell})$ and further define

$$\kappa_{j,k,\ell}(\omega) := \int_{\mathbb{R}^d} \chi_{j,k,\ell}(x, \omega) dx, \quad \omega \in \mathbb{R}^d.$$

Having in mind the asymptotic behaviour of the measure of the sets $V_{a,b}(x, \omega)$ (Lemma 4.7.3) we let

$$\mu_{j,k,\ell} := (j + k) \left(jk \cos \frac{\pi \ell}{2N(j,k)} \right)^{d-2}, (j, k, \ell) \in I, \ell \neq \pm N(j,k), j, k \geq 1,$$
$$\mu_{j,k,\pm N(j,k)} := j^{d-1} + k^{d-1} + 1, \quad j, k \in \mathbb{N}_0,$$

and if m is a moderate weight function on $\mathbb{R}^d \times \mathbb{R}^d$ then we denote

$$m_{j,k,\ell} := m(x_{j,k,\ell}, \omega_{j,k,\ell}).$$

Lemma 4.7.5. *(a) Let m be an invariant moderate weight function and assume $1 \leq p, q \leq \infty$. Equivalent norms on $(L_m^{p,q})_{SO(d)}^\flat(X_{a,b})$ and on $(L_m^{p,q})_{SO(d)}^\natural(X_{a,b})$ are given by*

$$\|(\lambda_{j,k,\ell})|(L_m^{p,q})_{SO(d)}^\flat\| \asymp \left(\int_{\mathbb{R}^d} \left(\sum_{(j,k,\ell) \in I} |\lambda_{j,k,\ell}|^p m_{j,k,\ell}^p \kappa_{j,k,\ell}(\omega) \right)^{q/p} d\omega \right)^{1/q},$$

$$\|(\lambda_{j,k,\ell})|(L_m^{p,q})_{SO(d)}^\natural\| \asymp \left(\int_{\mathbb{R}^d} \left(\sum_{(j,k,\ell) \in I} \left(|\lambda_{j,k,\ell}| m_{j,k,\ell} \mu_{j,k,\ell}^{-1} \right)^p \kappa_{j,k,\ell}(\omega) \right)^{q/p} d\omega \right)^{1/q}.$$

(b) In case $p = q$ we have the following simplification:

$$\|(\lambda_{j,k,\ell})|(L_m^p)_{SO(d)}^\flat\| \asymp \left(\sum_{(j,k,\ell) \in I} |\lambda_{j,k,\ell}|^p m_{j,k,\ell}^p \mu_{j,k,\ell} \right)^{1/p},$$

$$\|(\lambda_{j,k,\ell})|(L_m^p)_{SO(d)}^\natural\| \asymp \left(\sum_{(j,k,\ell) \in I} |\lambda_{j,k,\ell}|^p m_{j,k,\ell}^p \mu_{j,k,\ell}^{1-p} \right)^{1/p}.$$

Proof: (a) In order to determine the sequence spaces $Y_{\mathcal{A}}^\flat$ and $Y_{\mathcal{A}}^\natural$ we are allowed to choose freely the set $U = U_{a',b'}$, i.e., the constants $a', b' > 0$, see Lemma 4.3.1(a). Since $N(j,k) \asymp \frac{jk}{j+k}$ it is possible to choose a', b' such that $\alpha_{j,k}^\ell - \arcsin \frac{a'}{aj} - \arcsin \frac{b'}{bk} \geq \epsilon > 0$

for all $(j, k, \ell) \in I$, $\ell \neq \pm N(j, k)$ and such that the sets $V_{a', b'}(x_{j,k,\ell}, \omega_{j,k,\ell})$ are disjoint. By Lemma 4.7.3 there exist constants such that

$$C_1 |V_{a', b'}(x_{j,k,\ell}, \omega_{j,k,\ell})| \leq \mu_{j,k,\ell} \leq C_2 |V_{a', b'}(x_{j,k,\ell}, \omega_{j,k,\ell})| \tag{4.55}$$

for all $(j, k, \ell) \in I$, $j \neq 0$ or $k \neq 0$. Further, if $k = 0$ then

$$|V_{a', b'}(x_{j,0,0}, \omega_{j,0,0})| = |A(aj, a') \times B(0, b')| \asymp j^{d-1} + 1, \quad j \in \mathbb{N}_0,$$

and similarly for $j = 0$, $k \in \mathbb{N}_0$. Hence, (4.55) extends to all $(j, k, \ell) \in I$.

The component $\mathbb{T}$ of $\mathbb{H}_d$ is irrelevant for the computation of the sequence spaces. By disjointness of the sets $V_{a', b'}(x_{j,k,\ell}, \omega_{j,k,\ell})$ and by moderateness of m we obtain

$$\|(\lambda_{j,k,\ell})|(L_m^{p,q})_{SO(d)}^\flat\| = \left(\int_{\mathbb{R}^d} \left(\int_{\mathbb{R}^d} \left(\sum_{(j,k,\ell) \in I} |\lambda_{j,k,\ell}| \chi_{j,k,\ell}(x, \omega) m(x, \omega) \right)^p dx \right)^{q/p} d\omega \right)^{1/q}$$

$$\asymp \left(\int_{\mathbb{R}^d} \left(\sum_{(j,k,\ell)} |\lambda_{j,k,\ell}|^p m_{j,k,\ell}^p \, \kappa_{j,k,\ell}(\omega) \right)^{q/p} d\omega \right)^{1/q}.$$

The assertion for $(L_m^{p,q})_{SO(d)}^\natural$ is shown in the same way.

(b) The claim can be deduced from (a) since in case $p = q$ we may exchange integration and summation. Alternatively, it follows also from Lemma 4.3.1(e). $\qquad \square$

Now we are ready to specialize Theorem 4.6.3 to our specific situation. This results in the construction of radial Gabor frames. Recall also the operator Ω defined in (2.18), which essentially coincides with $\tilde{\pi}$.

Theorem 4.7.6. *Let w be a submultiplicative $SO(d)$-invariant weight and suppose $g \in (M_w^{1,1})_{rad}$. Then there exist constants $a_0, b_0 > 0$ such that for all positive $a < a_0, b < b_0$, $1 \leq p, q \leq \infty$ and all w-moderate $SO(d)$-invariant weights m the following holds:*

(i) The set of functions

$$\{g_{j,k,\ell}\}_{(j,k,\ell) \in I} := \left\{ \Omega \left(aj, bk, \sin \frac{\pi \ell}{2N(j,k)} \right) g, \ (j, k, \ell) \in I \right\}$$

forms a Banach frame of $(M_m^{p,q})_{rad}$ with corresponding sequence space $(L_m^{p,q})_{SO(d)}^\flat$.

(ii) The functions $\{g_{j,k,\ell}\}_{(j,k,\ell)}$ form an atomic decomposition of $(M_m^{p,q})_{rad}$ with corresponding sequence space $(L_m^{p,q})_{SO(d)}^\natural$. In particular, any $f \in (M_m^{p,q})_{rad}$ has an

expansion

$$f = \sum_{j=0}^{\infty}\sum_{k=0}^{\infty}\sum_{\ell=-N(j,k)}^{N(j,k)} \lambda_{j,k,\ell}(f)\,\Omega\left(aj, bk, \sin\frac{\pi\ell}{2N(j,k)}\right) g$$

with norm-convergence whenever $1 \le p,q < \infty$, and weak-* convergence otherwise. Moreover, we have the norm-equivalence

$$\|(\lambda_{j,k,\ell}(f))_{(j,k,\ell)\in I}|(L_m^{p,q})^{\natural}_{SO(d)}\| \asymp \|f|(M_m^{p,q})_{rad}\|.$$

(iii) There exists a 'dual frame' $\{e_{j,k,\ell}\}_{(j,k,\ell)\in I}$ with $e_{j,k,\ell} \in (M_w^{1,1})_{rad}$ in the sense of Theorem 4.6.3.

Proof: As already remarked the space $\mathbb{B}_w^{\mathcal{A}}$ coincides with $(M_w^{1,1})_{rad}$. Thus the Theorem follows essentially from Theorem 4.6.3. There is only a slight technical difficulty left. Namely, in order to make the left hand side of (4.27) less than 1 (which is possible by Lemma 4.4.10(b)), it might be necessary to choose a set $U = U_{a,b} \times V \subset \mathbb{H}_d$ for some proper subset $V \subsetneq \mathbb{T}$. If this is the case then we have to choose points $\tau_i \in \mathbb{T}$, $i = 1, \ldots, N$, such that $\cup_{i=1}^{N}\tau_i V = \mathbb{T}$. The points $(x_{j,k,\ell}, \omega_{j,k,\ell}, \tau_i)$, $(j,k,\ell) \in I$, $i = 1, \ldots, N$, form again a well-spread set in $\mathbb{H}_d$ and the norm of the corresponding sequence space is equivalent to $\sum_{i=1}^{N} \|(\lambda_{j,k,\ell,i})_{(j,k,\ell)\in I}|Y_{\mathcal{A}}^{\natural}(I)\|$ (or with $Y_{\mathcal{A}}^{\natural}(I)$ replaced by $Y_{\mathcal{A}}^{\flat}(I)$). By Theorem 4.6.3 the corresponding functions

$$g_{j,k,\ell,i} = \widetilde{\rho}(x_{j,k,\ell}, \omega_{j,k,\ell}, \tau_i)g = \tau_i\, e^{\pi i x_{j,k,\ell}\cdot\omega_{j,k,\ell}}\,\Omega\left(aj, bk, \sin\frac{\pi\ell}{2N(j,k)}\right)g$$

$$= \tau_i\, e^{\pi i x_{j,k,\ell}\cdot\omega_{j,k,\ell}} g_{j,k,\ell}$$

form a Banach frame or an atomic decomposition. Since the functions $g_{j,k,\ell,i}, i = 1, \ldots, N$, differ from each other only by a constant factor $c_i \in \mathbb{T}$, it is straightforward to see that it suffices to take only one of them, in order to have a Banach frame or an atomic decomposition. Also the assertion concerning the dual frame is still valid. □

Of course, also Theorems 4.6.1 and 4.6.2 are applicable, yielding only Banach frames or only atomic decomposition but with possibly larger constants a_0, b_0.

Finally, let us state explicitly the result for $L_{rad}^2 = M_{rad}^{2,2}$.

Theorem 4.7.7. *Assume $g \in (S_0)_{rad} = (M_0^{1,1})_{rad}$, $g \neq 0$. Then there exist constants $a_0, b_0 > 0$ such that for all $a < a_0, b < b_0$ the functions*

$$\widetilde{g}_{j,k,\ell} := \sqrt{\mu_{j,k,\ell}}\,\Omega\left(aj, bk, \sin\frac{\pi\ell}{2N(j,k)}\right)g, \qquad (j,k,\ell) \in I$$

form a (Hilbert-) frame for $L_{rad}^2(\mathbb{R}^d)$.

Proof: By Lemma 4.7.5(b) we have the norm equivalence

$$\|(\lambda_{j,k,\ell})_{(j,k,\ell)\in I}|(L^2)^\flat_{SO(d)}\| \asymp \left(\sum_{(j,k,\ell)\in I} |\lambda_{j,k,\ell}|^2 \mu_{j,k,\ell} \right)^{1/2}$$

and, hence, the result follows from Theorem 4.7.6 by renormalizing the frame elements. □

We remark that it remains open to obtain more explicit estimations of the parameters a, b in the previous theorem, when a particular function g is chosen, for example a Gaussian.

4.7.2 Radial Homogeneous Besov and Triebel-Lizorkin Spaces

Let us now discuss the function spaces, which are connected to wavelet analysis. First we give Triebel's definition of the homogeneous Besov and Triebel-Lizorkin spaces [Tri83b, Tri92].

Definition 4.7.1. Let $\phi \in \mathcal{S}(\mathbb{R}^d)$ be a function with supp $\phi \subset \{x \in \mathbb{R}^d, 1/2 \le |x| \le 2\}$ and such that $\sum_{j=-\infty}^{\infty} \phi_j \equiv 1$, where $\phi_j(x) := \phi(2^{-j}x)$.

(a) For $1 \le p, q \le \infty, s \in \mathbb{R}$ the homogeneous Besov spaces are defined by

$$\dot{B}^s_{p,q}(\mathbb{R}^d) := \left\{ f \in \mathcal{S}'(\mathbb{R}^d), \|f|\dot{B}^s_{p,q}\| = \left(\sum_{j=-\infty}^{\infty} 2^{sjq} \|\mathcal{F}^{-1}(\phi_j \hat{f})|L^p\|^q \right)^{1/q} < \infty \right\}.$$

(b) For $1 \le p < \infty, 1 \le q \le \infty, s \in \mathbb{R}$ the homogeneous Triebel-Lizorkin spaces are defined by

$$\dot{F}^s_{p,q}(\mathbb{R}^d) := \left\{ f \in \mathcal{S}'(\mathbb{R}^d), \|f|\dot{F}^s_{p,q}\| = \left\| \left(\sum_{j=-\infty}^{\infty} 2^{sjq} |\mathcal{F}^{-1}(\phi_j \hat{f})|^q \right)^{1/q} |L^p \right\| < \infty \right\}.$$

These are Banach spaces, which do not depend on the choice of ϕ. Since $\|f|\dot{B}^s_{p,q}\| = \|f|\dot{F}^s_{p,q}\| = 0$ if f is a polynomial they are considered modulo polynomials. If $p = q$ then it holds $\dot{B}^s_{p,p} = \dot{F}^s_{p,p}$. We note further that the definition of the Triebel-Lizorkin spaces for the case $p = \infty, q \ne \infty$ is somewhat more involved, see [Tri83b, p.46, p.50] and [FJ90, Section 5]. We abstain from discussing this case here since a characterization given below also includes it anyway.

There are also inhomogeneous counterparts, for which it holds $B_{p,q}^s(\mathbb{R}^d) = L^p(\mathbb{R}^d) \cap \dot{B}_{p,q}^s(\mathbb{R}^d)$ and $F_{p,q}^s(\mathbb{R}^d) = L^p(\mathbb{R}^d) \cap \dot{F}_{p,q}^s(\mathbb{R}^d)$ [Tri83b, p.242]. Many classical function spaces like Sobolev spaces, Hardy spaces, BMO, Hölder-Zygmund spaces etc. appear as particular instances of (homogeneous or inhomogeneous) Besov or Triebel-Lizorkin spaces, see [Tri83b, Tri92].

Important in our context is the following characterization of Besov and Triebel-Lizorkin spaces, which follows from Remark 18 (in connection with formula (95) for $N = 1$) and formula (74) in [Tri88], see also Sections 2.4.1 and 2.4.5 in [Tri92].

Theorem 4.7.8. *Let* $\phi \in \mathcal{S}(\mathbb{R}^d)$ *with* $0 \notin \operatorname{supp}\phi$ *and* $|\phi(x)| > 0$ *for* $1/2 \le |x| \le 2$.

(a) *Let* $1 \le p, q \le \infty, s \in \mathbb{R}$. *Then an equivalent norm on* $\dot{B}_{p,q}^s$ *is given by*

$$\left(\int_0^\infty t^{-sq} \|\mathcal{F}^{-1}(\phi(t\cdot)\hat{f})|L^p\|^q \frac{dt}{t} \right)^{1/q}.$$

(b) *Let* $1 \le p < \infty, 1 \le q \le \infty, s \in \mathbb{R}$. *Then an equivalent norm on* $\dot{F}_{p,q}^s$ *is given by*

$$\left\| \left(\int_0^\infty \int_{B(0,t)} t^{-sq} |\mathcal{F}^{-1}(\phi(t\cdot)\hat{f})(y + \cdot)|^q dy \frac{dt}{t^{d+1}} \right)^{1/q} |L^p \right\|,$$

where $B(0,t)$ *denotes the ball of radius* t *centered at* $0 \in \mathbb{R}^d$.

The convolution theorem shows a connection to the CWT. In fact, denoting $\psi(x) = \mathcal{F}^{-1}(\phi)(-x)$ we obtain

$$(\mathcal{F}^{-1}(\phi(t\cdot)\hat{f})(x) = f * \mathcal{F}^{-1}(\phi(t\cdot))(x) = t^{-d} f(T_x \psi(t^{-1}\cdot)) = t^{-d/2} f(T_x D_t \psi)$$
$$= t^{-d/2} \operatorname{CWT}_\psi f(x,t) = t^{-d/2} \operatorname{CWT}_\psi f(x,t,\operatorname{Id}) \quad \text{for all } f \in \mathcal{S}'(\mathbb{R}^d).$$

Thus, Theorem 4.7.8 shows that also the following expressions define equivalent norms on $\dot{B}_{p,q}^s$ and $\dot{F}_{p,q}^s$, respectively,

$$\|f|\dot{B}_{p,q}^s\| \asymp \left(\int_0^\infty t^{-q(s+d/2-d/q)} \|\operatorname{CWT}_\psi f(\cdot,t)|L^p\|^q \frac{dt}{t^{d+1}} \right)^{1/q}, \tag{4.56}$$

$$\|f|\dot{F}_{p,q}^s\| \asymp \left\| \left(\int_0^\infty \int_{B(0,t)} t^{-q(s+d/2)} |\operatorname{CWT}_\psi(y + \cdot,t)|^q dy \frac{dt}{t^{d+1}} \right)^{1/q} |L^p \right\|. \tag{4.57}$$

Let us now discuss how the Besov and Triebel-Lizorkin spaces fit into the coorbit space setting. As in Section 1.5.3 we take $\mathcal{G} = \mathbb{R}^d \rtimes (\mathbb{R}_+^* \times SO(d))$ and its representation $\pi(x,a,R) = T_x D_a U_R$ on $L^2(\mathbb{R}^d)$, so that the corresponding transform V_g is the CWT on $\mathbb{R}^d$. The equivalent norm (4.56) of $\dot{B}_{p,q}^s$ suggests to introduce the following function spaces on $\mathcal{G}$.

Definition 4.7.2. Let $1 \leq p, q \leq \infty, s \in \mathbb{R}$. Then $L_s^{p,q}(\mathcal{G})$ is the mixed norm space consisting of all measurable functions on $\mathcal{G}$, whose norm

$$\|F|L_s^{p,q}\| := \left(\int_0^\infty \left(\int_{\mathbb{R}^d} \int_{SO(d)} |F(x,t,R)|^p dRdx \right)^{q/p} t^{-sq} \frac{dt}{t^{d+1}} \right)^{1/q}$$

is finite.

Clearly, we have $L_0^{p,p} = L^p(\mathcal{G})$ and with the submultiplicative weight function

$$m_s(x,t,R) := t^{-s} = \Delta(x,t,R)^{s/d}, \qquad (x,t,R) \in \mathcal{G}, s \in \mathbb{R},$$

it holds $L_s^{p,p} = L_{m_s}^p(\mathcal{G}) =: L_s^p(\mathcal{G})$.

In order to treat also the $\dot{F}$-spaces we need to work with tent spaces on $\mathcal{G}$. These are slight modifications of the spaces introduced in [CMS85], see also [AM88, CV00, Sor92]. For $x \in \mathbb{R}^d$ we denote by $V(x) := \{(y,t) \in \mathbb{R}^d \times \mathbb{R}_+^*, |x-y| < t\}$ a cone of aperture 1, and for a ball $B \subset \mathbb{R}^d$ the tent $\widehat{B}$ over B is defined as the complement of $\cup_{x \notin B} V(x)$ in $\mathbb{R}^d \times \mathbb{R}_+^*$, i.e., $\widehat{B} = \{(x,t) \in \mathbb{R}^d \times \mathbb{R}_+^*, B(x,t) \subset B\}$ with $B(x,t)$ denoting a ball in $\mathbb{R}^d$ of radius t centered at x. For $1 \leq q < \infty$ let

$$A_q(F)(x,A) := \left(\int_{V(x)} |F(y,t,A)|^q dy \frac{dt}{t^{d+1}} \right)^{1/q}, \quad x \in \mathbb{R}^d, A \in SO(d),$$

$$A_\infty(F)(x,A) := \sup_{(y,t) \in V(x)} |F(y,t,A)|, \quad x \in \mathbb{R}^d, A \in SO(d)$$

and

$$C_q(F)(x,A) := \sup_{x \in B} \left(\frac{1}{|B|} \int_{\widehat{B}} |F(y,t,A)|^q dy \frac{dt}{t} \right)^{1/q},$$

where the supremum is taken over all balls B in $\mathbb{R}^d$ containing x.

Definition 4.7.3. For $1 \leq p < \infty, 1 \leq q \leq \infty$, the tent spaces are defined by

$$T^{p,q}(\mathcal{G}) := \{F \text{ measurable}, \|F|T^{p,q}\| := \|A_q(F)|L^p(\mathbb{R}^d \times SO(d))\| < \infty\}.$$

Moreover, for $1 \leq q < \infty$, we let

$$T^{\infty,q}(\mathcal{G}) := \{F \text{ measurable}, \|F|T^{\infty,q}\| := \|C_q(F)|L^\infty(\mathbb{R}^d \times SO(d))\| < \infty\}.$$

Further, for $s \in \mathbb{R}$ and $1 \leq p, q \leq \infty, (p,q) \neq (\infty,\infty)$, we denote

$$T_s^{p,q}(\mathcal{G}) := \{F \text{ measurable }, Fm_s \in T^{p,q}\}$$

with natural norm $\|F|T_s^{p,q}\| := \|Fm_s|T_s^{p,q}\|$.

As usual, we identify functions in $T_s^{p,q}$, which differ only on null sets. The tent spaces are Banach spaces. The subspace of $T^{p,q}$ consisting of those functions that do not depend on the component $SO(d)$ coincides with the original definition of tent spaces in [CMS85]. For further properties of $T_s^{p,q}$ we refer to [AM88, CMS85, CV00, Sor92]. We remark that our definition of $T^{p,\infty}$ differs from the original one in [CMS85], where the elements are additionally required to be continuous functions whose non-tangential limits exist. This additional requirement, however, would result in spaces that are no longer solid.

Lemma 4.7.9. *For $1 \leq p < \infty$ it holds $T_{s+d/p}^{p,p} = L_s^{p,p} = L_s^p$ with equivalent norms.*

Proof: Applying Fubini's theorem yields

$$\|F|T^{p,p}\|^p = \int_{SO(d)} \int_{\mathbb{R}^d} \int_0^\infty \int_{|y-x|\leq t} |F(y,t,R)|^p \frac{dydt}{t^{d+1}} dx dR$$
$$= \int_{SO(d)} \int_0^\infty \int_{\mathbb{R}^d} \int_{\mathbb{R}^d} \chi_{B(y,t)}(x) dx |F(y,t,R)|^p \frac{dydt}{t^{d+1}} dR$$
$$= |B(0,1)| \int_{SO(d)} \int_0^\infty \int_{\mathbb{R}^d} |F(y,t,R)|^p t^d \frac{dt}{t^{d+1}} dy dR = |B(0,1)| \|F|L_{-d/p}^p\|^p.$$

This also implies the assertion for general $s \in \mathbb{R}$. $\qquad\square$

In order to define the corresponding coorbit spaces we need to investigate whether $L_s^{p,q}$ and $T_s^{p,q}$ satisfy property (Y). Clearly, they both are solid BF-spaces. Furthermore, it was used in [Grö91] that they are also two-sided translation invariant. It seems, however, that there is no rigorous proof of this fact available in the literature. In particular, the right translation invariance of $T_s^{p,q}$ is not trivial.

Lemma 4.7.10. *Both $L_s^{p,q}$ and $T_s^{p,q}$ are two-sided translation invariant for all possible choices of p, q, s with*

$$\|L_{(y,r,S)}|L_s^{p,q}\| = r^{d(p^{-1}-q^{-1})-s}, \qquad \|R_{(y,r,S)}|L_s^{p,q}\| = r^{d/q+s},$$
$$\|L_{(y,r,S)}|T_s^{p,q}\| = r^{d/p-s}, \qquad \|R_{(y,r,S)}|T_s^{p,q}\| \leq Cr^{d/q+s} \max\{1, r^{-b}(1+|y|)^b\},$$

where $(y, r, S) \in \mathcal{G}$ and b is some constant depending on d, p and q.

Proof: It suffices to show the assertion for $s = 0$ since the general case follows then from the fact that m_s is submultiplicative (even a character), i.e., for $x \in \mathcal{G}$ it holds

$$\|L_x F|L_s^{p,q}\| = \|L_x (F L_{x^{-1}} m_s)|L_0^{p,q}\| = m_s(x)\|L_x(Fm_s)|L_0^{p,q}\|$$
$$\leq m_s(x)\|L_x|L_0^{p,q}\| \|F|L_s^{p,q}\|$$

and similarly for $T_s^{p,q}$ and for R_x.

(a) So let us treat $L^{p,q} = L_0^{p,q}$. With $(y, r, S) \in \mathcal{G}$ we obtain for $1 \leq p, q < \infty$ by simple substitutions

$$\|L_{(y,r,S)}F|L^{p,q}\|$$
$$= \left(\int_0^\infty \left(\int_{\mathbb{R}^d} \int_{SO(d)} |F(r^{-1}S^{-1}(x-y), r^{-1}t, S^{-1}A)|^p dA dx \right)^{q/p} \frac{dt}{t^{d+1}} \right)^{1/q}$$
$$= \left(\int_0^\infty \left(\int_{\mathbb{R}^d} \int_{SO(d)} |F(x,t,A)|^p r^d dA dx \right)^{q/p} r^{-d} \frac{dt}{t^{d+1}} \right)^{1/q} = r^{d(p^{-1}-q^{-1})}\|F|L^{p,q}\|.$$

Concerning right translation we get

$$\|R_{(y,r,S)}F|L^{p,q}\| = \left(\int_0^\infty \left(\int_{SO(d)} \int_{\mathbb{R}^d} |F(x+tAy, tr, AS)|^p dx dA \right)^{q/p} \frac{dt}{t^{d+1}} \right)^{1/q}$$
$$= \left(\int_0^\infty \left(\int_{SO(d)} \int_{\mathbb{R}^d} |F(x,t,A)|^p dx dA \right)^{q/p} r^d \frac{dt}{t^{d+1}} \right)^{1/q} = r^{d/q}\|F|L^{p,q}\|.$$

The results for $p = \infty$ or $q = \infty$ are obtained similarly.

(b) Let us now consider the spaces $T^{p,q}$. Left translation invariance is again rather easy, i.e., for $1 \leq p, q < \infty$ we obtain using simple substitutions

$$\|L_{(y,r,S)}F|T^{p,q}\|$$
$$= \left(\int_{SO(d)} \int_{\mathbb{R}^d} \left(\int_0^\infty \int_{|z-x|<t} |F(r^{-1}S^{-1}(z-y), r^{-1}t, S^{-1}A)|^q \frac{dzdt}{t^{d+1}} \right)^{p/q} dx dA \right)^{1/p}$$
$$= \left(\int_{SO(d)} \int_{\mathbb{R}^d} \left(\int_0^\infty \int_{|rSz+y-x|<rt} |F(z,t,A)|^q \frac{dzdt}{t^{d+1}} \right)^{p/q} dx dA \right)^{1/p}$$
$$= \left(\int_{SO(d)} \int_{\mathbb{R}^d} \left(\int_0^\infty \int_{|z-r^{-1}S^{-1}(x-y)|<t} |F(z,t,A)|^q \frac{dzdt}{t^{d+1}} \right)^{p/q} dx dA \right)^{1/p}$$
$$= \left(\int_{SO(d)} \int_{\mathbb{R}^d} \left(\int_0^\infty \int_{|z-x|<t} |F(z,t,A)|^q \frac{dzdt}{t^{d+1}} \right)^{p/q} r^d dx dA \right)^{1/p} = r^{d/p}\|F|T^{p,q}\|.$$

The cases $p = \infty, q < \infty$ and $p < \infty, q = \infty$ are handled in the same way.

The right translation invariance of $T^{p,q}$ requires some more effort. Denoting by $V^{(\alpha)}(x)$ the cone with aperture α, i.e., $V^{(\alpha)}(x) = \{(z,t), |x-z| \leq \alpha t\}$ we define

$$A_q^{(\alpha)}(F)(x,A) := \left(\int_{V^{(\alpha)}(x)} |F(z,t,A)|^q \frac{dzdt}{t^{d+1}} \right)^{1/q}, \qquad x \in \mathbb{R}^d, A \in SO(d).$$

For $1 \leq q < \infty$ we obtain

$$
\begin{aligned}
A_q(R_{(y,r,S)}(x, A) &= \left(\int_0^\infty \int_{|z-x|<t} |F(z + tAy, tr, AS)|^q \frac{dzdt}{t^{d+1}} \right)^{1/q} \\
&= \left(\int_0^\infty \int_{|z-r^{-1}tAy-x|<r^{-1}t} |F(z, t, AS)| r^d \frac{dzdt}{t^{d+1}} \right)^{1/q} \\
&\leq r^{d/q} \left(\int_0^\infty \int_{|z-x|<r^{-1}(1+|y|)t} |F(z, t, AS)| \frac{dzdt}{t^{d+1}} \right)^{1/q} = r^{d/q} A_q^{(r^{-1}(1+|y|))}(F)(x, AS)
\end{aligned}
$$

and a similar calculation yields an analogous estimation for $q = \infty$. Hereby, we used that $B(x + r^{-1}tAy, r^{-1}t) \subset B(x, r^{-1}(1 + |y|)t)$. It was shown in [CMS85, Proposition 4] (see also [CT75, Theorem 3.4] and [FS72, p.166] for $q = \infty$) that the definition of the tent spaces does not depend on the aperture α of the cone $V^{(\alpha)}(x)$, i.e., for $1 \leq p < \infty$ we have

$$
\| A_q^{(\alpha)}(F)(\cdot, A)|L^p(\mathbb{R}^d) \| \leq c(\alpha, p, q) \| A_q(F)(\cdot, A)|L^p(\mathbb{R}^d) \|.
$$

A deeper inspection of the proof in [CMS85] shows that $c(\alpha, p, q) \leq C \max\{1, \alpha^b\}$ with some exponent $b \geq 0$ depending on d, p and q, whose exact value is of minor importance in our context. Hence, for $1 \leq p, q < \infty$ we obtain (also using the invariance of the Haar measure of $SO(d)$)

$$
\begin{aligned}
\| R_{(y,r,S)}F|T^{p,q} \| &\leq \| A_q^{(r^{-1}(1+|y|))}(F)|L^p(\mathbb{R}^d \times SO(d)) \| \\
&\leq C \max\{1, r^{-b}(1 + |y|)^b\} r^{d/q} \| A_q(F)|L^p(\mathbb{R}^d \times SO(d)) \| \\
&= C \max\{1, r^{-b}(1 + |y|)^b\} r^{d/q} \| F|T^{p,q} \|.
\end{aligned}
$$

In order to treat the remaining case $p = \infty, 1 \leq q < \infty$ we use duality. Indeed it was proved in [CMS85, Theorem 1 and Section 8] that the dual of $T^{1,q'}$ (where as usual $1/q + 1/q' = 1$) may be identified $T^{\infty,q}$ for $1 < q < \infty$ via the bilinear pairing

$$
\langle F, G \rangle = \int_{SO(d)} \int_0^\infty \int_{\mathbb{R}^d} F(x, t, A) G(x, t, A) dx \frac{dt}{t} dA. \tag{4.58}
$$

(We note that the additional component $SO(d)$ does not change the result from [CMS85].) Moreover, denote by $CT^{1,\infty}$ the subspace of $T^{1,\infty}$ consisting of continuous functions with existing non-tangential limits. ($CT^{1,\infty}$ corresponds to the original definition of $T^{1,\infty}$ in [CMS85].) Clearly, also $CT^{1,\infty}$ is translation invariant with the same norm of the translation operator. It was proven in [CMS85] that the dual space of $CT^{1,\infty}$ coincides with the space CM of Carleson measures, and $T^{\infty,1}$ is a subspace of CM. In particular, $T^{\infty,1}$ can be realized as functionals on $CT^{1,\infty}$ via the pairing in (4.58).

A simple change of variable shows that

$$\langle R_{(y,r,S)}F, G\rangle = \langle F, R_{(-y,r^{-1},S^{-1})}G\rangle.$$

Thus for all $1 \leq q < \infty$ we have

$$\|R_{(y,r,S)}F|T^{\infty,q}\| = \sup_{G \in T^{1,q'}, \|G|T^{1,q'}\| \leq 1} |\langle F, R_{(-y,r^{-1},S^{-1})}\rangle|$$
$$\leq \|F|T^{\infty,q}\| \, \|R_{(-y,r^{-1},S^{-1})}|T^{1,q'}\|,$$

where for $q = 1$ we have to replace $T^{1,q'}$ by $CT^{1,\infty}$. Hence, the result for $T^{1,q'}$ implies also the one for $T^{\infty,q}$. $\square$

It seems that in the estimate for $\|R_{(x,r,S)}|T_s^{p,q}\|$ the term $(1 + |y|)$ can be omitted. At least, when $p = q$, this follows from the estimate for $L_s^p = T_{s+d/p}^{p,p}$. In the general case, however, our proof does not provide a stronger estimate. Fortunately, the estimate in Lemma 4.7.10 suffices for our purposes.

Suppose that w is the weight function associated to $Y = L_s^{p,q}$ or $Y = T_s^{p,q}$ by (4.19). It follows from the previous Lemma that there exist constants $C, \sigma_1, \sigma_2 \geq 0$ (depending on d, p, q, s) such that w can be estimated by

$$w(y, r, S) \leq C(1 + |y|)^{\sigma_1}(\max\{r, r^{-1}\})^{\sigma_2}. \tag{4.59}$$

Let us show that the associated space of admissible vectors $\mathbb{A}_w$ is non-empty. To do this we introduce the space $\mathcal{S}^\infty(\mathbb{R}^d) \subset \mathcal{S}(\mathbb{R}^d)$ consisting of those functions $f \in \mathcal{S}(\mathbb{R}^d)$ that have infinitely many vanishing moments, that is, for all $k \in \mathbb{N}$ there exists a constant C_k such that

$$|\hat{f}(\xi)| \leq C_k|\xi|^k \quad \text{for all } \xi \in \mathbb{R}^d.$$

Lemma 4.7.11. *Let w be a weight function that satisfies (4.59). Then it holds*

$$\mathcal{S}^\infty(\mathbb{R}^d) \subset \mathbb{B}_w \subset \mathbb{A}_w.$$

In particular, $\mathbb{A}_w$ is non-trivial.

Proof: Suppose $g \in \mathcal{S}^\infty(\mathbb{R}^d)$. It is well-known that for any $\alpha, \beta, \gamma \in \mathbb{N}$ there exists a constant $C_{\alpha,\beta,\gamma}$ such that

$$|\operatorname{CWT}_g g(x, t, R)| \leq C_{\alpha,\beta,\gamma} t^\alpha (1 + t)^{-\beta}(1 + |x|)^{-\gamma},$$

see e.g. Theorem 19.0.1 and page 116 in [Hol95]. Since $|(\operatorname{CWT}_g g)^\vee| = |\operatorname{CWT}_g g|$ it suffices to prove $\operatorname{CWT}_g g \in W(C_0, L_w^1)$ by Lemma 4.4.1. Let $U = B(0, a) \times [1/b, b] \times$

$SO(d)$. The control function satisfies

$$
\begin{aligned}
K(\mathrm{CWT}_g\, g, \chi_U, C_0)(x, t, R) &= \|(L_{(x,t,R)}\chi_U)\, \mathrm{CWT}_g\, g\|_\infty = \|\chi_U(L_{(x,t,R)^{-1}}\, \mathrm{CWT}_g\, g)\|_\infty \\
&= \sup_{(z,r,A)\in U} |\, \mathrm{CWT}_g\, g((x, t, R) \cdot (z, r, A))| = \sup_{(z,r,A)\in U} |\, \mathrm{CWT}(x + tRz, tr, RA)| \\
&\leq C_{\alpha,\beta,\gamma} \sup_{z\in B(0,a)} \sup_{r\in[1/b,b]} (tr)^\alpha (1 + tr)^{-\beta}(1 + |x + tRz|)^{-\gamma} \\
&\leq C' C_{\alpha,\beta,\gamma} t^\alpha (1 + t)^{-\beta}(1 + |x|)^{-\gamma}
\end{aligned}
$$

for all $\alpha, \beta, \gamma \in \mathbb{N}$. In the last line we used the moderateness of the function $x \mapsto (1 + |x|)^{-\gamma}$. Since the weight function w satisfies (4.59) we conclude (choosing α, β, γ large enough)

$$
\begin{aligned}
\|\, \mathrm{CWT}_g\, g|W^R(C_0, L^1_w)\| &= \|\, \mathrm{CWT}_g\, g|W(C_0, L^1_w)\| \\
&= \int_0^\infty \int_{\mathbb{R}^d} \int_{SO(d)} K(\mathrm{CWT}_g\, g, \chi_U, C_0)(x, t, R)\, w(x, t, R)\, dRdx\frac{dt}{t^{d+1}} < \infty.
\end{aligned}
$$

This means $g \in \mathbb{B}_w \subset \mathbb{A}_w$ and we are done. $\qquad\square$

In view of the previous lemma, we are enabled to define the coorbit spaces corresponding to the spaces $L^{p,q}_s$ and $T^{p,q}_s$.

Corollary 4.7.12. *It holds*

$$
\dot{B}^s_{p,q} = \mathrm{Co}L^{p,q}_{s+d/2-d/q} \qquad \text{and} \qquad \dot{F}^s_{p,q} = \mathrm{Co}T^{p,q}_{s+d/2}
$$

for all admissible values of p, q and s.

Proof: The reproducing formula for the CWT extends to $(\mathcal{S})'(\mathbb{R}^d)/\mathcal{P}$, where $\mathcal{P}$ denotes the space of all polynomials on $\mathbb{R}^d$, see e.g. Corollary 20.0.2 and page 117 in [Hol95] (note hereby that $(\mathcal{S}^\infty)'(\mathbb{R}^d)$ coincides with $\mathcal{S}'(\mathbb{R}^d)/\mathcal{P}$). Theorem 4.5.13(c) and the charaterizations (4.56) and (4.57) yield the claim. $\qquad\square$

We remark that $\mathrm{Co}T^{p,q}_{s+d/2}$ is also defined for $p = \infty, q < \infty$, so that this can be taken as definition of $\dot{F}^{\infty,q}_s$. Indeed, this definition is equivalent to the one found in [FJ90, p. 70].

Theorems 4.6.1–4.6.3 (with the trivial automorphism group) yield atomic decompositions and Banach frames for the spaces $\dot{B}^s_{p,q}$ and $\dot{F}^s_{p,q}$. These are of the same type as constructed by Frazier and Jawerth [FJ85, FJ88, FJ90, FJW91], see also [ACM03, Grö88, Grö91, Tri83b, Tri92].

In the sequel we will be interested in radial functions. Thus, we consider the symmetry group $\mathcal{A} := SO(d)$ acting on $\mathcal{G}$ by inner automorphisms, see Section 2.3.1. Clearly, $\mathcal{A}$

acts isometrically on the spaces $L_s^{p,q}$ and $T_s^{p,q}$, i.e., property (YA) is satisfied. Moreover, by Lemma 4.7.11 we have $\mathcal{S}_{rad}^\infty(\mathbb{R}^d) \subset \mathbb{B}_w^{\mathcal{A}} \subset \mathbb{A}_w^{\mathcal{A}}$ if w is the corresponding weight function (implying w satisfies (4.59)), in particular $\mathbb{A}_w^{\mathcal{A}}$ is non-trivial. Hence, the invariant coorbit spaces are well-defined and we have

$$\mathsf{Co}(L_{s+d/2-d/q}^{p,q})_{\mathcal{A}} = (\dot{B}_{p,q}^s)_{rad}(\mathbb{R}^d) = \{f \in \dot{B}_{p,q}^s, f \text{ is radial}\},$$
$$\mathsf{Co}(T_{s+d/2}^{p,q})_{\mathcal{A}} = (\dot{F}_{p,q}^s)_{rad}(\mathbb{R}^d) = \{f \in \dot{F}_{p,q}^s, f \text{ is radial}\}.$$

We are interested in atomic decompositions and Banach frames for these spaces. Let us first provide an invariant covering of $\mathcal{G}$ and the corresponding sequence spaces explicitly. For $\alpha > 0$ and $\beta > 1$ we define

$$U_{\alpha,\beta} := B(0,\alpha/2) \times [\beta^{-1/2}, \beta^{1/2}] \times SO(d) \subset \mathcal{G}.$$

Clearly, this set satisfies $U_{\alpha,\beta} = U_{\alpha,\beta}^{-1} = \mathcal{A}(U_{\alpha,\beta})$. In spirit of Section 3.2 we propose the following discrete set of points in $\mathcal{G}$. For some fixed unit vector $\eta \in S^{d-1}$ and constants $a > 0, b > 1$ we let

$$x_{j,k} := ((k-1/2)b^{-j}a\eta, b^{-j}, \mathrm{Id}) \in \mathcal{G}, \quad j \in \mathbb{Z}, k \in \mathbb{N},$$

giving

$$\mathcal{A}(x_{j,k}U_{\alpha,\beta}) = A((k-1/2)b^{-j}a, b^{-j}\alpha/2) \times [b^{-j}\beta^{-1/2}, b^{-j}\beta^{1/2}] \times SO(d), \quad (4.60)$$

where $A(r,h) = \{z \in \mathbb{R}^d, r - h \le |z| \le r + h\}$ denotes an annulus of width $2h$ and radius r. Obviously, the sets $\mathcal{A}(x_{j,k}U_{a,b}), j \in \mathbb{Z}, k \in \mathbb{N}$ cover $\mathcal{G}$ and $X_{a,b} = (x_{j,k})_{j \in \mathbb{Z}, k \in \mathbb{N}}$ is well-spread with respect to $\mathcal{A}$. We let further $\mu_{j,k} := |\mathcal{A}(x_{j,k}U_{a,b})|$. Elementary considerations show that there exist constants $C_1, C_2 > 0$ (possibly depending on a, b, d but not on j, k) such that

$$C_1 k^{d-1} \le \mu_{j,k} \le C_2 k^{d-1}. \quad (4.61)$$

From Corollary 4.6.4 we immediately deduce the following result concerning radial wavelet frames for $L_{rad}^2(\mathbb{R}^d)$. Recall that $\widetilde{\pi}$ is connected to the generalized translation τ by (2.12).

Theorem 4.7.13. *Let* $w(x,r,B) = w(r) = \max\{1, r^{d/2}\}$ *and assume* $g \in \mathbb{B}_w^{\mathcal{A}}$, *for instance* $g \in \mathcal{S}_{rad}^\infty(\mathbb{R}^d)$. *Then there exist constants* $a_0 > 0, b_0 > 1$ *such that for all* a, b *satisfying* $0 < a < a_0, 1 < b < b_0$ *the family*

$$\{k^{(d-1)/2}\widetilde{\pi}(x_{j,k})g\}_{j \in \mathbb{Z}, k \in \mathbb{N}} = \{k^{(d-1)/2}\tau_{b^{-j}(k-1/2)a}D_{b^{-j}}g\}_{j \in \mathbb{Z}, k \in \mathbb{N}}$$

forms a frame for $L_{rad}^2(\mathbb{R}^d)$. *Hereby,* τ *denotes the generalized translation (1.21).*

In order to treat also the $(\dot{B}_{p,q}^s)_{rad}$ and $(\dot{F}_{p,q}^s)_{rad}$ spaces, we first need to compute the sequence spaces associated to $L_s^{p,q}$ and $T_s^{p,q}$ as in Section 4.3. For $1 \le p, q \le \infty, s, u \in \mathbb{R}$ we introduce the following spaces of sequences on $\mathbb{Z} \times \mathbb{N}$,

$$l_{s,u}^{p,q} = \{(\lambda_{j,k})_{j\in\mathbb{Z},k\in\mathbb{N}}, \ \|(\lambda_{j,k})|l_{s,u}^{p,q}\| < \infty\},$$
$$t_{s,u}^{p,q} = \{(\lambda_{j,k})_{j\in\mathbb{Z},k\in\mathbb{N}}, \ \|(\lambda_{j,k})|t_{s,u}^{p,q}\| < \infty\}$$

with norms

$$\|(\lambda_{j,k})|l_{s,u}^{p,q}\| := \left(\sum_{j\in\mathbb{Z}} \left(\sum_{k\in\mathbb{N}} |\lambda_{j,k}|^p k^{pu} \right)^{q/p} b^{jqs} \right)^{1/q},$$

(with obvious modification for $p = \infty$ or $q = \infty$) and

$$\|(\lambda_{j,k})|t_{s,u}^{p,q}\| := \left\| \left(\sum_{j\in\mathbb{Z}} \sum_{k\in\mathbb{N}} |\lambda_{j,k}|^q k^{qu} b^{jqs} \chi_{[(k-1)b^{-j}a, kb^{-j}a]} \right)^{1/q} \Big| L^p(\mathbb{R}_+, \mu_d) \right\|, \ 1 \le p < \infty,$$

(with obvious modification for $q = \infty$). For $p = \infty, 1 \le q < \infty$ we let

$$\|(\lambda_{j,k})|t_{s,u}^{\infty,q}\| := \sup_{\ell\in\mathbb{Z}} \sup_{i\in\mathbb{N}} \left(b^\ell \sum_{j=\ell}^{\infty} \sum_{k=\lfloor a^{-1}b^{j-\ell}(i-2)\rfloor}^{\lceil a^{-1}b^{j-\ell}(i+2)\rceil} |\lambda_{j,k}|^q k^{qu} b^{j(qs-1)} \right)^{1/q}.$$

We may identify the sequence spaces corresponding to $L_s^{p,q}$ and $T_s^{p,q}$ with one of these spaces as shown in the next lemma.

Lemma 4.7.14. *For all admissible choices of p, q and s it holds*

$$\begin{array}{llll}
(L_s^{p,q})_{SO(d)}^\flat &=& l_{s+d/q-d/p,(d-1)/p}^{p,q}, & (L_s^{p,q})_{SO(d)}^\natural &=& l_{s+d/q-d/p,(1/p-1)(d-1)}^{p,q}, \\
(T_s^{p,q})_{SO(d)}^\flat &=& t_{s,0}^{p,q}, & (T_s^{p,q})_{SO(d)}^\natural &=& t_{s,-d+1}^{p,q}
\end{array}$$

with equivalence of norms.

Proof: By radiality we can always neglect the $SO(d)$ component in $L_s^{p,q}$ and $T_s^{p,q}$.
(a) By $A_{j,k}^{\alpha,\beta}$ we denote the annulus $A((k-1/2)b^{-j}a, b^{-j}\alpha/2)$. Further, let $\chi_{j,k}^{\alpha,\beta}$ be the characteristic function of the set $SO(d)(x_{j,k}U_{\alpha,\beta})$ and for simplicity we set $\chi_{j,k} = \chi_{j,k}^{a,b}$, see (4.60). Since the sets $\operatorname{supp}\chi_{j,k}$ are mutually disjoint apart from a set of measure

zero, we obtain for $1 \leq p, q < \infty$

$$
\begin{aligned}
\|(\lambda_{j,k})|(L_s^{p,q})_{SO(d)}^\flat\| &= \left(\int_0^\infty \left(\int_{\mathbb{R}^d} \sum_{j\in\mathbb{Z},k\in\mathbb{N}} |\lambda_{j,k}|^p \chi_{j,k}(x,t) dx \right)^{q/p} t^{-sq} \frac{dt}{t^{d+1}} \right)^{1/q} \\
&= \left(\sum_{j\in\mathbb{Z}} \left(\sum_{k\in\mathbb{N}} |\lambda_{j,k}|^p \int_{\mathbb{R}^d} \chi_{A_{j,k}^{a,b}}(x) dx \right)^{q/p} \int_{b^{-j-1/2}}^{b^{-j+1/2}} t^{-sq} \frac{dt}{t^{d+1}} \right)^{1/q} \\
&\asymp \left(\sum_{j\in\mathbb{Z}} \left(\sum_{k\in\mathbb{N}} |\lambda_{j,k}|^p b^{-dj} k^{d-1} \right)^{q/p} b^{j(sq+d)} \right)^{1/q} \\
&= \left(\sum_{j\in\mathbb{Z}} \left(\sum_{k\in\mathbb{N}} |\lambda_{j,k}|^p k^{d-1} \right)^{q/p} b^{jq(s+d(q^{-1}-p^{-1}))} \right)^{1/q}.
\end{aligned}
$$

The cases $p = \infty$ or $q = \infty$ are treated similarly. Essentially the same calculations also give the result for $(L_s^{p,q})_{SO(d)}^\natural$.

(b) Let us now treat $(T_s^{p,q})_{SO(d)}^\flat$. Again by disjointness of the sets $\operatorname{supp} \chi_{j,k}$ we obtain for $1 \leq q < \infty$

$$
\begin{aligned}
A_q\left(\sum_{j\in\mathbb{Z},k\in\mathbb{N}} |\lambda_{j,k}| \chi_{j,k} \right)(x) &= \left(\int_0^\infty \int_{|z-x|<t} \sum_{j\in\mathbb{Z},k\in\mathbb{N}} |\lambda_{j,k}|^q \chi_{j,k}(z,t) t^{-sq} \frac{dz dt}{t^{d+1}} \right)^{1/q} \\
&\asymp \left(\sum_{j\in\mathbb{Z}} \sum_{k\in\mathbb{N}} |\lambda_{j,k}|^q b^{jsq} \int_0^\infty \int_{|z-x|\leq t} \chi_{j,k}(z,t) dz \frac{dt}{t^{d+1}} \right)^{1/q}.
\end{aligned}
$$

If α, β are small enough then geometric considerations show that

$$
\begin{aligned}
C_1 \int_0^\infty \int_{|z-x|\leq t} \chi_{j,k}^{\alpha,\beta}(z,t) dz \frac{dt}{t^{d+1}} &\leq \chi_{[(k-1)b^{-j}a, kb^{-j}a]}(|x|) \\
&\leq C_2 \int_0^\infty \int_{|z-x|\leq t} \chi_{j,k}^{a,b}(z,t) dz \frac{dt}{t^{d+1}}
\end{aligned}
$$

for some constants $C_1, C_2 > 0$. Since by Lemma 4.3.1(a) the definition of the space $(T_s^{p,q})_{SO(d)}^\flat$ is independent of the choice of α, β, the assertion follows for $1 \leq p < \infty$ and $1 \leq q < \infty$. The case $p < \infty, q = \infty$ is handled similarly.

Now assume $p = \infty$ and $1 \leq q < \infty$. Let α, β such that the set $\operatorname{supp} \chi_{j,k}^{\alpha,\beta}$ are disjoint. We obtain

$$\|(\lambda_{j,k})|(T_s^{\infty,q})_{SO(d)}^\flat\|$$

$$= \sup_{x\in\mathbb{R}^d, t\in(0,\infty)} \left(\frac{1}{|B(x,t)|} \int_0^t \int_{B(x,t-r)} \left| \sum_{j\in\mathbb{Z}, k\in\mathbb{N}} \lambda_{j,k} \chi_{j,k}^{\alpha,\beta}(y,r) \right|^q r^{-sq} dy \frac{dr}{r} \right)^{1/q}$$

$$= C \sup_{\ell\in\mathbb{Z}} \sup_{t\in[b^{-\ell}\beta^{-1/2}, b^{-\ell}\beta^{1/2}]} \sup_{x\in\mathbb{R}^d} \left(t^{-d} \sum_{j\in\mathbb{Z}, k\in\mathbb{N}} |\lambda_{j,k}|^q \int_0^t \int_{B(x,t-r)} \chi_{j,k}^{\alpha,\beta}(y,r) r^{-sq} dy \frac{dr}{r} \right)^{1/q}$$

$$\asymp \sup_{\ell\in\mathbb{Z}} \sup_{x\in\mathbb{R}^d} \left(b^{\ell d} \sum_{j=\ell}^\infty \sum_{k\in\mathbb{N}} |\lambda_{j,k}|^q \int_{b^{-j}\beta^{-1/2}}^{b^{-j}\beta^{1/2}} |B(x, b^{-\ell}\beta^{1/2} - r) \cap A_{j,k}^{\alpha,\beta}| \frac{dr}{r} b^{jqs} \right)^{1/q}$$

The inner integral can be estimated from below and above by

$$D|B(x, \beta^{1/2}(b^{-\ell} - b^{-j})) \cap A_{j,k}^{\alpha,\beta}| \leq \int_{b^{-j}\beta^{-1/2}}^{b^{-j}\beta^{1/2}} |B(x, b^{-\ell}\beta^{1/2} - r) \cap A_{j,k}^{\alpha,\beta}| \frac{dr}{r} \qquad (4.62)$$

$$\leq D|B(x, b^{-\ell}\beta^{1/2} - b^{-j}\beta^{-1/2}) \cap A_{j,k}^{\alpha,\beta}|.$$

with $D = \int_1^\beta \frac{dr}{r}$. The intersection $B(x, b^{-\ell}\beta^{1/2} - b^{-j}\beta^{-1/2}) \cap A_{j,k}^{\alpha,\beta}$ is empty iff

$$\left| |(k-1/2) - b^j a^{-1}|x|| \right| > b^j a^{-1}(b^{-\ell}\beta^{1/2} - b^{-j}\beta^{-1/2}) + a^{-1}\alpha/2 =: \kappa_{j,\ell}$$

If the intersection is not empty then we observe that $B(x, b^{-\ell}\beta^{1/2} - b^{-j}\beta^{-1/2}) \cap A_{j,k}^{\alpha,\beta}$ approximately has the shape of a cylinder with height $b^{-j}\alpha/2$ and radius $b^{-\ell}\beta^{1/2} - b^{-j}\beta^{-1/2}$. Since $j \geq \ell$ we have $b^{-\ell}\beta^{1/2} - b^{-j}\beta^{-1/2} \asymp b^{-\ell}$. Thus, provided k is not near its extremal points, i.e.,

$$|(k-1/2) - b^j a^{-1}|x|| \leq \delta \kappa_{j,\ell}$$

for some fixed $\delta < 1$ then we have

$$|B(x, b^{-\ell}\beta^{1/2} - b^{-j}\beta^{-1/2}) \cap A_{j,k}^{\alpha,\beta}| \asymp b^{-\ell(d-1)} b^{-j}.$$

Moreover, if $j \geq l+1$ then this relation holds also for the left hand side in (4.62). Asymptotically, we have $\kappa_{j,k} \asymp \beta^{1/2} a^{-1} b^{j-\ell}$. Setting $|x| = b^{-\ell}(i+r)$ with $i \in \mathbb{N}$ and $|r| \leq 1$ the relation $|(k-1/2) - b^j a^{-1}|x|| \leq \delta\beta^{1/2} a^{-1} b^{j-\ell}$ implies

$$k - 1/2 \geq a^{-1} b^{j-\ell}(i + r - \delta\beta^{1/2}) \geq a^{-1} b^{j-\ell}(i - (\delta\beta^{1/2} + 1)) \qquad (4.63)$$

and

$$k - 1/2 \leq a^{-1} b^{j-\ell}(i + r + \beta^{1/2}) \leq a^{-1} b^{j-\ell}(i + (\delta\beta^{1/2} + 1)). \qquad (4.64)$$

By Lemma 4.3.1(a) the sequence space $(T_s^{\infty,q})_{SO(d)}^\flat$ is independent of the choice of α and β. Therefore, we are allowed to replace $\delta\beta^{1/2}+1$ by 2. Also, it is easy to see that the constant $-1/2$ in (4.63) and (4.64) can be neglected. Finally, we obtain

$$\|(\lambda_{j,k})|(T_s^{\infty,q})_{SO(d)}^\flat\| \asymp \sup_{\ell\in\mathbb{Z}}\sup_{i\in\mathbb{N}}\left(b^{\ell d}\sum_{j=\ell}^{\infty}\sum_{k=\lfloor a^{-1}b^{j-\ell}(i-2)\rfloor}^{\lceil a^{-1}b^{j-\ell}(i+2)\rceil}|\lambda_{j,k}|^q b^{-j-\ell(d-1)}b^{jqs}\right)^{1/q}$$

$$= \sup_{\ell\in\mathbb{Z}}\sup_{i\in\mathbb{N}}\left(b^\ell\sum_{j=\ell}^{\infty}\sum_{k=\lfloor a^{-1}b^{j-\ell}(i-2)\rfloor}^{\lceil a^{-1}b^{j-\ell}(i+2)\rceil}|\lambda_{j,k}|^q b^{j(qs-1)}\right)^{1/q}.$$

We note that for the lower bound one first has to replace $\sum_{j=\ell}^{\infty}$ by $\sum_{j=\ell+1}^{\infty}$. The substitution $\ell \mapsto \ell+1$, however, shows the final result then. The claim for the space $(T_s^{\infty,q})_{SO(d)}^\natural$ is shown in the same way.   □

Theorem 4.6.3 immediately yields the following result. Here, we denote by $w(Y)$ the weight function associated to a function space Y as in (4.19).

Theorem 4.7.15. *Let $1 \le p,q \le \infty$, $s \in \mathbb{R}$ and $w := \max\{w(L_s^{p,q}), w(T_s^{p,q})\}$. Further choose $g \in \mathbb{B}_w^{\mathcal{A}}$, for instance $g \in \mathcal{S}_{rad}^{\infty}(\mathbb{R}^d)$. Then there exist constants $a_0 > 0, b_0 > 1$ such that for all a,b satisfying $0 < a < a_0, 1 < b < b_0$ the following holds.*

(a) *The family*
$$\{g_{j,k}\}_{j\in\mathbb{Z},k\in\mathbb{N}} := \{\tau_{b^{-j}(k-1/2)a}D_{b^{-j}}g\}_{j\in\mathbb{Z},k\in\mathbb{N}}$$
forms a Banach frame for $(\dot{B}_{p,q}^s)_{rad}$ and $(\dot{F}_{p,q}^s)_{rad}$ with corresponding sequence spaces $l_{s+d/2-d/p,(d-1)/p}^{p,q}$ and $t_{s+d/2,0}^{p,q}$. In particular, we have the norm equivalences
$$\|f|(\dot{B}_{p,q}^s)_{rad}\| \asymp \|(\langle f,g_{j,k}\rangle)_{j,k}|l_{s+d/2-d/p,(d-1)/p}^{p,q}\|$$
and
$$\|f|(\dot{F}_{p,q}^s)_{rad}\| \asymp \|(\langle f,g_{j,k}\rangle)_{j,k}|t_{s+d/2,0}^{p,q}\|.$$

(b) *The family $\{g_{j,k}\}_{j\in\mathbb{Z},k\in\mathbb{N}}$ forms an atomic decomposition for $(\dot{B}_{p,q}^s)_{rad}$ and $(\dot{F}_{p,q}^s)_{rad}$ with corresponding sequence spaces $l_{s+d/2-d/p,(1/p-1)(d-1)}^{p,q}$ and $t_{s+d/2,-d+1}^{p,q}$.*

(c) *In addition, there exists a dual frame $\{e_{j,k}\}_{j\in\mathbb{Z},k\in\mathbb{N}}$ in the sense of Theorem 4.6.3.*

Finally, let us make some concluding remarks.

Remark 4.7.1.   (a) By Lemma 4.7.11 one may choose any function $g \in \mathcal{S}_{rad}^{\infty}(\mathbb{R}^d) \subset \mathbb{B}_w^{\mathcal{A}}$ in Theorem 4.7.15. Thus, we allow more flexibility than the results in [EF95], where the Fourier transform of g must have support in a subset of the unit ball $B(0,1)$.

(b) Our suggestion for the well-spread set $X = (x_{j,k})_{j \in \mathbb{Z}, k \in \mathbb{N}}$ is not the only possible choice, of course. It would be more in the spirit of [EF95] to determine the spatial lattice by the zeros of the corresponding Bessel functions. Also "completely irregular" sets (satisfying the well-spreadness assumption) would work. This shows again that Theorem 4.6.3 allows to construct more flexible decompositions than the ones in [EF95] derived by direct methods.

(c) In general, the constants a, b depend on p, q, s since the weight function w and, therefore, the estimate (4.25) does. However, fixing a weight w, the same constants a, b are valid for the whole class of spaces $\dot{F}_{p,q}^s$ and $\dot{B}_{p,q}^s$, whose associated weight functions $w(L_s^{p,q})$ and $w(T_s^{p,q})$ are dominated by w.

(d) In our discussion the weight functions m_s were only dependent on the dilation parameter. This restriction, however, is not necessary. Making the weight function also dependent on $x \in \mathbb{R}^d$ yields weighted Besov and Triebel-Lizorkin spaces. For example, a common choice would be $m_{s,u}(x, r, B) = (1 + |x|)^u r^{-s}$, $u, s \in \mathbb{R}$.

(e) Suppose we are given an orthonormal radial wavelet basis for $L_{rad}^2(\mathbb{R}^d)$, like the one constructed in Chapter 3. The question arises whether this is also an unconditional basis simultaneously for all $(\dot{B}_{p,q}^s)_{rad}$ and $(\dot{F}_{p,q}^s)_{rad}$ spaces (or at least a weak-$*$ basis if $p = \infty$ or $q = \infty$). Indeed, this is true if the "mother wavelet" ψ is contained in $\mathcal{S}_{rad}^\infty(\mathbb{R}^d)$. The proof of this fact is the same as the one found in [Grö88, Section 4]. In particular, this shows that the radial wavelets of Section 3.2 yield also decompositions of radial Besov and Triebel-Lizorkin spaces for suitable mother wavelets.

4.8 Notes and Open Problems

There is a striking relationship between our discretization method and sampling theory, i.e., the classical Shannon sampling theorem. Indeed, we proved in Theorems 4.6.2 and 4.6.3 that it is possible to recover an element $f \in \text{CoY}_\mathcal{A}$ by means of sample values of $\widetilde{V}_g f$. This in particular implies by the correspondence principle (Corollary 4.5.12(a)) that any function $F \in Y_\mathcal{A} * \widetilde{V}_g g$ (implying $F = \widetilde{V}_g f$ for some $f \in \text{CoY}_\mathcal{A}$) can be reconstructed from suitable sample values $F(x_i)$. Thus, the space $Y_\mathcal{A} * \widetilde{V}_g g$ has analogous properties as the Paley-Wiener space. Indeed, in [FG92b] Feichtinger and Gröchenig used similar techniques as in coorbit space theory to obtain irregular sampling theorems for band-limited functions on $\mathbb{R}^d$.

Feichtinger defined the modulation spaces on arbitrary locally compact Abelian groups G [Fei83c]. One may associate a Heisenberg group $\mathbb{H}(G)$ to G. This allows to carry out the corresponding coorbit space theory in this setting and one obtains results for

Gabor analysis on G. Also our extension involving automorphism groups still works. However, it remains open to investigate particular examples different from $G = \mathbb{R}^d$.

In [FG89b] Feichtinger and Gröchenig used their atomic decompositions in order to study embeddings of coorbit spaces by means of embeddings of the corresponding sequence spaces. In particular, they were able to show that an embedding of two coorbit spaces $\mathsf{Co}Y^{(1)} \subset \mathsf{Co}Y^{(2)}$ is compact if and only if the embedding of the corresponding sequence spaces is compact. The generalization to our invariant subspaces $\mathsf{Co}Y_{\mathcal{A}}$ seems to be straightforward. So we conjecture that $\mathsf{Co}Y_{\mathcal{A}}^{(1)}$ is compactly embedded into $\mathsf{Co}Y_{\mathcal{A}}^{(2)}$ if and only if $(Y^{(1)})_{\mathcal{A}}^{\natural}$ is compactly embedded into $(Y^{(2)})_{\mathcal{A}}^{\natural}$. The definiton of the spaces $Y_{\mathcal{A}}^{\natural}$ also involves the sequence $(|\mathcal{A}(x_i U)|)_{i \in I}$, which is unbounded in typical situations. Together with our conjecture this would cause the following phenomenon. Suppose $\mathsf{Co}Y^{(1)}$ is embedded into $\mathsf{Co}Y^{(2)}$ but not compactly. Then it may happen that the subspace $\mathsf{Co}Y_{\mathcal{A}}^{(1)}$ is compactly embedded into $\mathsf{Co}Y_{\mathcal{A}}^{(2)}$. Actually this was observed recently for the Besov and Triebel-Lizorkin spaces in [KLSS03, Skr03, SS00, ST01]. If $\mathcal{A}$ is an infinite closed subgroup of $O(d)$ then for certain choices of parameters, embeddings of $F_{p,q}^s$ and $B_{p,q}^s$ spaces become compact when restricted to subspaces of functions, which are invariant under $\mathcal{A}$. Coorbit space theory would give an abstract explanation for this phenomenon (provided our conjecture is true). Moreover, a counterpart of such results in the theory of modulation spaces seems to be not present in the literature. Our conjecture would provide an easy way to prove compactness of certain embeddings of modulation spaces of radial functions.

For other compactness criteria in coorbit spaces we refer to [DFG02].

The definition of the Besov, Triebel-Lizorkin and modulation spaces is also valid for $0 < p, q < 1$. In this case one obtains no longer Banach spaces, but quasi-Banach spaces, that is, the triangle inequality for the norm is replaced by $\|f+g\| \leq C(\|f\|+\|g\|)$ for some constant $C > 1$. One important motivation to consider such spaces comes from non-linear approximation. For example, if one approximates a function in $M^{p,p}$, $p < 2$ by N-terms of its Gabor expansion then the error of the best N-term approximation measured in the L^2-norm tends to zero like $N^{1/2-1/p}$, see [Grö01, Theorem 12.4.2]. Thus in order to achieve fast convergence, one should choose p as small as possible. However, if one restricts to $p \geq 1$, that is, to Banach spaces, the best possible rate is $N^{-1/2}$.

Although the proofs often become more technical, many results extend to these quasi-Banach spaces. In particular, the atomic decompositions of Besov and Triebel-Lizorkin spaces in [FJ90] hold for general $0 < p, q \leq \infty$, and also the radial decompositions in [EF95]. Some results concerning Gabor analysis on modulation spaces $M_s^{p,q}, 0 < p, q < 1$, can be found in [GS04]. However, it remains open to extend the general coorbit space theory to quasi-Banach spaces. One cannot expect that it is possible to treat also the quasi-Banach space case along the lines of the classical theory, since we extensively used the anti-dual $(\mathcal{H}^1_w)^\daleth_\mathcal{A}$. Indeed, a quasi-Banach space is no longer locally convex and, therefore, duality theory is no longer available, i.e., the dual space of a quasi-Banach space might be trivial. Nevertheless, a generalization of coorbit space theory to quasi-Banach spaces seems to be possible.

The radial modulation spaces $(M^{p,q}_m)_{rad}(\mathbb{R}^d)$, Besov spaces $(\dot{B}^s_{p,q})_{rad}(\mathbb{R}^d)$ and Triebel-Lizorkin spaces $(\dot{F}^s_{p,q})_{rad}(\mathbb{R}^d)$ may be identified with corresponding function spaces on $\mathbb{R}_+$, i.e., with function spaces on the Bessel-Kingman hypergroup of index $\frac{d-2}{2}$. The question arises whether it is possible to extend their definition to Bessel-Kingman hypergroups of arbitrary real-valued index $\nu > -1/2$. In case of the Besov spaces, this question can be answered positively (at least for $s > 0$). Indeed, the Besov-Hankel spaces introduced by Betancor and Rodríguez-Mesa [BRM98, BMRM01] can be interpreted as Besov spaces on the Bessel-Kingman hypergroup of arbitrary index ν. Moreover, if $\nu = \frac{d-2}{2}$ these spaces coincide with the radial Besov spaces $(\dot{B}^s_{p,q})_{rad}(\mathbb{R}^d)$ for $s > 0$, see [BMRM01, Theorema 2.1(ii)].

Recently, Leitner suggested a definition of the minimal character invariant Segal algebra $S_0 = M^{1,1}$ on arbitrary strong hypergroups [Lei]. In particular, his definition applies to Bessel-Kingman hypergroups of arbitrary index $\nu > -1/2$. Although this seems reasonable, it is not completely clear at the moment whether for $\nu = \frac{d-2}{2}$ Leitner's definition coincides with $(S_0)_{rad}(\mathbb{R}^d) = (M^{1,1})_{rad}(\mathbb{R}^d)$. Also it is open whether it is possible to extend his definition to arbitrary modulation spaces $M^{p,q}_s$ on hypergroups.

Recently, Dahlke, Steidl and Teschke generalized coorbit space theory into another direction [DST04a, DST04b]. There are interesting transforms arising from group representations that are not square-integrable in the usual sense, but modulo a (non-compact) subgroup H. This means that the parameter space of the transform $V_g f$ is a homogeneous space $\mathcal{G}/H$ and there exists a vector g such that $V_g f$ is contained in $L^2(\mathcal{G}/H)$ for all f. Examples include the continuous wavelet transform on the sphere introduced by Antoine and Vandergheynst [AV99, AV98], the windowed Fourier transform on the sphere developed by Torresani [Tor95], and a transform related to a representation of the affine Heisenberg group that "interpolates" between the STFT and the CWT [HL95, HL01, KT93, NH03, Tor91].

A convolution on $\mathcal{G}/H$ is no longer available. Nevertheless, one still has a reproducing formula at hand, which allows to carry out the corresponding coorbit space theory. Indeed, Dahlke et al. were able to define modulation spaces on the sphere as an application of coorbit space theory [DST04a].

Together with M. Fornasier the author was recently able to unify classical coorbit space theory and their two available generalizations into one theory [FR04]. Indeed, our starting point is the notion of continuous frame introduced by Ali, Antoine and Gazeau in [AAG93]. Let X be some locally compact space endowed with some Radon measure μ. A family $\{\phi_x\}_{x \in X}$ in a Hilbert space $\mathcal{H}$ indexed by X is called a continuous frame for $\mathcal{H}$ if there exist constants $C_1, C_2 > 0$ such that

$$C_1 \|f\|^2 \leq \int_X |\langle f, \phi_x \rangle|^2 d\mu(x) \leq C_2 \|f\|^2.$$

If X is a discrete set endowed with the counting measure then we obviously get a (Hilbert) frame in the usual sense. Observe further that the family $\{\pi(x)g, x \in \mathcal{G}\}$ forms a continuous frame if g is an admissible vector for a square-integrable representation π, and the same is true for the family $\{\widetilde{\pi}(\mathcal{A}x)g, \mathcal{A}x \in \mathcal{G}^{\mathcal{A}}\}$ from Chapter 2. Also square-integrable representations modulo subgroups fit into this setting.

Once again a reproducing formula is available and it is possible to develop the corresponding coorbit space theory. In this very general setting, some parts get even more technical. In particular, one has to work with two types of coorbit spaces. This is due to the fact that, in general, the (continuous) canonical dual frame does not coincide with the frame itself.

It is worth noting that most of the difficulties appear, when trying to apply this theory to non-classical examples like the continuous wavelet transform on the sphere. Typically it is a highly non-trivial problem to check the integrability conditions. Indeed, Dahlke et al. had to do some numerical computations to check the condition that allowed to define the modulation spaces on the sphere [DST04a].

Closely related to coorbit space theory is a recent concept invented by Gröchenig, called localization of frames [Grö04, CG03a, FG04]. Suppose we are given a (discrete) frame $\{f_i\}_{i \in I}$ for some Hilbert space $\mathcal{H}$. Whenever the matrix $(\langle f_i, f_j \rangle)_{i,j \in I}$ has off-diagonal decay, i.e., belongs to some suitable algebra of matrices over I, then the frame is called intrinsically localized. It is shown in [FG04, Theorem 3.8] that an intrinsically localized frame is automatically a Banach frame for a suitable class of Banach spaces.

Applied to our situation this would mean the following. Observe that the choice of the set U in Theorems 4.6.1–4.6.3 may depend on the weight function w, i.e., the choice of the sampling points $\{x_i\}_{i \in I}$. Thus, the frame elements $\{\widetilde{\pi}(x_i)g\}_{i \in I}$ depend on the space CoY. However, if one knows in addition that the frame $\{\widetilde{\pi}(x_i)g\}_{i \in I}$ for $\mathsf{CoL}_{\mathcal{A}}^2 = \mathcal{H}_{\mathcal{A}}$ is intrinsically localized (with respect to a suitable matrix algebra) then the frame expansion extends to all spaces $\mathsf{CoY}_{\mathcal{A}}$, i.e., one may choose the same frame elements $\{\widetilde{\pi}(x_i)g\}_{i \in I}$ simultaneously for all coorbit spaces.

Let us finally remark, that in [FR04] the concept of localization was extended to continuous frames. Indeed, it serves as a powerful tool to treat the question under which conditions the two types of generalized coorbit spaces associated to a continuous frame coincide.

Appendix

A   Integration on Spheres and on $SO(d)$

In this appendix we collect the necessary information for integrating on the sphere and on the special orthogonal group.

Denote by $S^{d-1} = \{x \in \mathbb{R}^d, |x| = 1\}$ the sphere of radius 1 in $\mathbb{R}^d$ and by $dS = dS^{d-1}$ its surface measure. The surface area of the sphere is $\int_{S^{d-1}} dS = |S^{d-1}| = \frac{2\pi^{d/2}}{\Gamma(d/2)}$ and dS is invariant under rotation, i.e.

$$\int_{S^{d-1}} f(R^{-1}x)dS(x) = \int_{S^{d-1}} f(x)dS(x) \quad \text{for all } R \in O(d), f \in L^1(S^{d-1}, dS).$$

If $d = 1$ then S^0 consists of two points and the integral is a sum. If $d = 2$ then S^1 is isomorphic to the torus $\mathbb{T} = \{z \in \mathbb{C}, |z| = 1\}$ and the integral is given by

$$\int_{S^1} f(x)dS(x) = \int_0^{2\pi} f(e^{it})dt.$$

In the sequel we describe a rule for integration on S^{d-1} for $d \geq 3$. Let $\eta \in S^{d-1}$ be a fixed vector. Then the set

$$S_\eta := \{x \in S^{d-1}, x \cdot \eta = 0\}$$

is isomorphic to S^{d-2} and

$$S^{d-1} = \{t\eta + \sqrt{1-t^2}\zeta, t \in [-1,1], \zeta \in S_\eta\}.$$

The following formula can be found in [Mül98, formula (1.41)]

$$\int_{S^{d-1}} f(x)dS(x) = \int_{-1}^{1} \int_{S_\eta} f(t\eta + \sqrt{1-t^2}\zeta)dS^{d-2}(\zeta)(1-t^2)^{\frac{d-3}{2}}dt. \qquad (A.1)$$

Thus an integral over S^{d-1} can be successively reduced to integrals over spheres of lower dimension.

Let us consider now integration on $SO(d)$. The corresponding measure dR is the Haar measure of $SO(d)$ normalized such that $|SO(d)| = \int_{SO(d)} dR = 1$. It is well-known that the sphere S^{d-1} is a homogeneous space with respect to $SO(d)$, i.e., it holds $S^{d-1} = SO(d)/SO(d-1)$ (where $SO(d-1)$ is identified with the subgroup of $SO(d)$ that fixes a special point $\eta \in S^{d-1}$, for example the north pole). Thus, a function f on the sphere can be interpreted as a function f' on $SO(d)$ that is left invariant under $SO(d-1)$, i.e., $f'(RS) = f'(R)$ for all $R \in SO(d), S \in SO(d-1)$. If $\eta \in S^{d-1}$ is fixed then this identification can be done via

$$f'(R) = f(R\eta), \quad R \in SO(d). \tag{A.2}$$

One immediately verifies that with this definition f' is left invariant under the fixgroup $G_\eta := \{R \in SO(d), R\eta = \eta\} \cong SO(d-1)$. Furthermore, the invariant measure on the homogeneous space $SO(d)/SO(d-1)$ coincides with the surface measure on S^{d-1} (possibly up to multiplication with a constant).

We recall the following theorem concerning invariant measures on homogeneous spaces, which is also known as Weil's formula [Fol95, Theorem 2.49].

Theorem A.1. *Suppose that $\mathcal{G}$ is a locally compact group with Haar measure μ and H a compact subgroup of $\mathcal{G}$ with Haar measure ν. Then there exists an invariant measure ρ on $\mathcal{G}/H$ which is unique up to a constant factor. If this factor is suitably chosen we have*

$$\int_{\mathcal{G}} f(x)d\mu(x) = \int_{\mathcal{G}/H} \int_H f(x\xi)d\nu(\xi)d\rho(xH).$$

From this theorem we conclude that for the function f' on $SO(d)$ given by (A.2) it holds

$$\int_{SO(d)} f'(R)dR = \int_{SO(d)} f(R\eta)dR = |S^{d-1}|^{-1} \int_{S^{d-1}} f(x)dS(x). \tag{A.3}$$

The normalization is due to the fact that we agreed to choose the Haar measure on $SO(d)$ such that $|SO(d)| = 1$.

For a general function on $SO(d)$, Weil's formula implies that one can compute its integral by successive integrations over special orthogonal groups of lower dimension. In the special case $d = 2$ we have the identification $SO(2) \cong S^1 \cong [0, 2\pi)$. Indeed, we may parametrize $SO(2)$ by means of the matrices

$$A(\theta) := \begin{pmatrix} \cos\theta & \sin\theta \\ -\sin\theta & \cos\theta \end{pmatrix} \tag{A.4}$$

and the Haar measure is given by $\int_{SO(2)} f(R)dR = (2\pi)^{-1} \int_0^{2\pi} f(A(\theta))d\theta$.

For $d \geq 3$ the following explicit rule is derived from Weil's formula in [SD80, p.171]. Denote $e_1 = (1, 0, \ldots, 0)^T \in \mathbb{R}^d$ and identify $SO(d-1)$ with the subgroup of $SO(d)$ which fixes e_1. Further, let

$$A(\theta) := \begin{pmatrix} \cos\theta & \sin\theta & 0 & \cdots & 0 \\ -\sin\theta & \cos\theta & 0 & \cdots & 0 \\ 0 & 0 & 1 & \cdots & 0 \\ \vdots & \vdots & \vdots & \ddots & \vdots \\ 0 & 0 & 0 & \cdots & 1 \end{pmatrix}, \qquad \theta \in [0, \pi). \tag{A.5}$$

With this notation it holds [SD80, formula (3.11)]

$$\int_{SO(d)} f(R)dR = \frac{|S^{d-2}|}{|S^{d-1}|} \int_{SO(d-1)} \int_{SO(d-1)} \int_0^\pi f(RA(\theta)S) \sin^{d-2}\theta d\theta dRdS. \tag{A.6}$$

Let us consider the case $d = 3$. The subgroup of $SO(3)$ fixing e_1 is isomorphic to $SO(2)$ and may be parametrized by means of the matrices

$$B(\phi) = \begin{pmatrix} 1 & 0 & 0 \\ 0 & \cos\phi & \sin\phi \\ 0 & -\sin\phi & \cos\phi \end{pmatrix}. \tag{A.7}$$

Formula (A.3) together with the explicit form of the Haar measure on $SO(2)$ yields

$$\int_{SO(3)} f(R)dR = \frac{1}{8\pi^2} \int_0^{2\pi} \int_0^{2\pi} \int_0^\pi f(B(\phi)A(\theta)B(\psi)) \sin\theta d\theta d\phi d\psi. \tag{A.8}$$

We remark that the values (θ, ϕ, ψ) in the decomposition $R = B(\phi)A(\theta)B(\psi)$ of an arbitrary $R \in SO(3)$ are the Euler angles and (A.8) is the corresponding integral formula.

B Other Approaches to Gabor Transforms on Hypergroups

In Chapter 2 we discussed the STFT of radial functions on $\mathbb{R}^d$. In particular, we derived a formula (2.19) for the corresponding integral transform on $\mathbb{R}_+$, which can be interpreted as the STFT on the Bessel-Kingman hypergroup of index $\alpha = \frac{d-2}{2}$. However, there are also some alternative approaches for the STFT on (special classes of) commutative hypergroups present in the literature. We discuss some of their features in this appendix. It turns out that all of these alternative approaches have some severe drawbacks. Thus it seems that our approach for the radial STFT (2.19) can be considered as the most natural one, at least for the Bessel-Kingman hypergroups of index $\frac{d-2}{2}$. However, it remains open to adapt our approach to general hypergroups.

B.1 The Approach of Trimèche and Dachraoui

Suppose $\mathcal{K}$ is a commutative hypergroup with dual objects $\mathcal{X}^b(\mathcal{K}), \widehat{\mathcal{K}}, \mathsf{S}$. Consider the hypergroup translation $T_y, y \in \mathcal{K}$, see (1.52), and the "modulation" operator

$$m_\alpha f(x) = \alpha(x)f(x), \quad \alpha \in \mathcal{X}^b(\mathcal{K}).$$

Since α is bounded by definition of $\mathcal{X}^b(\mathcal{K})$, the operator m_α is continuous on $L^2(\mathcal{K})$, i.e., $\|m_\alpha f\|_2 \leq \|f\|_2$. Now for $g, f \in L^2(\mathcal{K})$ consider the transform

$$W_g f(x, \alpha) = \langle f, m_\alpha T_x g \rangle = \widehat{(f T_x g)}(\alpha), \quad x \in \mathcal{K}, \alpha \in \widehat{\mathcal{K}}, \tag{B.1}$$

which, at first glance, seems to be the natural analogue of the STFT on $\mathbb{R}^d$ (1.64).

In [Dac01] Dachraoui introduced this transform for the special case of the Bessel-Kingman hypergroups and in [Dac03] for the Laguerre hypergroups. Also the windowed spherical Fourier transform introduced by Trimèche in [Tri97] fits into this definition. Indeed, let $\mathcal{G}$ be a locally compact group and K a compact subgroup of $\mathcal{G}$ such that $(\mathcal{G}, K)$ is a Gelfand pair, i.e., the double coset hypergroup $\mathcal{K} := \mathcal{G}//K$ is commutative (see also page 27). The characters α in the dual $\widehat{\mathcal{K}}$ are called the spherical functions of $(\mathcal{G}, K)$. Such an α can be interpreted as a function on $\mathcal{G}$, which is biinvariant under K. The elements of S are the spherical functions, which are positive definite on $\mathcal{G}$. Now, for a function $g \in L^2(\mathcal{G}//K)$ Trimèche defines [Tri97, formula (3.II.2)] the windowed spherical Fourier transform of $f \in L^2(\mathcal{G}//K)$ as

$$\Phi_g f(y, \alpha) := (\overline{\alpha} f) * \overline{g_0}(y), \quad y \in \mathcal{G}, \alpha \in \mathsf{S},$$

where $*$ denotes the convolution on $\mathcal{G}$ and $g_0(x) = g(x^{-1})$. Since the convolution of two biinvariant functions as again biinvariant, $y \mapsto \Phi_g f(y, \alpha)$ can be interpreted as a function on $L^2(\mathcal{G}//K)$, and $\Phi_g f(y, \alpha)$ may be rewritten as

$$
\Phi_g f(KyK, \alpha) = \int_{\mathcal{G}//K} f(KxK) \overline{T_{Ky^{-1}K} g(KxK) \alpha(KxK)} dm(KxK)
$$
$$
= \langle f, \alpha T_{Ky^{-1}K} g \rangle_{L^2(\mathcal{G}//K)},
$$

where $\mathcal{T}$ is the translation on the hypergroup $\mathcal{G}//K$ generated by (1.51). Hence, also the definition of Trimèche is essentially a special case of (B.1).

In case of the Bessel-Kingman hypergroup of index $\nu = \frac{d-2}{2}$, which corresponds to radial analysis in $\mathbb{R}^d$ (and to the Gelfand-pair setting with $\mathcal{G} = \mathbb{R}^d \rtimes SO(d)$ and $K = SO(d)$) there is a connection to the classical STFT on $\mathbb{R}^d$.

Theorem B.1. *Let $f, g \in L^2_{rad}(\mathbb{R}^d)$ with corresponding functions $\mathsf{f}, \mathsf{g} \in L^2(\mathbb{R}_+, \mu_d)$. Then it holds*

$$
W_{\mathsf{g}} \mathsf{f}(r, \lambda) = P_{rad}^{(x)} \operatorname{STFT}_g f(x, \omega) = \frac{1}{|S^{d-1}|} \int_{S^{d-1}} \operatorname{STFT}_g f(|x|\xi, \omega) dS(\xi),
$$

where $x, y \in \mathbb{R}^d, r = |x|, \lambda = |\omega|$.

Proof: Using polar decomposition and Fubini's theorem we obtain

$$
\frac{1}{|S^{d-1}|} \int_{S^{d-1}} \operatorname{STFT}_g f(|x|\xi, \omega) dS(\xi)
$$
$$
= \frac{1}{|S^{d-1}|} \int_{S^{d-1}} \int_{\mathbb{R}^d} f(t) \overline{g(t - |x|\xi)} e^{-2\pi i t \cdot \omega} dt \, dS(\xi)
$$
$$
= \frac{1}{|S^{d-1}|^2} \int_0^\infty \mathsf{f}(s) \int_{S^{d-1}} \int_{S^{d-1}} \overline{g(s\eta - |x|\xi)} dS(\xi) e^{-2\pi i s \eta \cdot \omega} dS(\eta) d\mu_d(s)
$$
$$
= \int_0^\infty \mathsf{f}(s) \overline{\tau_r \mathsf{g}(s)} \mathcal{B}_d(s\lambda) d\mu_d(s) = W_{\mathsf{g}} \mathsf{f}(r, \lambda).
$$

Hereby, we used the definitions (1.21) and (1.34) of the generalized translation τ and the spherical Bessel functions $\mathcal{B}_d$, respectively. $\qquad \square$

Let us now investigate some properties of W_g. Denote by π the Plancherel measure on $\widehat{\mathcal{K}}$, see Theorem 1.4.8. The next theorem establishes the continuity of $W_g : L^2(\mathcal{K}) \to L^2(\mathcal{K} \times \widehat{\mathcal{K}}, m \otimes \pi)$.

Theorem B.2. *Suppose $f, g \in L^2(\mathcal{K})$. Then it holds*

$$
\int_{\mathcal{K}} \int_{\widehat{\mathcal{K}}} |W_g f(x, \alpha)|^2 d\pi(\alpha) dm(x) \leq \|f\|_2^2 \|g\|_2^2.
$$

Proof: Using the Plancherel theorem 1.4.8, commutativity of $\mathcal{K}$, Fubini's theorem and boundedness of T_y on $L^2(\mathcal{K})$ (Theorem 1.4.4) we obtain

$$\int_{\mathcal{K}}\int_{\hat{\mathcal{K}}} |(\widehat{f\overline{T_xg}})(\alpha)|^2 d\pi(\alpha)dm(x) = \int_{\mathcal{K}}\int_{\mathcal{K}} |f(y)\overline{T_xg}(y)|^2 dm(y)dm(x)$$

$$= \int_{\mathcal{K}} |f(y)|^2 \int_{\mathcal{K}} |T_yg(x)|^2 dm(x)dm(y) \leq \int_{\mathcal{K}} |f(y)|^2 \int_{\mathcal{K}} |g(x)|^2 dm(x)dm(y)$$

$$= \|f\|_2^2\|g\|_2^2.$$

$\square$

Next, we provide an inversion formula.

Theorem B.3. *Suppose* $f, g, \gamma \in L^2(\mathcal{K})$ *such that* $\langle T_y\gamma, T_yg \rangle \neq 0$ *for almost all* $y \in \mathcal{K}$. *Then it holds*

$$f(y) = \frac{1}{\langle T_y\gamma, T_yg \rangle} \int_{\mathcal{K}}\int_{\hat{\mathcal{K}}} W_gf(x,\alpha)(m_\alpha T_x\gamma)(y)d\pi(\alpha)dm(x) \quad a.e.. \quad (B.2)$$

Proof: Let $h \in L^2(\mathcal{K})$. Using the Levitan-Plancherel theorem 1.4.8 we obtain

$$\int_{\mathcal{K}}\int_{\hat{\mathcal{K}}} W_gf(x,\alpha)\langle m_\alpha T_x\gamma, h \rangle d\pi(\alpha)dm(x) = \int_{\mathcal{K}}\int_{\hat{\mathcal{K}}} \overline{(\widehat{f\overline{T_xg}})(\alpha)}\overline{(\widehat{h\overline{T_x\gamma}})(\alpha)}d\pi(\alpha)dm(x)$$

$$= \int_{\mathcal{K}}\int_{\mathcal{K}} (f\overline{T_xg})(z)(\overline{h}T_x\gamma)(z)dm(z)dm(x)$$

$$= \int_{\mathcal{K}} f(z)\overline{h(z)} \int_{\mathcal{K}} T_z\gamma(x)\overline{T_zg(x)}dm(x)dm(z) = \int_{\mathcal{K}} f(z)\langle T_z\gamma, T_zg \rangle\overline{h(z)}dm(z).$$

Hereby, we used also the commutativity of $\mathcal{K}$ and Fubini's theorem. Since h was arbitrary we conclude that (B.2) holds. $\square$

Special cases of this theorem were proved in [Dac01, Dac03]. We remark that the condition on γ, g is satisfied if $g = \gamma$ is a positive function. Moreover, in case that $\mathcal{K}$ is a group (for example $\mathcal{K} = \mathbb{R}^d$) we have $\langle T_yg, T_y\gamma \rangle = \langle g, \gamma \rangle$ and we get the well-known inversion formula for the STFT (1.66). However, for a general hypergroup the expression $\langle T_y\gamma, T_yg \rangle$ is not constant with respect to $y \in \mathcal{K}$. In fact, it may happen that $y \mapsto \|T_yg\|$ is not bounded away from zero and this leads to instabilities in the reconstruction. Indeed, in case of the Bessel-Kingman hypergroups of index $\nu > -1/2$ (which includes the situation of radial functions in $\mathbb{R}^d, d \geq 2$) we have the following result. Let us first introduce the dilation operator

$$D_af(r) = a^{-(\nu+1)}f(a^{-1}r), \quad a > 0, r \in \mathbb{R}_+.$$

On $L^2(\mathcal{K}_\nu) = L^2(\mathbb{R}_+, \mu_{(\nu)})$ it is isometric, i.e., $\|D_af\|_2 = \|f\|_2$.

Theorem B.4. *Let $\mathcal{K}_\nu$ denotes the Bessel-Kingman hypergroup of index $\nu > -1/2$.*

(a) For all $\mathsf{f}, \mathsf{g} \in L^2(\mathcal{K}_\nu)$ we have

$$\lim_{a \to \infty} \|W_{\mathsf{g}}(D_a \mathsf{f})| L^2(\mathbb{R}_+ \times \mathbb{R}_+, \mu_{(\nu)} \otimes \mu_{(\nu)})\| = 0$$

(b) For all $\mathsf{f}, \mathsf{g} \in L^2(\mathcal{K}_\nu)$ we have

$$\lim_{a \to 0} \|W_{\mathsf{g}}(D_a \mathsf{f})| L^2(\mathbb{R}_+ \times \mathbb{R}_+, \mu_{(\nu)} \otimes \mu_{(\nu)})\| = \|\mathsf{f}\|_2 \|\mathsf{g}\|_2.$$

<u>**Proof:**</u> The proof of Theorem B.2 showed that

$$\|W_{\mathsf{g}}(D_a \mathsf{f})| L^2(\mathbb{R}_+ \times \mathbb{R}_+, \mu_{(\nu)} \otimes \mu_{(\nu)})\|^2 = \int_0^\infty |D_a \mathsf{f}(r)|^2 \|\tau_r \mathsf{g}\|_2^2 d\mu_{(\nu)}(r).$$

With the substitution $r \mapsto ar$ we obtain

$$\int_0^\infty |D_a \mathsf{f}(r)|^2 \|\tau_r \mathsf{g}\|_2^2 d\mu_{(\nu)}(r) = \int_0^\infty |\mathsf{f}(r)|^2 \|\tau_{ar} \mathsf{g}\|_2^2 d\mu_{(\nu)}(r).$$

Hence, we have to investigate the behaviour of $\|\tau_{ar} \mathsf{g}\|_2^2$ when a tends to zero or to infinity. The Plancherel-Levitan theorem 1.4.8 together with Theorem 1.4.7(c) yields

$$\|\tau_{ar} \mathsf{g}\|_2^2 = \int_0^\infty |\mathcal{B}_{(\nu)}(ar\lambda)|^2 |\widehat{\mathsf{g}}(\lambda)|^2 d\mu_{(\nu)}(\lambda)$$

Further, it is well known (see e.g. [Fol92, Theorem 5.1]) that there exists a constant C_α such that

$$\mathcal{B}_{(\nu)}(r) \leq \min\{1, C_\alpha r^{-(\nu+1/2)}\}.$$

Hence, $\lim_{a \to \infty} \mathcal{B}_{(\nu)}(ar) = 0$ for all $r > 0$ and with Lebesgue's dominated convergence theorem we get

$$\lim_{a \to \infty} \int_0^\infty |\mathcal{B}_{(\nu)}(ar\lambda)|^2 |\widehat{\mathsf{g}}(\lambda)|^2 d\mu_{(\nu)}(\lambda) = 0$$

for all $r > 0$. Using Lebesgue's dominated convergence theorem once more we finally obtain

$$\lim_{a \to \infty} \int_0^\infty |\mathsf{f}(r)|^2 \|\tau_{ar} \mathsf{g}\|_2^2 d\mu_{(\nu)}(r) = 0.$$

We further note that $\lim_{a \to 0} \mathcal{B}_{(\nu)}(ar) = 1$ for all $r \geq 0$ and by similar arguments as above we obtain

$$\lim_{a \to 0} \int_0^\infty |\mathsf{f}(r)|^2 \|\tau_{ar} \mathsf{g}\|_2^2 d\mu_{(\nu)}(r) = \|\mathsf{f}\|_2^2 \|\mathsf{g}\|_2^2.$$

This shows (b). $\qquad\qquad\qquad\qquad\qquad\qquad\qquad\qquad\qquad\qquad\qquad\qquad$ $\square$

From part (a) it follows immediately that the reconstruction operator (see Theorem B.3)

$$W_\gamma^\bullet : L^2(\mathcal{K}_\nu \times \widehat{\mathcal{K}_\nu}) \supset \mathcal{D}(W_\gamma^\bullet) \to L^2(\mathcal{K}_\nu),$$

$$W_\gamma^\bullet(F)(s) = \frac{1}{\langle \tau_s g, \tau_s \gamma \rangle} \int_0^\infty \int_0^\infty F(r,\lambda) m_\lambda \tau_r \gamma(s) d\mu_{(\nu)}(\lambda) d\mu_{(\nu)}(r)$$

is **unbounded** (for any choice of $\gamma \in L^2(\mathcal{K}_\nu)$), which means that inversion of $W_g f$ is unstable. This is a very severe drawback of W_g and, thus, we have a strong argument not to consider the transform W_g as a suitable analogue of the STFT.

In case $\nu = \frac{d-2}{2}$ the transform W_g coincides with the windowed spherical Fourier transform for the Gelfand pair $(\mathbb{R}^d \rtimes SO(d), SO(d))$. Trimèche claimed [Tri97, Proposition 3.II.1] that there is a Plancherel (Moyal) formula for general windowed spherical Fourier transforms. Part (a) of Theorem B.4 shows, however, that Trimèche's Proposition is false! (His "proof" uses Proposition 3.II.1(i) in [Tri97], which is easily seen to be false.)

Furthermore, there is no analogue to the fundamental identity of time-frequency analysis (1.67) for W_g. Indeed, for a general hypergroup and in particular, for the Bessel-Kingman hypergroups, the operators $m_\lambda \tau_r$ do not obey a canonical commutation relation like the classical modulation and translation operators (1.63). This means that

$$m_\lambda \tau_r \neq C(\lambda, r) \tau_r m_\lambda \qquad \text{for all } r, \lambda > 0$$

and for any choice of $C(\lambda, r) \in \mathbb{C}$. This is mainly due to the fact that in general $\tau_r(fg) \neq \tau_r(f)\tau_r(g)$. Indeed, the crucial ingredient for the proof of the fundamental identity of time-frequency analysis is exactly a commutation relation like (1.63) so that there is **no** function $C(r,\lambda)$ (independent of g, f) such that $W_g f(r,\lambda) = C(r,\lambda) W_{\widehat{g}} \widehat{f}(\lambda, r)$. We remark that these rough arguments can be made precise by computing $W_g f$ and $W_{\widehat{g}} \widehat{f}$ for special choices of f, g.

B.2   The Approach of Czaja and Gigante

In [CG03b] Czaja and Gigante proposed the following approach. Assume that $\mathcal{K}$ is a strong commutative hypergroup, which means that also $\widehat{\mathcal{K}}$ is a hypergroup. This implies, in particular, that the support S of the Plancherel measure π coincides with $\mathcal{K}$ and π is the Haar measure on $\widehat{\mathcal{K}}$. Of course, Pontryagin hypergroups are strong. Thus, Bessel-Kingman hypergroups provide examples of strong hypergroups. Although not stated explicitly it is assumed further in [CG03b] that π is σ-finite, i.e., $\widehat{\mathcal{K}}$ is σ-compact. Denote by $\mathcal{T}$ the translation on $\mathcal{K}$ and by τ the translation on $\widehat{\mathcal{K}}$. For a character $\alpha \in \widehat{\mathcal{K}}$ and a function $g \in L^2(\mathcal{K})$ the modulation $\mathcal{M}_\alpha$ is defined by

$$\mathcal{M}_\alpha g := \left(\sqrt{\tau_\alpha(|\widehat{g}|^2)} \right)^\vee.$$

Hereby, $\hat{}$ denotes the Fourier transform (1.56) on $\mathcal{K}$ and $\vee$ the inverse Fourier transform (1.57). If $\mathcal{K}$ is a group and $\hat{g}$ positive then $\mathcal{M}_\alpha g = \overline{\alpha} g$ is a usual modulation. However, on a general hypergroup $\mathcal{M}_\alpha$ is not a multiplication operator by a character any more. Indeed, $\mathcal{M}_\alpha$ is not even a linear operator!

Now, for $f, g \in L^2(\mathcal{K})$ the generalized continuous Gabor transform is defined by

$$\Psi_g f(y, \alpha) := \langle f, \mathcal{T}_{\bar{y}} \mathcal{M}_\alpha g \rangle = \int_{\mathcal{K}} f(x) \overline{(\mathcal{T}_{\bar{y}} \mathcal{M}_\alpha g)(x)} dm(x).$$

Czaja and Gigante were able to prove a Plancherel (Moyal) formula and an inversion formula for Ψ_g.

Theorem B.5. *[CG03b, Theorem 3.4] Let $g \in L^2(\mathcal{K})$ with $\|g\|_2 = 1$. Then for all $f \in L^2(\mathcal{K})$ it holds*

$$\|\Psi_g f | L^2(\mathcal{K} \times \widehat{\mathcal{K}}, m \otimes \pi)\| = \|f | L^2(\mathcal{K})\|.$$

Theorem B.6. *[CG03b, Theorem 3.6] Let $g \in L^2(\mathcal{K})$ with $\|g\|_2 = 1$ and $(\widehat{\mathcal{K}}_n)_n$ a sequence of subsets $\widehat{\mathcal{K}}_n \subset \widehat{\mathcal{K}}$ with $\widehat{\mathcal{K}}_n \subset \widehat{\mathcal{K}}_{n+1}$, $\pi(\widehat{\mathcal{K}}_n) < \infty$ and $\cup_{n=1}^\infty \widehat{\mathcal{K}}_n = \widehat{\mathcal{K}}$ (recall that π is assumed to be σ-finite). Then for any $n \in \mathbb{N}$ the function*

$$f^n(x) = \int_{\widehat{\mathcal{K}}_n} \int_X \Psi_g f(y, \alpha) \mathcal{T}_{\bar{y}} \mathcal{M}_\alpha g(x) dm(y) d\pi(\alpha)$$

belongs to $L^2(\mathcal{K})$ and satisfies

$$\lim_{n \to \infty} \|f^n - f\|_2 = 0.$$

Despite these theorems the author believes that also the transform Ψ_g is not a suitable analogue for the Gabor transform on $\mathbb{R}^d$. One reason is that the modulation operator $\mathcal{M}_\alpha$ is non-linear. Moreover, if $\mathcal{K}$ is a Pontryagin hypergroup then one may also define the transform Ψ_g on $\widehat{\mathcal{K}}$. The question arises whether an analogue of the fundamental identity of time-frequency analysis (1.67) holds. However, bearing the formula for $\mathcal{M}_\alpha$ in mind there is no hope that such an identity might hold in general.

Symbols

$\mathcal{A}$	compact automorphism group
$\mathbb{A}_w^{\mathcal{A}}$	space of invariant analyzing vectors, p. 111
$A(r,h)$	annulus of radius r and width h, p. 147
$\mathcal{B}_d$	spherical Bessel function, p. 19
$\mathbb{B}_w^{\mathcal{A}}$	'better' space of invariant analyzing vectors, p. 120
$\dot{B}_{p,q}^s$	homogeneous Besov space, p. 154
$(\dot{B}_{p,q}^s)_{rad}$	its subspace of radial distributions
$B(x,r)$	closed ball of radius r centered at x
$\mathbb{C}$	complex numbers
$C(X)$	continuous functions on some locally compact space X
$C^b(X)$	bounded continuous functions
$C_0(X)$	continuous functions vanishing at infinity
$C_c(X)$	continuous functions with compact support
$\mathsf{Co}Y$	coorbit space with respect to Y, p. 119
$\mathsf{Co}Y_{\mathcal{A}}$	subspace of $\mathsf{Co}Y$ of invariant elements, p. 119
CWT_ψ	continuous wavelet transform, p. 38
dS, dS^{d-1}	surface measure on the sphere S^{d-1}, p. 171
Δ	modular function of $\mathcal{G}$
D_a	dilation operator, p. 38
$\mathcal{D}(K)$	admissible vectors, i.e. domain of Duflo-Moore operator K, p. 41
$\mathcal{D}_{\mathcal{A}}(K)$	admissible invariant vectors, p. 51
ϵ_x	Dirac measure at x
$\epsilon_{\mathcal{A}x}$, $\epsilon_{(r)}$	invariant Dirac measure, p. 11; "radial" Dirac measure, p. 16
$\mathcal{F}$	Fourier transform, p. 18
$\dot{F}_{p,q}^s$	homogeneous Triebel-Lizorkin space, p. 154
$(\dot{F}_{p,q}^s)_{rad}$	its subspace of radial distributions
$\mathcal{G}$	locally compact, σ-compact group
$\mathcal{G}^{\mathcal{A}}$	space of orbits under $\mathcal{A}$, p. 9
$G_U^{\#}$	U-oscillation of a function G, p. 110
Γ	Gamma function
$\mathcal{H}$	Hilbert space

$\mathcal{H}_\mathcal{A}$	subspace of $\mathcal{H}$ of invariant elements, p. 48
$(\mathcal{H}_w^1)_\mathcal{A}$	$(\mathcal{H}_w^1)_\mathcal{A} = \mathrm{Co}(L_w^1)_\mathcal{A}$, p. 112
$(\mathcal{H}_w^1)_\mathcal{A}^{\urcorner}$	anti-dual of $(\mathcal{H}_w^1)_\mathcal{A}$, p. 113
Id	identity matrix, identity operator
K	Duflo-Moore operator, p. 41
$\mathcal{K}$	hypergroup, orbit hypergroup $\mathcal{G}^\mathcal{A}$
L_x	left translation, p. 9
$\mathcal{L}_x$	generalized left translation, p. 11
L^p, L_w^p	Lebesgue space, weighted Lebesgue space, p. 2
$L_{rad}^p(\mathbb{R}^d)$	radial square-integrable functions on $\mathbb{R}^d$
$L_s^{p,q}$	mixed norm space, p. 138 or p. 156
L_{loc}^1	locally integrable functions, p. 3
l^p, l_w^p	Lebesgue sequence space, weighted Lebesgue sequence space
m	Haar measure on $\mathcal{G}^\mathcal{A}$, $\mathcal{K}$, p. 9
M, $M(X)$	complex regular Borel measures on X, p. 2
M_ξ	modulation operator, p. 36
$M_s^{p,q}$, $(M_s^{p,q})_{rad}$	modulation space, its subspace of radial distributions, p. 139
μ	Haar measure on $\mathcal{G}$
μ_d	measure on $\mathbb{R}_+$: $d\mu_d(r) = |S^{d-1}|r^{d-1}dr$, p. 14
$\mathbb{N}$, $\mathbb{N}_0$	natural numbers, natural numbers including 0
$O(d)$	orthogonal group in dimension d
Ω	generalized combined translation and modulation, p. 58
p_K	semi-norms defining the topology of L_{loc}^1, p. 3
π	unitary irreducible representation of $\mathcal{G}$ on $\mathcal{H}$
$\widetilde{\pi}(x)$	operator on invariant elements, p. 49
$P_\mathcal{A}$, P_{rad}	projection onto invariant, resp. radial, functions, p. 6, p. 14
$\mathbb{R}$, $\mathbb{R}^d$	real numbers, d-dimensional Euclidean space
$\mathbb{R}_+$	positive half line $[0,\infty)$
$\mathbb{R}_+^*$	multiplicative group $(0,\infty)$
R_x	right translation, p. 9
$\mathcal{R}_x$	generalized right translation, p. 11
supp	support of a function, measure, distribution
σ	unitary representation of $\mathcal{A}$ on $\mathcal{H}$, p. 47
S^{d-1}, $|S^{d-1}|$	unit sphere in $\mathbb{R}^d$, its surface area, p. 171
$SO(d)$	special orthogonal group in dimension d
$\mathcal{S}(\mathbb{R}^d)$	Schwartz space, p. 3
$\mathcal{S}'(\mathbb{R}^d)$	tempered distributions, p. 3
$\mathcal{S}^\infty(\mathbb{R}^d)$	Schwartz functions with infinitely many vanishing moments, p. 160
$\mathcal{S}_{rad}, \mathcal{S}'_{rad}, \mathcal{S}_{rad}^\infty$	their subspaces of radial functions, resp. distributions
S_0, $(S_0)_{rad}$	Feichtinger algebra, its subspace of radial functions, p. 139
STFT_g	short time Fourier transform, p. 36

τ_r	generalized translation for radial function, p. 15
$T^{(k)}$	$T^{(k)} = \tau_{k/2}$, p. 69
T_x	translation operator on $\mathbb{R}^d$, p. 16
$\mathcal{T}_x$	generalized translation, p. 11
$T_s^{p,q}$	tent spaces, p. 156
$U_R, U_{\mathcal{A}}$	rotation operator, p. 38; action of $\mathcal{A}$ on functions on $\mathcal{G}$, p. 4
V_g	(abstract) wavelet transform, resp. voice transform, p. 41
$\widetilde{V}_g$	restriction of V_g to $\mathcal{H}_{\mathcal{A}}$, p. 49
$(\widetilde{V}_g f)_U^{\#}$	U-oscillation of $\widetilde{V}_g f$, p. 110
$V(x)$	a cone in $\mathbb{R}^d \times (0, \infty)$ centered at x, p. 156
$W(B, Y)$	Wiener amalgam space on $\mathcal{G}$, p. 105
$W_{\mathcal{A}}(B, Y)$	its subspace of elements, which are invariant under $\mathcal{A}$
(Y), (YA)	property (Y), p. 93; property (YA), p. 95
$Y, Y_{\mathcal{A}}$	function space on $\mathcal{G}$, its subspace of invariant functions
$Y_{\mathcal{A}}^{\natural}, Y_{\mathcal{A}}^{\flat}$	sequence spaces associated to Y, p. 100
$\mathbb{Z}$	integers
$\sim$	$F(x) \sim G(x) \iff \lim_{x \to \infty} \frac{F(x)}{G(x)} = 1$
$\asymp$	$A \asymp B \iff C_1 A \leq B \leq C_2 A$, p. 4
$\lvert x \rvert$	Euclidean norm of $x \in \mathbb{R}^d$
$x \cdot y$	scalar product of $x, y \in \mathbb{R}^d$
$\lfloor x \rfloor$	biggest number $k \in \mathbb{Z}$ such that $k \leq x$
$\lceil x \rceil$	smallest number $k \in \mathbb{Z}$ such that $k \geq x$
$\lVert \cdot \lvert B \rVert$	norm of the Banach space B
$\lVert \cdot \lvert B \rVert$	norm of an operator from a Banach space B into itself
$\lVert \cdot \lvert B_1 \to B_2 \rVert$	norm of an operator from B_1 into B_2, p. 4
$\langle \cdot, \cdot \rangle$	inner product in a Hilbert space,
	duality between some vector space and its dual
$\widehat{f}$	Fourier transform of f, p. 18 or p. 32
$\widehat{f}, H_d(f)$	Hankel transform of f, p. 19
$\overline{F}$	complex conjugate of F
F^*	$F^*(x) = \Delta(x^{-1}) \overline{F(x^{-1})}$
$F^{\vee}$	$F^{\vee}(x) = F(x^{-1})$
$F^{\triangledown}$	$F^{\triangledown}(x) = \overline{F(x^{-1})}$
$w^{\bullet}$	$w^{\bullet}(x) = w(x) + \Delta(x^{-1}) w(x^{-1})$
$\lvert U \rvert$	(Haar) measure of a set U
U°	interior of a set U
$\overline{U}$	closure of a set U

Bibliography

[AAG93]   S.T. Ali, J.P. Antoine, and J.P. Gazeau. Continuous frames in Hilbert spaces. *Annals of Physics*, 222:1–37, 1993.

[AAG00]   S.T. Ali, J.P. Antoine, and J.P. Gazeau. *Coherent States, Wavelets and Their Generalizations*. Springer, New York, 2000.

[AAR99]   E.A. Andrews, R. Askey, and R. Ranjan. *Special Functions*. Cambridge University Press, 1999.

[ACM03]   A. Aldroubi, C. Cabrelli, and U.M. Molter. Wavelets on irregular grids with arbitrary dilation matrices, and frame atoms for $L^2(\mathbb{R}^d)$. *Preprint*, 2003.

[AM88]   J. Alvarez and M. Milman. Interpolation of tent spaces and applications. In *Function Spaces and Applications, Proc. US-Swed. Semin., Lund/Swed.*, volume 130 of *Lect. Notes Math.*, pages 11–21. 1988.

[AV98]   J.P. Antoine and P. Vandergheynst. Wavelets on the n-sphere and related manifolds. *J. Math. Phys.*, 39:3987–4008, 1998.

[AV99]   J.P. Antoine and P. Vandergheynst. Wavelets on the 2-sphere: A group-theoretical approach. *Appl. Comput. Harmon. Anal.*, 7:262–291, 1999.

[BH94]   W.R. Bloom and H. Heyer. *Harmonic Analysis of Probability Measures on Hypergroups*. de Gruyter, Berlin, New York, 1994.

[BJR98]   C. Benson, J. Jenkins, and G. Ratcliff. The spherical transform of a Schwartz function on the Heisenberg group. *J. Funct. Anal.*, 154:379–423, 1998.

[BMRM01]   J.J. Betancor, J.M. Méndez, and L. Rodríguez-Mesa. Espacios de Besov asociados a la transformación de Hankel. *Margarita mathematica, Univ. La Rioja*, pages 661–675, 2001.

[Böl03] H. Bölcskei. Orthogonal frequency division multiplexing based on offset QAM. In *Feichtinger, H.G. (ed.) et al., Advances in Gabor Analysis*, pages 321–352. Birkhäuser, Boston, 2003.

[Bon86] F.F. Bonsall. Decomposition of functions as sums of elementary functions. *Quart. J. Math. Oxford*, 37(2):129–136, 1986.

[BR96] C. Benson and G. Ratcliff. A classification of multiplicity free actions. *J. Algebra*, 181:152–186, 1996.

[BRM98] J.J. Betancor and L. Rodríguez-Mesa. On the Besov-Hankel spaces. *J. Math. Soc. Japan*, 50:781–788, 1998.

[BT96] D. Bernier and K.F. Taylor. Wavelets from sqare-integrable representations. *SIAM J. Math. Anal.*, 27(2):594–608, 1996.

[CCS02] P.G. Casazza, O. Christensen, and D.T. Stoeva. Frame expansions in separable Banach spaces. *Preprint*, 2002.

[CG03a] E. Cordero and K. Gröchenig. Localization of frames II. *Preprint, to appear in Appl. Comput. Harmon. Anal.*, 2003.

[CG03b] W. Czaja and G. Gigante. Continuous Gabor transform for strong hypergroups. *J. Fourier Anal. Appl.*, 9:321–339, 2003.

[Chr03] O. Christensen. *An Introduction to Frames and Riesz Bases*. Birkhäuser, Boston, 2003.

[CMS85] R.R. Coifman, Y. Meyer, and E.M. Stein. Some new function spaces and their applications to harmonic analysis. *J. Funct. Anal.*, 62:304–335, 1985.

[CT75] A.P. Calderón and A. Torchinsky. Parabolic maximal functions associated with a distribution. *Adv. in Math.*, 16:1–64, 1975.

[CV00] W.S. Cohn and I.E. Verbitsky. Factorization of tent spaces and Hankel operators. *J. Funct. Anal.*, 175:308–329, 2000.

[Dac01] A. Dachraoui. Weyl-Bessel transforms. *J. Comput. Appl. Math.*, 133:263–276, 2001.

[Dac03] A. Dachraoui. Weyl transforms associated with Laguerre functions. *J. Computat. Appl. Math.*, 153:151–162, 2003.

[Dau92] I. Daubechies. *Ten Lectures on Wavelets*. SIAM, 1992.

[DFG02]   M. Dörfler, H.G. Feichtinger, and K. Gröchenig. Compactness criteria in function spaces. *Colloq. Math.*, 94(1):37–50, 2002.

[DGM86]   I. Daubechies, A. Grossmann, and Y. Meyer. Painless nonorthogonal expansions. *J. Math. Phys.*, 27:1271–1283, 1986.

[Dix82]   J. Dixmier. *C*-Algebras*. North-Holland, Amsterdam – New York – Oxford, 1982.

[DM76]   M. Duflo and C.C. Moore. On the regular representation of a nonunimodular locally compact group. *J. Funct. Anal.*, 21:209–243, 1976.

[DS52]   R.J. Duffin and A.C. Schaefer. A class of nonharmonic Fourier series. *Trans. Amer. Math. Soc.*, 72:341–366, 1952.

[DST04a]   S. Dahlke, G. Steidl, and G. Teschke. Coorbit spaces and Banach frames on homogeneous spaces with applications to the sphere. *Adv. Comput. Math.*, 21:147–180, 2004.

[DST04b]   S. Dahlke, G. Steidl, and G. Teschke. Weighted coorbit spaces and Banach frames on homogeneous spaces. *J. Fourier Anal. Appl.*, 10:507–539, 2004.

[Dun73]   C.F. Dunkl. The measure algebra of a locally compact hypergroup. *Trans. Amer. Math. Soc.*, 179:331–348, 1973.

[Edw65]   R.E. Edwards. *Functional Analysis*. Holt, Rinehart and Winston, 1965.

[EF95]   J. Epperson and M. Frazier. An almost orthogonal radial wavelet expansion for radial distributions. *J. Fourier Anal. Appl.*, 1:311–353, 1995.

[EG67]   W.R. Emerson and F.P. Greenleaf. Covering properties and Følner conditions for locally compact groups. *Math. Zeitschr.*, 102:370–384, 1967.

[Fei79]   H.G. Feichtinger. Gewichtsfunktionen auf lokalkompakten Gruppen. *Sitzungsber. Öster. Akad. Wissensch.*, 1979.

[Fei81a]   H.G. Feichtinger. A characterization of minimal homogeneous Banach spaces. *Proc. Am. Math. Soc.*, 81(1):55–61, 1981.

[Fei81b]   H.G. Feichtinger. Banach spaces of distributions of Wiener's type and interpolation. In *Functional Analysis and Approximation, Proc. Conf., Oberwolfach 1980, ISNM 60*, pages 153–165. 1981.

[Fei81c]   H.G. Feichtinger. On a new Segal algebra. *Monatsh. Math.*, 92:269–289, 1981.

[Fei83a]   H.G. Feichtinger. A new family of functional spaces on the Euclidean
            n-space. 1983.

[Fei83b]   H.G. Feichtinger. Banach convolution algebras of Wiener's type. In *Func-
            tions, Series, Operators, Proc. int. Conf., Budapest 1980, Vol. I, Colloq.
            Math. Soc. Janos Bolyai 35*, pages 509–524. 1983.

[Fei83c]   H.G. Feichtinger. Modulation spaces on locally compact groups. Technical
            report, Vienna, 1983.

[Fei89]   H.G. Feichtinger. Atomic characterizations of modulation spaces through
            Gabor-type representations. *Rocky Mt. J. Math.*, 19(1):113–126, 1989.

[Fei90]   H.G. Feichtinger. Generalized amalgams, with applications to Fourier
            transform. *Can. J. Math.*, 42(3):395–409, 1990.

[FG88]   H.G. Feichtinger and K. Gröchenig. A unified approach to atomic decom-
            positions via integrable group representations. In *Function Spaces and
            Applications, Proc. US-Swed. Semin., Lund/Swed.*, volume 1302 of *Lect.
            Notes Math.*, pages 52–73. Springer, 1988.

[FG89a]   H.G. Feichtinger and K. Gröchenig. Banach spaces related to integrable
            group representations and their atomic decompositions. Part I. *J. Funct.
            Anal.*, 21:307–340, 1989.

[FG89b]   H.G. Feichtinger and K. Gröchenig. Banach spaces related to integrable
            group representations and their atomic decompositions. Part II. *Monatsh.
            Math.*, 108:129–148, 1989.

[FG92a]   H.G. Feichtinger and K. Gröchenig. Gabor wavelets and the Heisenberg
            group: Gabor expansions and short time Fourier transform from the group
            theoretical point of view. In *Chui, C.K. (ed.): Wavelets: a tutorial in
            theory and applications*, pages 359–397. 1992.

[FG92b]   H.G. Feichtinger and K. Gröchenig. Iterative reconstruction of multivariate
            band-limited functions from irregular sampling values. *SIAM J. Math.
            Anal.*, 23(1):244–261, 1992.

[FG92c]   H.G. Feichtinger and K. Gröchenig. Non-orthogonal wavelet and Gabor
            expansions, and group representations. In *Ruskai, Mary Beth (ed.) et
            al., Wavelets and their Applications*, pages 353–375. Jones and Bartlett
            Publishers, Boston, 1992.

[FG04]   M. Fornasier and K. Gröchenig. Intrinsic localization of frames. *Preprint,
            to appear in Constructive Approximation*, 2004.

[Fil00]   F. Filbir. *Approximationsprozesse auf kommutativen Hypergruppen.* Shaker Verlag, 2000.

[FJ85]   M. Frazier and B. Jawerth. Decomposition of Besov spaces. *Indiana Univ. Math. J.*, 34:777–799, 1985.

[FJ88]   M. Frazier and B. Jawerth. The ϕ-transform and applications to distribution spaces. In *Cwikel, M. (ed.) et al., Function Spaces and Applications*, volume 1302 of *Lecture Notes in Math.*, pages 223–246. Springer, 1988.

[FJ90]   M. Frazier and B. Jawerth. A discrete transform and decompositions of distribution spaces. *J. Funct. Anal.*, 93:34–170, 1990.

[FJW91]   M. Frazier, B. Jawerth, and G. Weiss. *Littlewood-Paley Theory and the Study of Function Spaces.* AMS, 1991.

[FN03]   H.G. Feichtinger and K. Nowak. A first survey of Gabor multipliers. In *Feichtinger, H.G. (ed.) et al., Advances in Gabor Analysis*, pages 99–128. Birkhäuser, Basel, 2003.

[Fol92]   G.B. Folland. *Fourier Analysis and its Applications.* Brooks/Cole, 1992.

[Fol95]   G.B. Folland. *A Course in Abstract Harmonic Analysis.* CRC Press, 1995.

[FR04]   M. Fornasier and H. Rauhut. Continuous frames, function spaces, and the discretization problem. *Preprint*, 2004.

[FS72]   C. Fefferman and E.M. Stein. H^p spaces of several variables. *Acta Math.*, 129:137–193, 1972.

[FS97]   G.B. Folland and A. Sitaram. The uncertainty principle, a mathematical survey. *J. Fourier Anal. Appl.*, 66:207–238, 1997.

[FZ98]   H.G. Feichtinger and G. Zimmermann. A Banach space of test functions for Gabor analysis. In *Feichtinger, H.G. (ed.) et al., Gabor Analysis and Algorithms. Theory and Applications.*, pages 123–170. Birkhäuser, Boston, MA, 1998.

[GB85]   L.C. Grove and C.T. Benson. *Finite Reflection Groups.* Springer, 1985.

[GMP85]   A. Grossmann, J. Morlet, and T. Paul. Transforms associated to square integrable group representations. I. General results. *J. Math. Phys.*, 26(10), 1985.

[GMP86]   A. Grossmann, J. Morlet, and T. Paul. Transforms associated to square
           integrable group representations. II. Examples. *Ann. Inst. Poincaré Phys.*
           Théor., 45(3), 1986.

[Grö88]   K. Gröchenig. Unconditional bases in translation and dilation invariant
           function spaces on $\mathbb{R}^n$. In *Sendov, B. (ed.) et al., Constructive Theory of*
           Functions, Proc. Conf. Varna 1987, pages 174–183. 1988.

[Grö91]   K. Gröchenig. Describing functions: Atomic decomposition versus frames.
           Monatsh. Math., 112:1–41, 1991.

[Grö01]   K. Gröchenig. *Foundations of Time-Frequency-Analysis.* Birkhäuser,
           Boston, 2001.

[Grö04]   K. Gröchenig. Localization of frames, Banach frames, and the invertibility
           of the frame operator. *J. Fourier Anal. Appl.*, 10(2):105–132, 2004.

[GS04]   Y.V. Galperin and S. Samarah. Time-frequency analysis on modulation
           spaces $M_m^{p,q}$, $0 < p, q \leq \infty$. *Appl. Comput. Harmon. Anal.*, 16:1–18, 2004.

[GV64]   I.M. Gelfand and N.J. Vilenkin. *Verallgemeinerte Funktionen (Distribu-*
           tionen) IV. VEB, Berlin, 1964.

[Her92]   P. Hermann. Induced representations of hypergroups. *Math. Z.*, 211:687–
           699, 1992.

[Her95]   P. Hermann. Representations of double coset hypergroups and induced
           representations. *Manuscripta Math.*, 88:1–24, 1995.

[Hin00]   J. Hinz. Hypergroup actions and wavelets. In *H. Heyer (ed.) et al., Infi-*
           nite Dimensional Harmonic Analysis (Japanese-German Symposium Ky-
           oto 1999), pages 167–176. Gräbner Verlag, 2000.

[HL95]   J.A. Hogan and J.D. Lakey. Extensions of the Heisenberg group by dila-
           tions and frames. *Appl. Comput. Harmon. Anal.*, 2:174–199, 1995.

[HL01]   J.A. Hogan and J.D. Lakey. Embeddings and uncertainty principles for
           generalized modulation spaces. In *Benedetto, J.J. (ed.) et al.: Modern*
           Sampling Theory, pages 73–105. Birkhäuser, Basel, 2001.

[Hol95]   M. Holschneider. *Wavelets, An Analysis Tool.* Oxford University Press,
           1995.

[Hör83]   L. Hörmander. *The Analysis of Linear Partial Differential Operators I.*
           Springer, 1983.

[HW89]   C.E. Heil and D.F. Walnut. Continuous and discrete wavelet transforms. *SIAM Review*, 31(4):628–666, 1989.

[Jew75]   R.I. Jewett. Spaces with an abstract convolution of measures. *Adv. Math.*, 18:1–101, 1975.

[Kin65]   J.F.C. Kingman. Random walks with spherical symmetry. *Acta Math.*, 109:11–53, 1965.

[KLSS03]   T. Kühn, H. Leopold, W. Sickel, and L. Skrzypczak. Entropy numbers of Sobolev embeddings of radial Besov spaces. *J. Approx. Theory*, 121:244–268, 2003.

[KT93]   C. Kalisa and B. Torrésani. *N*-dimensional affine Weyl-Heisenberg wavelets. *Ann. Inst. H. Poincaré Phys. Théor.*, 59:201–236, 1993.

[Lei]   M. Leitner. Character invariant Segal algebras on hypergroups. unpublished manuscript.

[LMR98]   A.K. Louis, P. Maaß, and A. Rieder. *Wavelets*. Teubner, Stuttgart, 1998.

[LOR02]   R. Lasser, J. Obermaier, and H. Rauhut. Harmonic analysis on discrete generalized hypergroups. Preprint, 2002.

[LT77]   J. Lindenstrauss and L. Tzafriri. *Classical Banach spaces I*. Springer, 1977.

[LT79]   J. Lindenstrauss and L. Tzafriri. *Classical Banach spaces II*. Springer, 1979.

[Mal88]   S. Mallat. *A Wavelet Tour of Signal Processing*. Academic Press, San Diego, CA, 1988.

[Mal89]   S.G. Mallat. Multiresolution approximation and wavelet orthonormal bases of $L^2(\mathbb{R})$. *Trans. Amer. Math. Soc.*, 315:69–87, 1989.

[MB79]   P. Milnes and J.V. Bondar. A simple proof of a covering property of locally compact groups. *Proc. AMS*, 73:117–118, 1979.

[Mic51]   E. Michael. Topologies on spaces of subsets. *Trans. Amer. Math. Soc.*, 71:151–182, 1951.

[Mos72]   R.D. Mosak. The L^1- and C^*-algebras of $[FIA]_B^-$ groups, and their representations. *Trans. Amer. Math. Soc.*, 163:277–310, 1972.

[Mül98]   C. Müller. *Analysis of Spherical Symmetries in Euclidean Spaces*. Springer, 1998.

[MZ55]   D. Montgomery and L. Zippin. *Topological Transformation Groups*. John Wiley & Sons, 1955.

[NH03]   B. Nazaret and M. Holschneider. An interpolation family between Gabor and wavelet transformations. Applications to differential calculus and construction of anisotropic Banach spaces. In *Albeverio, S. (ed.) et al., Nonlinear Hyperbolic Equations, Spectral Theory, and Wavelet Transformations*, pages 363–394. Birkhäuser, Basel, 2003.

[NT97]   M.M. Nessibi and K. Trimèche. Inversion of the Radon transform on the Laguerre hypergroup by using generalized wavelets. *J. Math. Anal. Appl.*, 208:337–363, 1997.

[OW96]   N. Obata and N.J. Wildberger. Generalized hypergroups and orthogonal polynomials. *Nagoya Math. J.*, 142:67–93, 1996.

[Rau01]   H. Rauhut. An uncertainty principle for periodic functions. Diploma Thesis, TU München, 2001.

[Rau03a]   H. Rauhut. Banach frames in coorbit spaces consisting of elements which are invariant under symmetry groups. *Preprint*, 2003.

[Rau03b]   H. Rauhut. Wavelet transforms associated to group representations and functions invariant under symmetry groups. *Preprint, to appear in Int. J. Wavelets Multiresolut. Inf. Process.*, 2003.

[Ric60]   C.E. Rickart. *Banach Algebras*. van Nostrand, 1960.

[Rös95a]   M. Rösler. Convolution algebras which are not necessarily positivity preserving. *Cont. Math.*, 183:299–318, 1995.

[Ros95b]   K.A. Ross. Signed hypergroups – a survey. *Cont. Math.*, 183:319–329, 1995.

[Rös97]   M. Rösler. Radial wavelets and wavelets on hypergroups. Preprint, 1997.

[RR03]   H. Rauhut and M. Rösler. Radial multiresolution in dimension three. *Preprint, to appear in Constr. Approx.*, 2003.

[Rud87]   W. Rudin. *Real and Complex Analysis*. McGraw-Hill, Singapore, third edition, 1987.

[Sch71]   H.H. Schaefer. *Topological Vector Spaces*. Springer, 1971.

[SD80]   W. Schempp and B. Dreseler. *Einführung in die harmonische Analyse*. Teubner, Stuttgart, 1980.

[Skr03]   L. Skrzypczak. Function spaces in presence of symmetries: Compactness of embeddings, regularity and decay of functions. In *D. Haroske (ed.) et al., Function Spaces, Differential Operators and Nonlinear Analysis. The Hans Triebel Anniversary Volume.*, pages 453–466. Birkhäuser, Basel, 2003.

[Sor92]   J. Soria. Weighted tent spaces. *Math. Nachr.*, 155:231–256, 1992.

[Spe75]   R. Spector. Aperçu de la théorie des hypergroupes. In *Analyse harmonique sur les groupes de Lie*, volume 497 of *Lecture Notes in Math.* Springer, Berlin, 1975.

[SS00]   W. Sickel and L. Skrzypczak. Radial subspaces of Besov and Lizorkin-Triebel classes: extended Strauss lemma and compactness of embeddings. *J. Fourier Anal. Appl.*, 6:639–662, 2000.

[ST01]   L. Skrzypczak and B. Tomasz. Compactness of embeddings of the Trudinger-Strichartz type for rotation invariant functions. *Houston J. Math.*, 27:633–647, 2001.

[Tor91]   B. Torrésani. Wavelets associated with representations of the affine Weyl-Heisenberg group. *J. Math. Phys.*, 32:1273–1279, 1991.

[Tor95]   B. Torrésani. Position-frequency analysis for signals defined on spheres. *Signal Process.*, 43:341–346, 1995.

[Tri83a]   H. Triebel. Modulation spaces on the Euclidean n-space. *Z. Anal. Anwendungen*, 5(2):443–457, 1983.

[Tri83b]   H. Triebel. *Theory of Function Spaces.* Birkhäuser, 1983.

[Tri88]   H. Triebel. Characterizations of Besov-Hardy-Sobolev spaces: A unified approach. *J. Approx. Theory*, 52(2):162–203, 1988.

[Tri92]   H. Triebel. *Theory of Function Spaces II.* Birkhäuser, 1992.

[Tri95]   K. Trimèche. Wavelets on hypergroups. In *Ross, K.A. (ed.) et al., Harmonic Analysis and Hypergroups.* Birkhäuser, Boston, 1995.

[Tri97]   K. Trimèche. *Generalized Hypergroups and Wavelets.* Gordon and Breach, Amsterdam, 1997.

[Vre79]   R.C. Vrem. Harmonic analysis on compact hypergroups. *Pacific J. Math.*, 1:239–251, 1979.

[Wer95]   D. Werner. *Funktionalanalysis.* Springer, 1995.

[Woj97]   P. Wojtaszczyk. *A Mathematical Introduction to Wavelets.* Cambridge University Press, Cambridge, 1997.

[Yos80]   K. Yosida. *Functional Analysis.* Springer, 1980.